Attractors, Bifurcations, and Chaos

Nonlinear Phenomena in Economics

Springer
Berlin
Heidelberg
New York
Barcelona
Hong Kong
London
Milan
Paris
Singapore
Tokyo

Tönu Puu

Attractors, Bifurcations, and Chaos

Nonlinear Phenomena in Economics

With 186 Figures
and 2 Tables

Prof. Tönu Puu
Umeå University
Department of Economics
SE-90187 Umeå
Sweden

ISBN 3-540-66862-4 Springer-Verlag Berlin Heidelberg New York

Library of Congress Cataloging-in-Publication Data
Die Deutsche Bibliothek – CIP-Einheitsaufnahme
Attractors, bifurcations, and chaos: nonlinear phenomena in economics; with 2 tables / Tönu Puu. – Berlin; Heidelberg; New York; Barcelona; Hong Kong; London; Milan; Paris; Singapore; Tokyo: Springer, 2000
ISBN 3-540-66862-4

Hardcover-Design: Erich Kirchner, Heidelberg

SPIN 10724779 42/2202-5 4 3 2 1 0 – Printed on acid-free paper

Preface

The present book relies on various editions of my earlier book "*Nonlinear Economic Dynamics*", first published in 1989 in the Springer series "*Lecture Notes in Economics and Mathematical Systems*", and republished in three more, successively revised and expanded editions, as a Springer monograph, in 1991, 1993, and 1997. The first three editions were focused on applications. The last was considerably different, by including some chapters with mathematical background material (ordinary differential equations and iterated maps), so as to make the book more self-contained and suitable as a textbook for economics students of dynamical systems. To the same pedagogical purpose, the number of illustrations were expanded. The author also prepared some of the software used in producing the illustrations for use by the readers on the PC, by making the programs interactive and providing them with a user interface. Simulations are essential, when dealing with nonlinear systems, where closed form solutions do not exist. Even theoretical science then becomes experimental. (The software prepared for that book can still be acquired directly from the author at the address `tonu.puu@econ.umu.se`.)

The present book has been so much changed, that I felt it reasonable to give it a new title. There are two new mathematics chapters (on partial differential equations and on catastrophe theory), making the mathematical background material fairly complete. There is also an account of the recently emergent method of critical lines and absorbing areas for non-invertible maps added to the chapter on maps.

As for the application part, the inclusion of partial differential equations made it possible to discuss topics of pattern formation in the space economy, which is another interest of the author, stemming from collaborative work over many years with Professor Martin J. Beckmann of Brown University and the TU München, last given account of in my other Springer monograph "*Mathematical Location and Land Use Theory*" also dating from 1997.

The critical line method is put to use in most application models, for drawing attractors and for better understanding their bifurcations. A short account of collaborative work in progress with the Urbino group, aiming at an extension of the critical line method to critical surfaces in three dimensions, is also included in the application to oligopoly. In writing these parts the author is very much indebted for the expert advice from Professor Laura Gardini of the University of Urbino.

The mathematical background chapters should provide enough of up to date methodological tools for any economics student of nonlinear dynamics, no matter what the particular field of aimed application is. A unique feature in this book, among literature for economics students, is a fairly complete account of the perturbation methods.

As for the applications presented in this volume, most of the material on business cycles stems from various editions of my previous dynamics book, though much new information has been added using the critical line method. The same holds true for the discussion of oligopoly.

Completely new in this book is the attempt to understand economic development from the viewpoint of increasing diversity, in contrast to the normal focus on quantitative growth. Coexistence of different attractors provides for divergence in development, due to minor differences in the dynamical process itself, for alternative histories and futures if we so wish. The question is whether this result of nonlinearity should not be at least as intriguing for economics as unpredictability due to deterministic chaos. Among other things the cherished relation between steady states and optimality becomes problematic once they are no longer unique.

Linearity for dynamical processes goes together with convexity of the background structure. Expressed more explicitly, unique (optimal) steady states and predictability, so comfortable for the belief in a well ordered reality, belong together, and they have to leave the stage together, once we admit that things are not so simple.

The author gratefully acknowledges the financial support over many years of his research by the *Swedish Research Council for the Humanities and the Social Sciences*, a support which definitely has been a necessary condition for the earlier stages of this work. The sponsorship has now shifted to the *Swedish Transport & Communications Research Board*, to which the author is equally indebted.

Umeå in October 1999
Tönu Puu

Contents

1 Introduction

1.1 Dynamics Versus Equilibrium Analysis

Dynamic analysis in economics is as old as economics itself. A glance at the subject index in Joseph Schumpeter's monumental "*History of Economic Analysis*" is sufficient to convince you about this. Even dynamic mathematical models are fairly old. Cournot duopoly dates back to 1838 and the cobweb model of price adjustments to 1887.

Throughout the history of economics there has been a competition between the dynamic and the equilibrium outlooks. As an alternative to a truly causal or recursive dynamics there is always the concept of an equilibrium balance of forces, even in intertemporal perspective.

In general the equilibrated forces are results of optimising behaviour, and of correct (self-fulfilling) anticipations of the actions of others. Therefore the epistemological polarity - causal versus teleological explanation - is involved.

Certain controversies in the history of economics reflect this polarity of different philosophical outlooks. One example is provided by those who objected the original Marshallian concept of market equilibrium on the grounds that price could not be determined by production cost and utility in consumption at once. Such objections need not be ascribed to mathematical ignorance, i.e. inability to understand the mathematics of a simultaneous system of equations, as they in fact were.

On the whole, equilibrium analysis has been dominant during the history of economics, especially in various classical periods. Maybe it is natural that a period of scientific consolidation emphasizes the harmony in a balance of forces, whereas more turbulent periods of scientific renewal favour the genuinely dynamic outlook.

A basic theme in Schumpeter's history is the alternation between periods of renewal and what he names ”classical situations”. Implicit is that we tend to overvalue the consolidation periods and undervalue those of renewal. The term "classical situation" roughly corresponds to what Thomas Kuhn called "normal science". Schumpeter identified three such classical periods in the history of economics, and described them in the following somewhat ironic words: ”*... there is a large expanse of common ground ... and a feeling of repose, both of which create an expression of finality - the finality of a Greek temple that spreads its perfect lines against a cloudless sky*”.

Schumpeter would certainly have agreed that today we have a fourth classical situation, if there ever was one, with a complete dominance of general equilibrium theory, as formulated by above all Debreu. The concept of "commodity" not only represents all physically different brands, it also specifies availability at different moments and in different locations. When an intertemporal equilibrium, based on "rational" expectations (once called "perfect foresight" by a more appropriate name) is established, economic evolution merely becomes a cinema performance of a film produced beforehand, and a dynamic process over time just represents a very specific sequence of coordinate values of the intertemporal equilibrium point.

How different this outlook is from that represented in Paul Samuelson's "*Foundations of Economic Analysis*" ten years before Debreu, where equilibrium analysis was regarded as the simplified study of stationary processes alone! The claim to higher generality is a sign of dominance as good as any.

1.2 Linear Versus Nonlinear Modelling

When the flourishing field of economic growth theory finally collapsed under the attacks from "The Club of Rome" and the "Limits to Growth" movement, this was the end to the latest real outburst of economic dynamics. Although several economists managed to deliver convincing counterattacks on the *ad hoc* character of the computer simulated models on which the doomsday scenarios were based, economic dynamics never recovered. "Endogenous growth" just seems to be a last sign of life of an essentially dead research area.

The basic reason for this must have been the inherent limitations of linear dynamics then in use. A linear dynamic model can either produce explosive or damped movement. The latter option was used in the study of stability in dynamised market equilibrium models, whereas the former became the basis of growth theory.

As for the damped case, we may ask whether a model is at all dynamic if it can only explain the progressive decay of any exogenously introduced deviation from eternal equilibrium.

As for the explosive case, it is not only in contradiction to even the most casual observation of real processes of change in nature or society, it also involves the following problem of scientific procedure:

In general we may assume that functions relating the variables studied are nonlinear but smooth. Hence they can in certain intervals be approximated by linear relations. But then we must be sure that the approximation is not used in a model which produces change so large that the bounds of the approximation interval are violated.

When Hooke's law, representing the restoring force of a spring as proportionate to its elongation, was used in models of mechanical vibrations, it was granted that the variables were kept in the interval for which the approximation was reasonable. In economic growth theory, in contrast, the linear relations were used to produce change so large that the bounds of linear approximation would be automatically violated, no matter how liberally they were set. The problem with explosive linear models is hence not just one of realism, but also one of scientific procedure.

There remains a neutral boundary case in the linear models, of neither explosion, nor damping, which was once used to produce standing business cycles, but such boundary cases need so specific relation between the parameters to hold that it has zero probability. In terms of modern systems theory it is structurally unstable.

Apart from what has already been said, there is another serious drawback of linear modelling: The uniqueness of attractors, which in the perspective of dynamics also means uniqueness of development paths. Nonlinearity admits the coexistence of multiple attractors, and hence of development paths, which means that a process, due to minor divergences in the process itself can take completely different courses.

In general, this is seen as an improvement by scientists, though many economists think differently, unique optimal trajectories being a cherished idea, as if society was a rocket and reaching the moon one single, unique as well as obvious, target. This square way of thinking, more than anything else, may be what repels the general enlightened audience from current mainstream economics.

1.3 Perturbation Methods

A problem is how to get ahead with theory once we decide to rid us of the assumption of linearity. A linear relation is completely specified by a limited set of parameters, a nonlinear one has an infinite number of degrees of freedom.

The precedence of Duffing, van der Pol, and Lord Rayleigh, however, has demonstrated how much can be achieved by modifying the linear terms with higher order terms in the reverse direction, and from the mathematics of catastrophe theory we are in general encouraged to work with truncated Taylor series, without bothering about analyticity and remainders. So even if we want to do something more than apply the scant supply of general theorems for nonlinear systems, we have a sound basis for taking the next step in approximation - from linear to polynomials of a higher degree.

Once we specify the nonlinearities, we have the whole set of tools called "perturbation methods" at our disposal. As so much else in the theory of dynamical systems they go back to Poincaré. The general idea is to study what happens when a linear system, such as an oscillator whose solution we know, is just modified a little by introducing a small nonlinearity, and deriving the solution as a power series in the smallness parameter.

The solution process for the nonlinear system is so reduced to solving a sequence of linear differential equations, and even finishing the solution process before the computations become too messy often gives accurate information. The perturbation methods have hardly ever been used in economics, although the "final" monograph is as old as Stoker's dating from 1950.

Later work by Hayashi in 1964, comparing the perturbation approach with the outcome of extensive simulation on an analog computer, demonstrated its usefulness for cases where the nonlinearities were no longer vanishingly small. Moreover, the perturbation methods were found equally useful for the opposite case, where the nonlinearity is very large instead, so that it dominates at all times except for very fast transitions through phase space. For an up to date account on perturbation methods, which are extensively used in the following, we refer to Jordan and Smith or Kevorkian and Cole.

In addition to perturbation methods, computer simulation of the systems is used throughout the book, and the whole study is indeed focused on the phase portraits. In a science, such as economics, which, as far as formal models are concerned, for so long has resorted on general existence proofs, simulation may make the impression of an inconclusive, heuristic, and hence inferior scientific procedure.

Such an attitude would, however, declare all experimental science as inferior. As matter of fact we now know that dynamical systems that can be solved in closed form constitute, not only a sparse subset, but a nontypical subset of all systems. Accordingly, experiment, i.e., simulation, along with general consideration of the geometry of phase space, is our only way of getting ahead with the systems which are really significant.

It may be a good idea to use some computer software for simulating dynamical systems along with the reading of this book. The Nusse and Yorke package, sold in bookstores with the manual, contains many excellent procedures along with a representative selection of popular dynamical systems, both maps and differential equations. The latest edition also includes a welcome possibility to compile one's own systems, which was incomparably more messy before. For those who possess a general programming language software and are able to cook their own simulation procedures, the book by Parker and Chua can be recommended for their recipes, which approximately correspond to those implemented in the Nusse and Yorke package.

1.4 Structural Stability

Another important development in systems dynamics, particularly related to geometrical aspects, is the qualitative theory of differential equations, also called the "generic" theory. In a remarkable series of articles Peixoto around 1975 explored the qualitative features of solutions to sets of first order differential equations under the mild assumption that the flow portrait determined by them was structurally stable.

The remarkable thing was that, although nonlinear differential equations in general can produce so rich solution patterns as to even defy attempts of scientific classification, the sole assumption of structural stability reduces the singularities to a finite set of isolated points, each of which locally looks like the singularities of a linear system. A global theorem on the instability of ("homoclinic" and "heteroclinic") saddle connections was added and rendered amazingly informative qualitative descriptions of global flow portraits, given only that they were determined by structurally stable dynamical systems.

About this, as well as about most of the other topics in modern dynamics, the book of "cartoons" by Abraham and Shaw cannot be too highly recommended.

1.5 Chaos and Fractals

The most spectacular development in modern systems theory undoubtedly was the discovery of chaos. We used to think that evolution processes were either well understood, deterministic, and predictable, or else not well understood, stochastic, and unpredictable. Though Poincaré already discovered that things were not that simple, it was shocking for a broader audience to learn that the most simple nonlinear, though completely deterministic systems, could produce unpredictable series of events.

It, of course, lies in the essence of determinism that if we know the initial conditions with infinite exactitude, we can always determine the orbit. But, whenever there is some inexactness, however small, in the specification of initial conditions, as always is the case in practice, then the tendency to separate nearby trajectories inherent in chaotic systems leads to complete unpredictability.

The practical importance of chaos is high in fields such as meteorology and economics, as it calls in doubt the very sense of the costly business of long run forecasting.

The strange attractors to which chaotic orbits tend have an intriguing structure, being neither curves, nor surfaces, nor solids, and they have dissipated to the general audience much more than the idea of deterministic chaos itself has. The concept of a fractal, an object of fractional dimension, was originally developed by Mandelbrot as an alternative to Euclidean geometry, and pretended to be better suited to represent natural objects than the traditional geometrical shapes. The essential feature is inexhaustible minute detail and self similarity at various scales of enlargement. The fractal images produced today, of planets, landscapes, flowers, clouds, and smoke, are indeed so deceptively "realistic" to the eye that they can be used in science fiction movies, and the reverse of their production as iterative (dynamic) processes has found practical use in the science of image compression.

There is an aesthetic attraction in the fractal images produced today (see for instance the work by Barnsley, Field/Golubitsky, Mandelbrot, Peitgen/Jürgens/Saupe) which is not entirely irrelevant for science. Even if we do not subscribe to the polemical opinions of some great scientists, such as Dirac, Poincaré, or Weyl, that beauty is a more important driving force in the development of science than even logic or observation, it is undeniable that aesthetics and images have extreme importance for scientific intuition, imagination and understanding. Fractals as the fingerprints of chaos will hence be given their due share of space in the following.

There are certain dangers with all new captivating scientific ideas: The temptation for vague analogy formation with a new paradigm, such as chaos, is great, and typically it is the logistic iteration (as it was the cusp catastrophe) that is applied in economics. The source of inspiration being the mathematical model itself, it is not surprising that in terms of economic substance such models are not very deeply rooted in basic economic principles. This need not necessarily be so.

Ironically, one of the earliest chaotic models, studied by Cartwright and Littlewood around 1940, was the forced nonlinear oscillator. The nonlinear oscillator itself was very similar in structure to the multiplier-accelerator model of the business cycle, with nonlinearities due to the Hicksian floor and roof, and the forcing is easily obtained by the influence of cyclically changing exports to a world market. But this model, unlike the quadratic map, or the Lorenz and Rössler systems is seldom discussed in the popular books. We have to resort to a monograph by Levi, or to Guckenheimer and Holmes to find a complete account.

1.6 The Choice of Topics Included

This book does not aim at being a survey. The author has included a set of models from micro- and macroeconomics which were developed by himself, rather than those developed by others. However, these models are with very few exceptions only slight modifications of models that have a certain age and are well rooted in economic theory. There is a pedagogical point in such a strategy - to show that firmly rooted economic principles need no, or almost no, stretching at all to produce bifurcations, chaos and other extravagant features of nonlinear dynamics.

The dynamics of the Cournot duopoly model already mentioned, dating from 1838, is an obvious item to be included. Rand hinted at many possible nonlinearities in the model that might lead to interesting dynamics. The author, however, thinks that he found the simplest conditions, isoelastic demand and constant marginal costs, under which this occurs. The advantage of these assumptions is that it becomes possible to solve for the reaction functions explicitly.

Another microeconomic model included is a search algorithm for the optimum by a monopolist facing a marginal revenue curve which has two profit maximizing intersections with the marginal cost curve. This was shown by

Joan Robinson in 1933 to occur whenever the market was a composite of two groups of consumers with different demand elasticities.

The macroeconomic part of the book mainly deals with business cycle models of the multiplier-accelerator type. They are used with a nonlinear investment function, such as suggested by Hicks and Goodwin around 1950. We treat them both in the format of discrete time, as originally suggested by Samuelson and Hicks, and in continuous time as suggested by Goodwin and Phillips.

We also propose a model with continuous space, resulting in a case of partial differential equations. The spatial interaction arises from nothing more than a linear interregional trade multiplier, which fits well in with the basically Keynesian philosophy in the multiplier-accelerator models. There are two more models included, where space plays an essential role: The migration model by Hotelling from 1921, and the interregional trade model by Beckmann from 1952.

Despite the dominant traditional layout in most of the models presented, the concluding Chapters are a bit different as they concentrate on the multiplicity of attractors and of development routes, and focus such issues as economic development from the viewpoint of increasing diversity, with the Darwinian development tree invoked as an analogy, and catastrophe theory being used as a tool for analysis. Such analogy too is an old idea in economics, for instance considered by Marx already.

Though mainstream economists still prefer to believe that the economic system is so rational that we can do with uniqueness, optimality, and equilibrium, a growing number of researchers, however, appear to think that present day instabilities and turbulence in the economy and in the society warrant dynamic studies expressly focusing nonlinearities.

A selection of recent monographs on the topic are included in the reference list, though, as usual in a very active research field, much of the development is only represented in journals and in special conference proceedings. For an excellent comprehensive survey we refer to Rosser.

2 Differential Equations: Ordinary

2.1 The Phase Portrait

There is no more useful tool for the study of differential equations, in particular if they are in two dimensions, than the phase portrait. Many important systems both in physics and in economics in fact live in two dimensions. All second order systems are two dimensional. To this category belong all the oscillators, exemplified by the mathematical pendulum, or by the Samuelson-Hicks business cycle model if put in continuous time. It should be remembered that a second order differential equation, as characteristic of an oscillator, can always be put in the style of two coupled first order equations.

Take the general equation:

$$\ddot{x} = g(x, \dot{x}) \tag{2.1}$$

We can always define a new variable $y = \dot{x}$, and we have the system:

$$\dot{x} = y \tag{2.2}$$

$$\dot{y} = g(x, y) \tag{2.3}$$

The reverse procedure, of course, also works. A more general statement of our object of study is the system:

$$\dot{x} = f(x, y) \tag{2.4}$$

$$\dot{y} = g(x, y) \tag{2.5}$$

The variables x, y are the natural axes for the phase portrait. From the connection with second order systems we often deal with one variable and its time derivative, its rate of change, but this need not always be so.

Except for the fact that many interesting systems in applications are two dimensional, two dimensions strikes a nice balance between the triviality of one dimension and the mess of three dimensions and higher. In one dimension there are nothing but points and intervals, having no geometrical shape, in three dimensions things are already difficult to visualize. Nevertheless it is true that two-dimensional pictures are useful even for certain illustrations of dynamic processes in higher dimension.

By the emphasis on phase portraits we focus on general qualitative properties of differential equations and their solutions, rather than on closed form solutions. There are, after all, very sparse classes of equations, such as the linear, the exact, and the separable equations, which are at all possible candidates for closed form solution. Much ingenuity has to be spent and wasted on attaining such solutions, which it is now known, are possible only with sparse and, even worse, atypical sets of differential equations.

There are several things to be seen in a phase portrait: If we divide equations (2.4)-(2.5) we get:

$$\frac{dy}{dx} = \frac{g(x,y)}{f(x,y)} \tag{2.6}$$

This derivative represents a field direction in the phase diagram. Equation (2.6) allows us to determine the direction for any point in the phase diagram. The only problem that can arise is if $f(x,y) = g(x,y) = 0$ holds. It may seem that already a zero denominator alone would cause problems, but in that case we could just shift the axes and study dx/dy instead. In this way the denominator becomes a numerator, and the derivative zero, which means a horizontal slope, or, in the original setup, a vertical slope. Only at the singular point, where $f(x,y) = g(x,y) = 0$ holds true, is it impossible to define a field direction.

Drawing short line segments in the phase diagram based at a lattice of regularly spaced points, we get a first impression of how the solution curves go. As a matter of fact (2.6) is the tangent vector to the local solution curve. If we treat (2.6) as a differential equation and solve it we get the integral curves in the phase diagram.

Fig. 2.1. Field directions and solution curves for the pendulum.

Note that even if we accomplish this, we have not yet solved for the variables as functions of time, but having got that far we have reduced the order of the differential equation. If it was a second order oscillator then we would have arrived at a relation between the variable and its time derivative, i.e., a first order differential equation. For this reason such solutions are called first integrals. They originally arose in conservative mechanical systems and represented the division of a constant total energy between potential and kinetic.

Fig. 2.1 illustrates the direction field for the pendulum $\ddot{x} + \sin x = 0$, or the equivalent pair of first order equations $\dot{x} = y$, $\dot{y} = -\sin x$. The gray direction field lines represent the slope $dy / dx = -(\sin x) / y$. Even without the selection of solution curves drawn, the direction field would make it possible to sketch the first integral curves. In the chosen example of the pendulum these curves are easily obtained as the closed form solutions $y = \pm\sqrt{K + \cos x}$.

On the horizontal axis in Fig. 2.1 we have x, the angular deflection of the pendulum from its rest position, on the vertical we have y, the angular veloc-

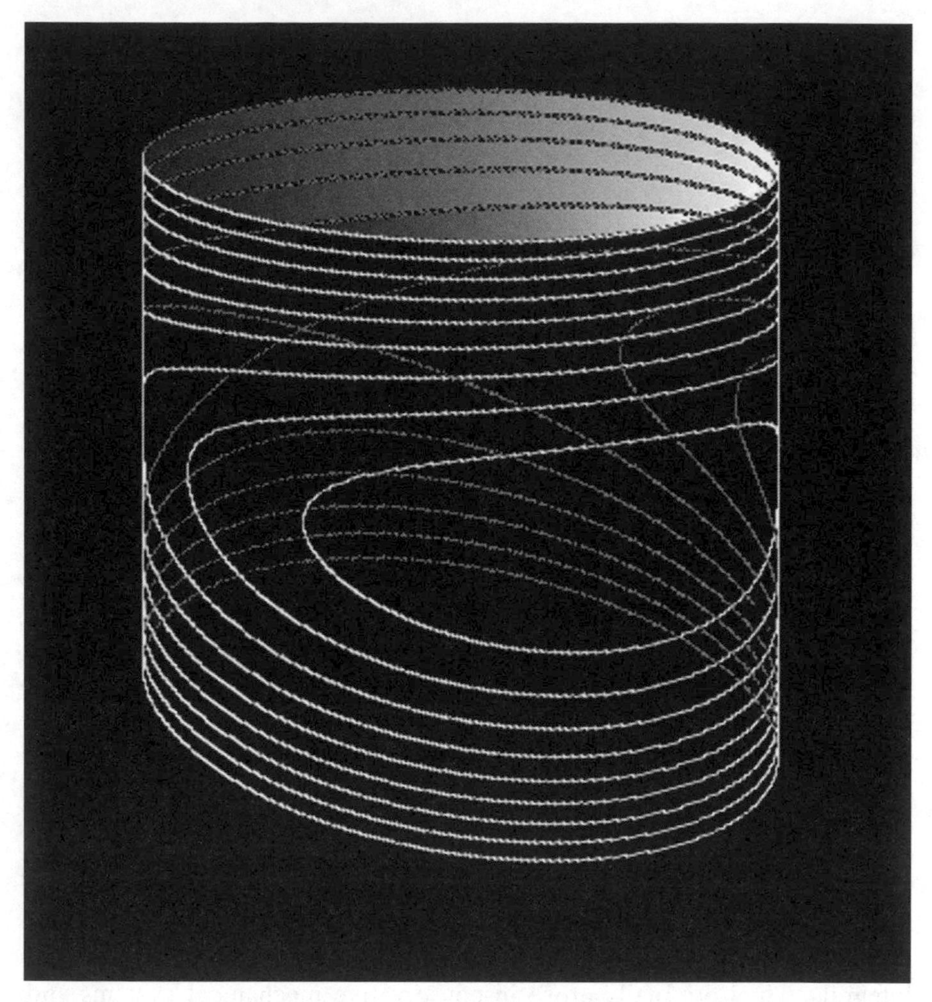

Fig. 2.2. Phase portrait of the pendulum wrapped around a cylindre.

ity of motion. In elementary text books the approximation $\sin x \approx x$ is often used, and then the equation for the pendulum is transformed into that of the simple harmonic oscillator $\ddot{x} + x = 0$. The solutions of the latter are all concentric closed orbits.

We see that this in the more general case only holds for small movements at low energy around the rest positions. This is reasonable because a variable and its sine are equal for small values, but deviate considerably for large.

The movement along these closed orbits in Fig. 2.1 is clockwise, as always with oscillators of this type. This is because we on the vertical axis have the rate of change of the variable on the horizontal. Accordingly, in the upper half, where the rate of change is positive, motion goes right, in the lower part, where the rate is negative, motion goes left.

The closed orbits are confined within a pair of intersecting sinoid curves. As a matter of fact they do not intersect. The seeming intersection points represent the unstable equilibria when the pendulum is positioned with the bob right up. (Of course we must think of the bob as fixed to a rigid rod and not to a string.) The sinoid curves are split in sections by these unstable rest points, and the sections represent critical motion when the pendulum is released from its upper unstable position, accelerates, and then retards again, exactly so as to come to rest at the unstable top equilibrium anew. This critical motion would by the way take infinite time.

What about the wavy lines outside the critical curves? As a matter of fact they represent periodic motion too, in the case when the energy is so large that the pendulum turns over again and again. When the bob gets up, the motion is retarded, but never so much that it stops and turns back as in the closed orbits. It passes the top and then gains force again. As there is no friction damping in the model this can go on for ever and ever. We may ask how those wavy lines can be closed orbits. The answer is that we have displayed the picture in terms of the horizontal scale from -2π to 2π, which is twice a main branch of a trigonometric function. We could extend the picture horizontally without limit, by adding new copies of the diagram, but we have too much already. There are two intersection points representing the unstable rest position of the bob. They should in fact be identified.

This is achieved by wrapping the diagram around a cylinder, so that the intersection points come over each other. Then, of course, the stable equilibrium point with the bob downwards, of which there are three copies in the figure, also come over each other. In this way the wavy curves too become closed, but around the cylinder. This is shown in Fig. 2.2.

Accordingly, we have seen the following things in the phase portrait: the direction fields, the first integral solution curves, and the points of equilibrium, the stable and unstable. This last characteristic is not quite true, because the rest position with the bob down is not stable either, at least not asymptotically stable. Any tiny push will set it in perpetual motion. It turns stable only once we add friction.

Adding a small friction term to the pendulum model changes the equation into $\ddot{x} + \sin x + \delta \dot{x} = 0$. The effect on the direction field is a slight change to $dy / dx = -(\sin x) / y - \delta$, but the effect on the solution curves, or the phase portrait as it is usually called, is enormous. This is all illustrated in Fig. 2.3.

Fig. 2.3. Direction field and phase portrait with friction added.

We see that the stable equilibria really become asymptotically stable, with the closed orbits replaced by inward spirals. The unstable equilibria remain saddle points, but there are no longer any trajectories between them. The outgoing trajectories from saddle points end up at the spiral attractors, the ingoing trajectories issue from somewhere in outer space.

The singular points, called centres, which are surrounded by closed orbits, are structurally unstable phenomena, and so are the trajectories going out from one saddle and in to another, called heteroclinic trajectories. We remember that in our original case there was just one saddle, so the trajectories actually go out from that saddle and in to itself. These particular types of heteroclinic trajectories are called homoclinic. We see that, centres and homoclinic/heteroclinic trajectories, being structurally unstable, i.e., swept away by any change of the friction coefficient δ from zero, however slight, the entire phase portrait is qualitatively converted into something different.

Arnol'd in fact warns us to make structurally unstable models. Scientific modelling is always subject to mistakes, and we could not afford to have models such that their conclusions turn out completely wrong should the assumptions not all be true to the last tiny detail. According to Arnol'd, friction should be included in a model of the pendulum, no matter whether we knew of its existence as an empirical fact or not, just as strategy of scientific prudence.

There is even more to the difference between Figs. 2.1 and 2.3. There is no longer necessarily a point in wrapping the picture around a cylinder, even if it can again be done. The stable spiral attractors are in fact all different. The initial energy may be so large that the pendulum first makes one, two, three, or one million full turns before finally starting its motion back and forth around the stable equilibrium point. Even if it physically is the same rest point, there is a point in distinguishing in the phase portrait between the different cases in terms of how many full turns the pendulum first makes.

The ingoing trajectories to the saddle points then get a new significance, as watersheds between the attractors to the left and right of them. All the stable equilibria, competing for the ultimate state of the system, have their basins of attraction, and the ingoing saddle trajectories furnish the basin boundaries.

If we again wrap the phase portrait around a cylinder, then the ingoing saddle trajectories just become one pair of spirals around the cylinder, but they separate nothing, as both sides then are the same basin of the single attractor. The simple reason is that if we cut a cylinder along its height, in a helicoid motion or not, then it still remains in one piece. The difference between the styles of presentation is just a matter of taste, whether we want to distinguish between the rest positions depending on the number of primary full turns or not.

Let us now turn our interest to the singular points or the equilibria of the system (2.4)-(2.5). They are defined by:

$$f(x,y) = 0 \tag{2.7}$$

$$g(x,y) = 0 \tag{2.8}$$

For the example case of a pendulum without friction we for instance have $f(x,y) = y = 0$ and $g(x,y) = -\sin x = 0$, i.e. $x = n\pi$ and $y = 0$, where n is any integer. For even integers we deal with centres, for odd with saddle points. The singular points are the really interesting elements of the phase portrait. The trajectories or solution curves everywhere else are topologically equivalent to a set of parallel straight lines and hence not very exciting. The theorem on the existence and uniqueness of solutions for differential equations guarantees that, except when (2.7)-(2.8) holds, the system (2.4)-(2.5) has a unique solution in the neighbourhood of any point. There hence passes just one trajectory through each point in the phase portrait. This also makes it possible to accommodate two initial values for a solution curve, for both variables, as indeed it should in a second order system.

2.2 Linear Systems

Let us so start looking at the singular points for the linear system:

$$\dot{x} = ax + by \tag{2.9}$$

$$\dot{y} = cx + dy \tag{2.10}$$

It is obvious that if we put the right hand sides equal to zero we get the unique solution $x = y = 0$. This uniqueness is always true with linear systems. The normal procedure for solving the system (2.9)-(2.10) is to try exponentials $x = \alpha \exp \lambda t$, $y = \beta \exp \lambda t$. Substituting into (2.9)-(2.10) we thus get:

$$\lambda \alpha \exp \lambda t = a\alpha \exp \lambda t + b\beta \exp \lambda t \tag{2.11}$$

$$\lambda \beta \exp \lambda t = c\alpha \exp \lambda t + d\beta \exp \lambda t \tag{2.12}$$

Dividing through by $\exp \lambda t$ and rearranging, (2.11)-(2.12) can be written:

$$(a - \lambda)\alpha + b\beta = 0 \tag{2.13}$$

$$c\alpha + (d - \lambda)\beta = 0 \tag{2.14}$$

Due to elementary algebra such a homogeneous system of equations has a nonzero solution for α, β if and only if the determinant is zero, i.e. if

$$\begin{vmatrix} a - \lambda & b \\ c & d - \lambda \end{vmatrix} = 0 \tag{2.15}$$

or, written out,

$$\lambda^2 - (a+d)\lambda + (ad - bc) = 0 \tag{2.16}$$

holds. This condition is called the characteristic equation, and it serves to determine the values of λ, called the eigenvalues of the system. Before continuing we should note that (2.16) can also be written in a different form.

Defining the coefficient matrix:

$$\begin{bmatrix} a & b \\ c & d \end{bmatrix} \tag{2.17}$$

we have its trace and determinant:

$$\mathrm{Tr} = a + d \tag{2.18}$$

$$\mathrm{Det} = ad - bc \tag{2.19}$$

With these expressions (2.18)-(2.19), equation (2.16) becomes:

$$\lambda^2 - \mathrm{Tr}\lambda + \mathrm{Det} = 0 \tag{2.20}$$

This quadratic equation is readily solved and yields the two eigenvalues:

$$\lambda_1, \lambda_2 = \frac{\mathrm{Tr}}{2} \pm \frac{1}{2}\sqrt{\mathrm{Tr}^2 - 4\mathrm{Det}} \tag{2.21}$$

From (2.21) we see that the trace equals the sum of the eigenvalues, $\mathrm{Tr} = \lambda_1 + \lambda_2$, whereas the determinant equals their product, $\mathrm{Det} = \lambda_1\lambda_2$.

Once we have determined the eigenvalues, we can consider equations (2.13)-(2.14) anew, for each of the eigenvalues *i*=1,2:

$$(a - \lambda_i)\alpha_i + b\beta_i = 0 \tag{2.22}$$

$$c\alpha_i + (d - \lambda_i)\beta_i = 0 \tag{2.23}$$

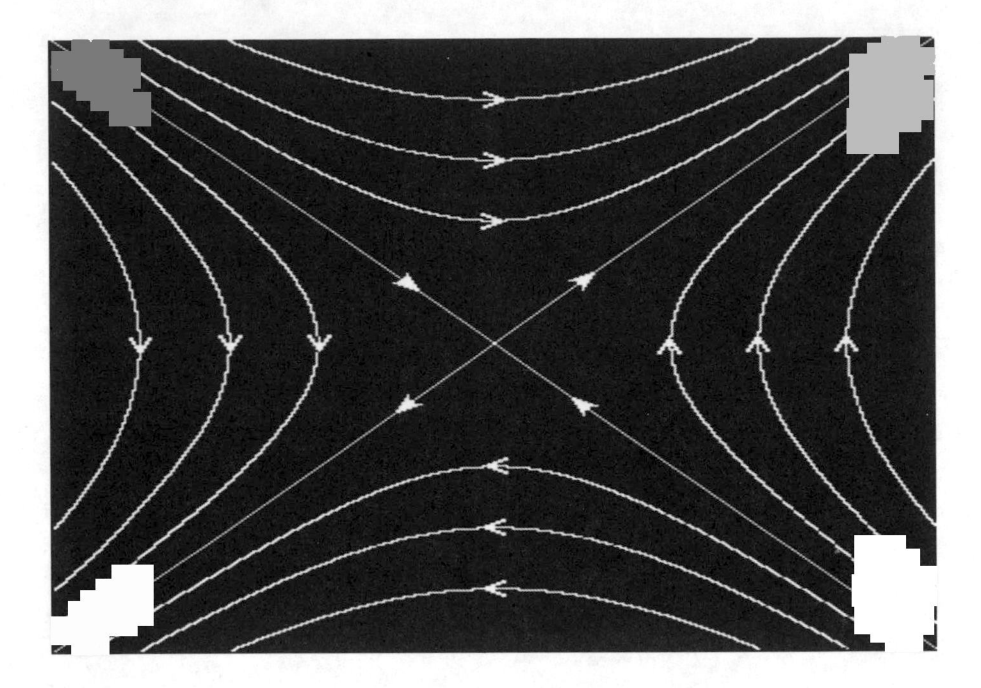

Fig. 2.4. Saddle point.

The solutions for α_i, β_i, called the characteristic vectors, have one degree of freedom due to the fact that the coefficient matrix has not full rank, as indeed it should not if we are to at all get a solution. One of the components may hence be chosen arbitrarily, and the second can then be computed from either equation, because it makes no difference which one we use.

The ratio β_i / α_i, however, gives a characteristic direction, associated with the eigenvalue λ_i, and the set of both ratios imply a coordinate transformation corresponding to the eigenvectors, such that the system is diagonalized.

For the classification of singularities we, in addition to the trace and the determinant of (2.17) also need the discriminant, i.e., the expression under the root sign in (2.21):

$$\Delta = \mathrm{Tr}^2 - 4\,\mathrm{Det} \tag{2.24}$$

It is now true that, if $\Delta > 0$, then the two eigenvalues are real, and the attempted solutions in terms of exponentials work. They do for both eigenvalues, as does any weighted sum of such solutions. Thus:

Fig. 2.5. Stable node.

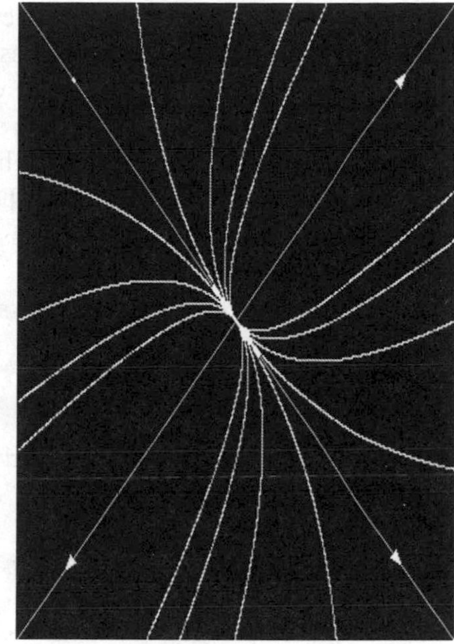

Fig. 2.6. Unstable node.

$$x = A\exp\lambda_1 t + B\exp\lambda_2 t \tag{2.25}$$

$$y = C\exp\lambda_1 t + D\exp\lambda_2 t \tag{2.26}$$

As a matter of fact these coefficients are determined by the initial conditions, but they are only two and the coefficients are four. To determine them we need the characteristic directions which provide exactly the additional constraints we need.

As a rule the eigenvalues are different. Given they are real we see from (2.21) that they have the same sign if Det>0, and opposite signs if Det<0. This is so because the square root of the discriminant contains the trace squared, and if a negative determinant increases the expression under the root even more it is bound to dominate, thus producing roots of opposite sign.

If the roots have opposite sign, then the singularity is a saddle point, because there is one direction, associated with the negative eigenvalue, in which the solution converges at the point, and another, associated with the positive eigenvalue, in which the solution diverges from it. In Fig. 2.4 we see the saddle.

In the case both eigenvalues have the same sign we deal with a node, stable if both eigenvalues are negative, unstable if both eigenvalues are positive. The former is hence the only stable case dealt with until now, whereas the latter is even more unstable than the saddle, having two unstable directions.

As we see from (2.21) the sign of the trace makes the difference between a stable and an unstable node. Given Det>0 and Tr<0 we have a stable node, given Det>0 and Tr>0 we have an unstable node. Both types are seen in Figs. 2.5-2.6.

All the time up to now we have assumed that the discriminant is positive, so that the eigenvalues are real. If $\Delta < 0$, then the eigenvalues become:

$$\lambda_1, \lambda_2 = \frac{\mathrm{Tr}}{2} \pm i \frac{1}{2} \sqrt{4\mathrm{Det} - \mathrm{Tr}^2} \tag{2.27}$$

where as usual $i = \sqrt{-1}$.

Let us define $\rho = \mathrm{Tr}/2$ and $\omega = \sqrt{4\mathrm{Det} - \mathrm{Tr}^2}/2$. Then we can write the eigenvalues (2.21) as:

$$\lambda_1, \lambda_2 = \rho \pm i\omega \tag{2.28}$$

i.e., as a pair of complex conjugate numbers. According to a basic identity in the algebra of complex numbers we however have:

$$\exp(\rho t \pm i\omega t) = \exp \rho t (\cos \omega t \pm i \sin \omega t) \tag{2.29}$$

Accordingly the solutions (2.25)-(2.26) can be written:

$$x = \exp \rho t (A \cos \omega t + B \sin \omega t) \tag{2.30}$$

$$y = \exp \rho t (C \cos \omega t + D \sin \omega t) \tag{2.31}$$

where the imaginary numbers in the solutions cancel out against suitably conjugate complex coefficients.

The coefficients can also be put in the form of an amplitude and a phase lead, so that we get:

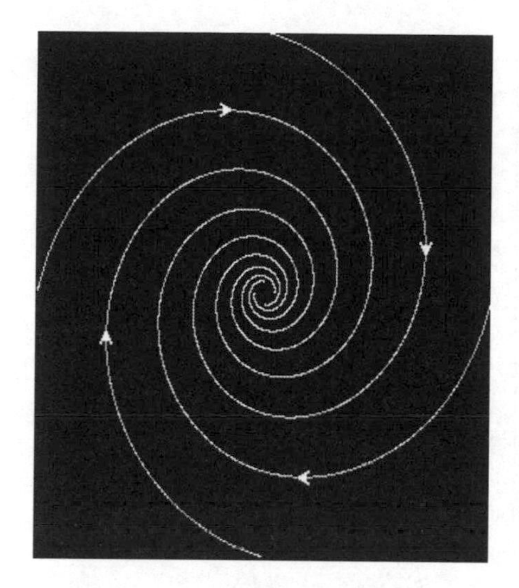

Fig. 2.7. Stable spiral.

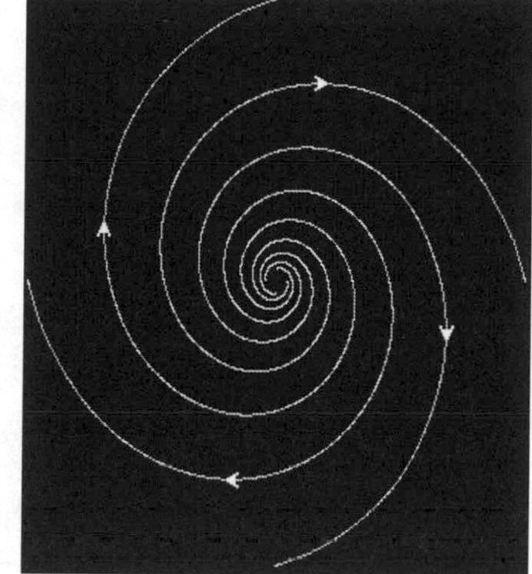

Fig. 2.8. Unstable spiral.

$$x = F \exp \rho t \cos(\omega t + \phi) \tag{2.32}$$

$$y = G \exp \rho t \cos(\omega t + \psi) \tag{2.33}$$

Thus we see that in the case of a negative discriminant we have a spiralling movement, defined by the trigonometric functions, with a contraction or expansion of the spiral defined by the exponential depending on whether the coefficient ρ is negative or positive. It has the same sign as the trace according to (2.27) and (2.28), so the sign of the trace again differentiates between stability and instability. The stable and unstable spirals are shown in Figs. 2.7-2.8.

In the special case when the trace is zero we have the boundary case of a centre, first encountered in the frictionless pendulum.

The thoughtful reader notes that we have again two coefficients too many in comparison to the initial conditions. But, again similar considerations as in the case of real eigenvalues reduce the degrees of freedom appropriately. We do not concentrate on these computational details, because our purpose is to get a characterization of the singular points, not to actually obtain closed form solutions.

We have thus seen that the linear system of differential equations has one unique singular point, or point of equilibrium. The types found are three: Nodes, saddles, and spirals. The nodes and spirals are of two types, stable and unstable, or attractors and repellors.

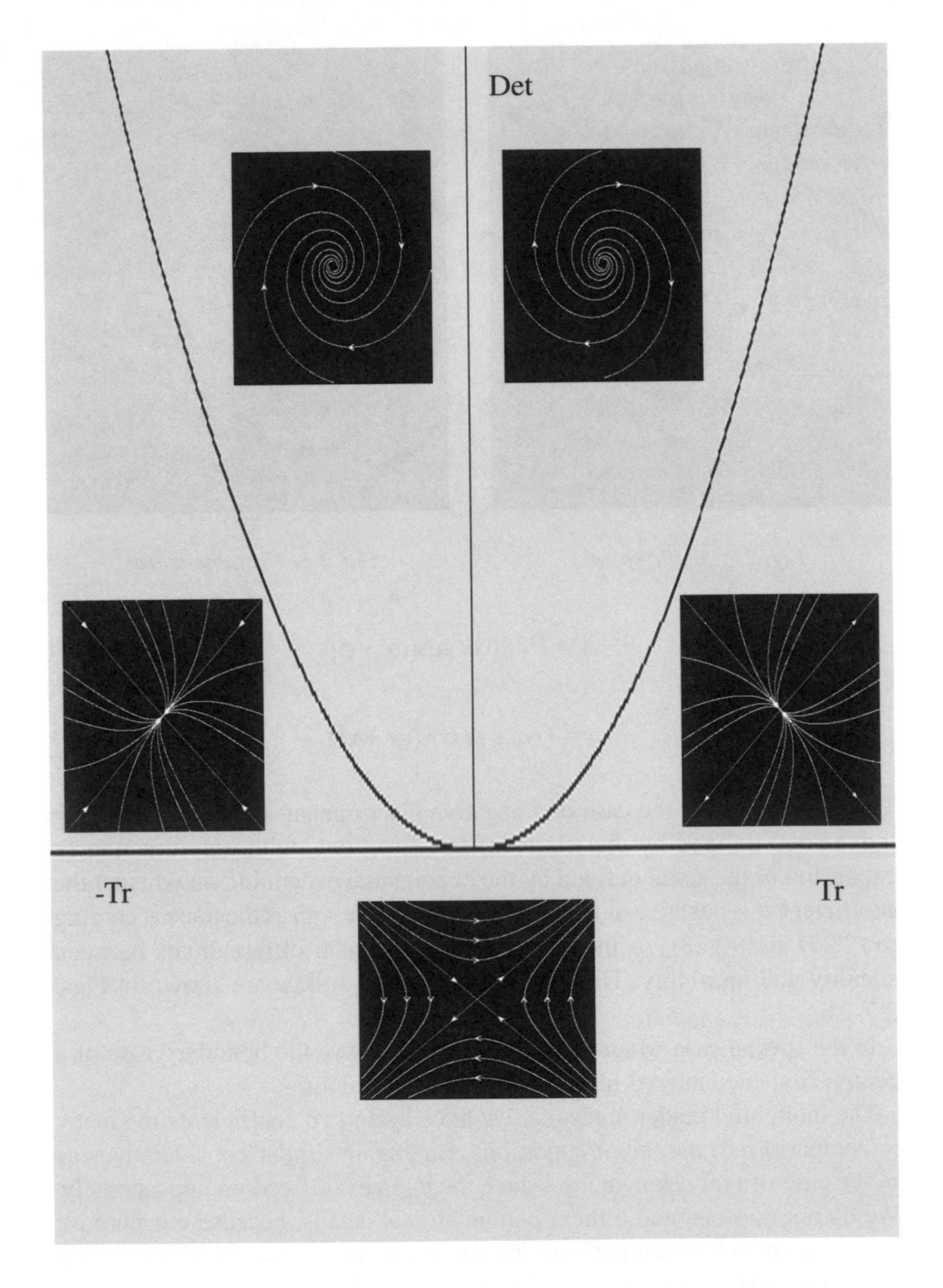

Fig. 2.9. Determinant versus trace for linear system, and parabola of zero discriminant.

The difference between stability and instability is made by the sign of the trace, if negative we have stability, if positive instability. The sign of the discriminant determines whether we have a node or a spiral.

The saddle, by its very nature, is unstable, but as we have seen it can become a watershed between different basins of attraction if there are several attractors. We just saw that in linear systems there are not, but in general there is such coexistence of attractors. The distinguishing feature of a saddle is a negative determinant.

We also dealt with the centre as a special case when the trace is zero. There are more such cases, like lines of accumulating singularities, but they all arise only for special isolated values of the trace, determinant, or discriminant.

There are various names for such isolated cases, such as structurally unstable, or nongeneric, and the general idea is that they can be ignored as they are unlikely to arise on mere probabilistic grounds due to their association with specific parameter values. We met the idea in discussing Arnol'd's point above, and we will dwell on it anew.

Fig. 2.9 summarises the information. We have the trace on the horizontal axis, and the determinant on the vertical. Accordingly, saddles occur below the horizontal zero line for the determinant. Stability for nodes and spirals is to the left of the vertical line at zero trace, instability to the right.

The discriminant, according to equation (2.24), is zero along the parabola $\text{Det} = \text{Tr}^2 / 4$, also shown in Fig. 2.9. Above it there are spirals, below it nodes. The location of the various (generic, or structurally stable) singular points in terms of this determinant/trace diagram is shown by the small inset phase portraits.

That is indeed very much to know about the linear system. The question now is: How much do we know about a general nonlinear system! The answer is very disappointing: We know nothing at all! There may be an infinity of singularities, even accumulating along lines or over areas, and the singularities may look like anything, for instance being composites of any number of parabolic and hyperbolic sectors.

That is, if we do not restrict the systems in any way. The really fantastic thing is that Peixoto in 1977 managed to show that the reasonable and really nonrestrictive assumption of structural stability makes things almost as simple for the general nonlinear system as they are for linear systems, especially for differential equations in the plane.

There may now be several singularities, but they are finite in number. Moreover, they belong to the five generic types already encountered in the discussion of the linear systems, nothing else.

He also proved the global result that there may be no heteroclinic or homoclinic trajectories in a structurally stable system, such as we encountered in the case of the undamped pendulum.

These results are so fantastic that we have to put the concept of structural stability down in more detail.

2.3 Structural Stability

To define structural stability formally for a system such as (2.4)-(2.5), which we repeat here for convenience:

$$\dot{x} = f(x, y) \tag{2.34}$$

$$\dot{y} = g(x, y) \tag{2.35}$$

we need two other concepts: topological equivalence, and perturbation.

The concept of perturbation is used to define what we mean by a small change to a model. Consider, along with (2.34)-(2.35) another slightly different pair of differential equations:

$$\dot{x} = F(x, y) \tag{2.36}$$

$$\dot{y} = G(x, y) \tag{2.37}$$

such that in any point x, y

$$|F - f| < \varepsilon, \qquad |G - g| < \varepsilon \tag{2.38}$$

and

$$\left|\frac{\partial F}{\partial x} - \frac{\partial f}{\partial x}\right| < \varepsilon, \qquad \left|\frac{\partial F}{\partial y} - \frac{\partial f}{\partial y}\right| < \varepsilon \tag{2.39}$$

$$\left|\frac{\partial G}{\partial x} - \frac{\partial g}{\partial x}\right| < \varepsilon, \qquad \left|\frac{\partial G}{\partial y} - \frac{\partial g}{\partial y}\right| < \varepsilon \tag{2.40}$$

Fig. 2.10. Topological equivalence.

hold. The system (2.36)-(2.37) differs from (2.34)-(2.35) so little that both the corresponding right hand sides, and their first derivatives differ by less than a small number ε. The inclusion of just the first derivatives but no higher ones is a technical point that is not restrictive, and it makes the concept of perturbation work for our purposes in a way that corresponds to intuition.

Next on the programme is to explain the concept of topological equivalence. Any system of differential equations, such as (2.34)-(2.35) or (2.36)-(2.37), defines a set of solution curves in the plane, i.e. the flow or phase portrait we have been discussing.

In general the phase portraits defined by systems such as (2.34)-(2.35) or (2.36)-(2.37) will look different, but it is possible that they can still be similar in a topological sense. We can imagine any one of the portraits drawn on a sheet of rubber and continuously deformed into the other by stretching as long as it does not lead to tearing. If their relation is one of such rubber sheet equivalence the flow portraits are said to be topologically equivalent.

Fig. 2.11. Topological nonequivalence. Splitting singularity.

Fig. 2.10 illustrates a case of topologically equivalent flows, where the flow, organized around a node and a saddle point, on the bottom picture has been smoothly transformed into the top picture.

There is, of course, also a formal definition. Topological equivalence means that any of the flow portraits can be mapped onto the other, by a continuous one-to-one mapping, so that trajectories are mapped on trajectories, and singular points on singular points, the directions of the trajectories and the characters of the singular points being preserved.

In Figs. 2.11 and 2.12 we see flows on the top and bottom sheets that, in contrast to Fig. 2.10, are not topologically equivalent. In Fig. 2.11 there is one complex type of saddle with six sectors, a so called monkey saddle, on the bottom, and two separate ordinary saddles, with four sectors each on the top. There is no way to map two singularities onto one single of a different type by a continuous coordinate transformation. Similarly, in Fig. 2.12, on the bottom, there is one single trajectory that joins two saddle points. On the top this trajectory has split in two.

Fig. 2.12. Topological nonequivalence. Splitting trajectory.

We are now ready for the definition of structural stability. A model, such as represented by (2.34)-(2.35), is structurally stable, provided that a small perturbation of the model only leads to a change of its phase portrait which is related to the original one by topological equivalence.

Trajectories may be deformed, rotated and translated in space, as in Fig. 2.10, but they do not split, reverse direction, or disappear, as in Fig. 2.12. Singular points may likewise be moved and deformed, but they retain their character, and do not split or disappear either, as they do in Fig. 2.11. Accordingly, the bottom picture of Fig. 2.10 is structurally stable, whereas those of Figs. 2.11-2.12 are structurally unstable.

To make things precise we should say that the system leading to the monkey saddle flow in the bottom picture of Fig. 2.11 arises from the system of differential equations $\dot{x} = x^2 - y^2, \dot{y} = -2xy$, whereas the transformations to the top pictures of Figs. 2.11-2.12 arise through the addition of small constants: $\dot{x} = x^2 - y^2 + \delta, \dot{y} = -2xy + \varepsilon$. The monkey saddle splits in two different saddles if only one of the constants δ, ε becomes nonzero, whereas the heteroclinic trajectory joining them splits if both become nonzero.

The reward in terms of information, issuing from the assumption of structural stability is rich: There are finitely many singular points, and they have the same elementary type as those of linear systems - for our purpose they are nodes, spirals, and saddles. Points such as the monkey saddle in Fig. 2.11 are ruled out. Moreover, there is the global result that saddle points can never be directly connected by a heteroclinic saddle connection. Accordingly, the saddle connection in Fig. 2.12 is ruled out.

2.4 Limit Cycles

This however is not all that can occur in two dimensions. Suppose we have the system:

$$\dot{x} = \left(1 - \sqrt{x^2 + y^2}\right)x + y \tag{2.41}$$

$$\dot{y} = \left(1 - \sqrt{x^2 + y^2}\right)y - x \tag{2.42}$$

It is easiest to understand this system if we transform it to polar coordinates:

$$x = \rho \cos\theta \tag{2.43}$$

$$y = \rho \sin\theta \tag{2.44}$$

In the new coordinates the system (2.41)-(2.41) reads:

$$\dot{\rho} = (1 - \rho)\rho \tag{2.45}$$

$$\dot{\theta} = -1 \tag{2.46}$$

Written in this form the system is suitable for a very easy qualitative analysis: We see that in the angular coordinate there is a constant rotational motion everywhere in the phase diagram, in the negative, i.e., clockwise sense. As for the radius coordinate, its rate of change is negative whenever the radius is larger than 1, positive whenever the radius is smaller than 1. So, there is asymptotic approach to unit radius both from below and from above, and we expect the combination to be a spiralling motion towards a closed orbit at unit distance from the origin of phase space.

The system is so simple that it can be integrated right away. The second equation is linear, the first is of the logistic type which can be solved by separation of variables and expansion in partial fractions. Even though our purpose is not computational, it is instructive to follow the details. Hence we first write (2.45) as:

$$\frac{d\rho}{(1-\rho)\rho} = dt \tag{2.47}$$

which has the variables separated. Next, we note that (2.47) can be expanded into:

$$\frac{d\rho}{\rho} + \frac{d\rho}{(1-\rho)} = dt \tag{2.48}$$

which can be integrated right away and yields:

$$\ln \rho - \ln(1-\rho) = t - \ln(-K) \tag{2.49}$$

Of course K is a constant of integration, and it is entered by its logarithm and with the minus signs for convenience. Exponentiating (2.49) we get:

$$\frac{\rho}{1-\rho} = -\frac{\exp t}{K} \tag{2.50}$$

or, solving for ρ ,

$$\rho = \frac{1}{1 - K\exp(-t)} \tag{2.51}$$

Fig. 2.13. Approch to limit cycle.

This is the well known solution to the logistic equation. As for the angular coordinate we may write the solution down immediately:

$$\theta = \phi - t \tag{2.52}$$

where ϕ is another integration constant, an arbitrary phase lead.

Equations (2.51)-(2.52) indeed corroborate what we inferred from general considerations about the phenomena in the phase diagram.

We now easily find the solution of (2.41)-(2.42) in our original coordinates:

$$x = \frac{\cos(\phi - t)}{1 - K\exp(-t)} \tag{2.53}$$

$$y = \frac{\sin(\phi - t)}{1 - K\exp(-t)} \tag{2.54}$$

It is easy to see that there is an asymptotic behaviour of the system (2.53)-(2.54). With increasing time the second term in the denominator simply vanishes and we have $x = \cos(\phi - t)$, $y = \sin(\phi - t)$. Taking squares and adding, we get $x^2 + y^2 = 1$, which defines the unit circle which is the limit cycle.

Note that Fig. 2.13 has a certain likeness to both Figs. 2.1 and 2.3. There is a closed orbit as in Fig. 2.1, but only one, not a concentric family. The rest of the curves are spirals, as in Fig. 2.3. They, however, do not spiral towards a fixed point, but towards the limit cycle.

The limit cycle thus is a new sort of attractor present in two dimensions along with the fixed points. Unlike the case of a centre with a family of concentric orbits, the limit cycle is a structurally stable object in the phase space. Accordingly, it too is admitted along with the various fixed points in structurally stable systems according to Peixoto's Characterisation Theorem.

There is a powerful tool in connection with the limit cycles, which normally goes under the name of the Poincaré-Bendixon Theorem. The theorem specifies conditions for the existence of limit cycles. In brief it says the following:

If we can define an annular region in phase space such that the direction field is oriented inwards towards the annular region everywhere on its boundary, and if there are no fixed points in the annular region, then there exists a limit cycle in it. Such an annular region can be the area enclosed by for instance two ellipses. Let us check the use of the Poincaré-Bendixon Theorem on our exemplificatory case (2.41)-(2.42). Let us delimit the area by two circles with radii r and R respectively. The lower case radius is smaller than the upper case.

A simple way of checking the direction of a vector field in relation to a boundary curve is to check the sign of the projection onto the normal of the boundary curve. If it is positive the direction is outward, if it is negative it is inward. For the boundary circles we deal with, the unit normals are just the vectors $(x / r, y / r)$ and $(x / R, y / R)$. From (2.41)-(2.42) we then have:

$$(\dot{x}, \dot{y}) \cdot \left(\frac{x}{r}, \frac{y}{r} \right) = (1 - r) r \tag{2.55}$$

$$(\dot{x}, \dot{y}) \cdot \left(\frac{x}{R}, \frac{y}{R} \right) = (1 - R) R \tag{2.56}$$

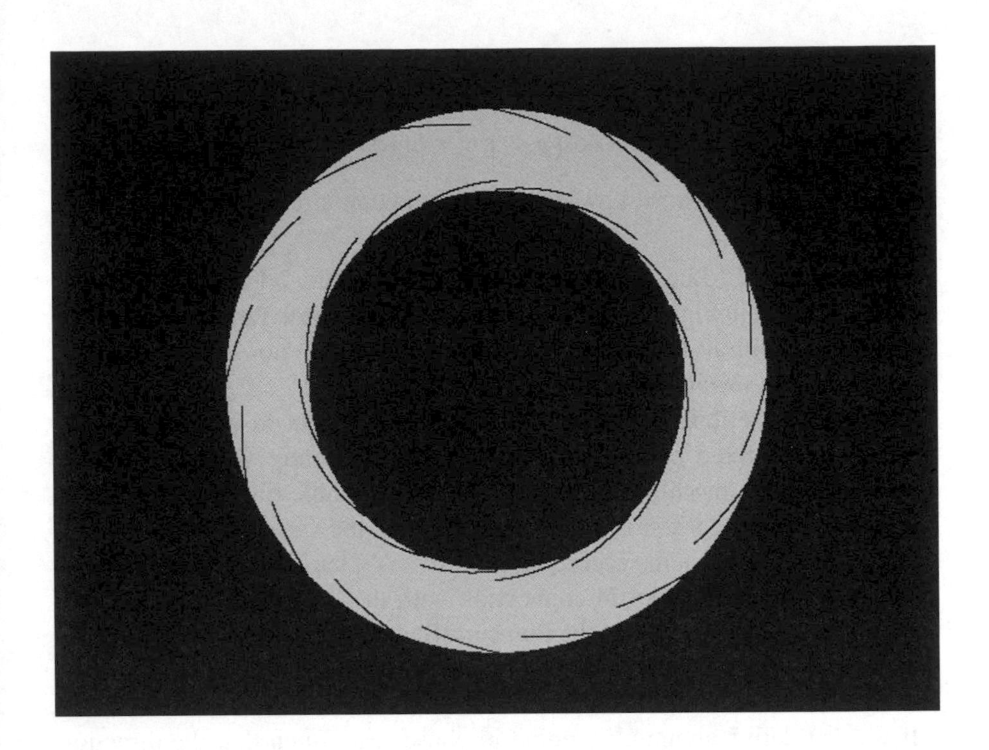

Fig. 2.14. The Poincaré-Bendixon Theorem.

If we now choose any $r < 1$ and any $R > 1$, then we note that (2.55) is positive and (2.56) is negative. Accordingly, the direction field is inward on the outer boundary and outward on the inner, as shown in Fig. 2.14 and as required by the theorem.

We thus have the annular region, and the correct direction for the vector field, so what about fixed points? It is easy to see that (2.41)-(2.42) only has one fixed point: $x = y = 0$. This is the origin of coordinate space, and not included in the annular region. Hence the annulus contains no fixed point, so there is a limit cycle.

In our case we could even use the theorem to find the limit cycle. The conditions for the direction field hold according to (2.55)-(2.56) no matter how close we let the boundary circles creep to the limit cycle. We can hence let the annulus shrink to a circle! In general, of course, things are not that simple. Neither is it so simple to find the suitable annular region. The tricky part usually is to define the annular region itself.

2.5 The Hopf Bifurcation

Consider the system (2.41)-(2.42). The first term is responsible for the fact that the system gains energy inside the unit circle $x^2 + y^2 = 1$, and loses it outside. Hence the system becomes damped far out in the phase diagram, but explosive close to the origin. This in fact makes the limit cycle attractive. It also makes the fixed point at the origin unstable. The linear last terms of opposite sign just make the system rotating.

Suppose now that we reverse the sign of thc first term in (2.41)-(2.42):

$$\dot{x} = -\left(1 - \sqrt{x^2 + y^2}\right)x + y \tag{2.57}$$

$$\dot{y} = -\left(1 - \sqrt{x^2 + y^2}\right)y - x \tag{2.58}$$

Then we see that the system instead loses energy inside the unit circle, whereas it gains energy outside it. We would hence expcct the fixed point to have become attractive, and the system to be explosivc outside the unit circle. That circle would hence seem to separate the interior attraction basin of the origin from the outside where the system explodes. The limit cycle seems to have become unstable and work as a basin boundary just as a stable trajectory to a saddle point.

It is equally easy to solve (2.57)-(2.58) as it was to solve (2.41)-(2.42). Reverting to polar coordinates we have:

$$\dot{\rho} = -(1 - \rho)\rho \tag{2.59}$$

$$\dot{\theta} = -1 \tag{2.60}$$

The solution of (2.60) is (2.52) as before, whereas for (2.59) we note that, if we just replace *dt* by *-dt* everywhere in (2.47)-(2.48) and *t* by *-t* in (2.50)-(2.51), then nothing else needs to be altered in the derivation or reasoning. Hence the counterparts of (2.53)-(2.54) are:

$$x = \frac{\cos(\phi - t)}{1 - K\exp(t)} \tag{2.61}$$

$$y = \frac{\sin(\phi - t)}{1 - K\exp(t)} \tag{2.62}$$

We see that the denominator no longer vanishes with time, but grows without limit. Accordingly, if $K < 0$ then the system approaches the origin in a spiralling motion. If $K > 0$ then the system explodes in finite time, more precisely when $t = -\ln K$. Taking squares of (2.61)-(2.62) for $t = 0$ and adding we get:

$$x_0^2 + y_0^2 = \frac{\cos^2\phi + \sin^2\phi}{(1-K)^2} = \frac{1}{(1-K)^2} \tag{2.63}$$

where x_0, y_0 denote the initial conditions. We hence see that the sign of K indeed depends on whether we start inside or outside the unit circle.

Equations (2.61)-(2.62) hence corroborate all that was conjectured. Reversing the signs of the first terms of (2.41)-(2.42) thus makes the stable limit cycle unstable, shifting its function from that of an attractor to that of a basin boundary. At the same time the unstable fixed point at the origin is made stable. Accordingly the attractor shifts from being a limit cycle to being a fixed point.

There is a special name for such shifts or bifurcations, to use the technically correct term: Hopf Bifurcation. As a matter of fact the coefficient in (2.41)-(2.42) need not make a dramatic shift all the way from +1 to -1.

A general way to write the systems (2.41)-(2.42) and (2.57)-(2.58) is:

$$\dot{x} = k\left(1 - \sqrt{x^2 + y^2}\right)x + y \tag{2.64}$$

$$\dot{y} = k\left(1 - \sqrt{x^2 + y^2}\right)y - x \tag{2.65}$$

with a parameter k. The bifurcation occurs when the parameter shifts by any tiny step from positive to negative. The parameter k thus by passing from negative to positive values makes the system jump from a fixed point to a limit cycle. The reverse change makes the system jump from limit cycle to fixed point again.

We recall from our discussion of the linear system that the intermediate case with $k = 0$ is the structurally unstable case of a centre.

It is also worthwhile noting that the jump is from a fixed point at origin to a cycle with unit radius. Such a bifurcation is called a hard bifurcation. The limit cycle could also have sneaked in with a zero radius from the beginning, only growing in size with the parameter. Such a bifurcation where initially the cycle with zero radius cannot be distinguished from the fixed point is called soft. The Hopf Bifurcation will be encountered repeatedly in the sequel.

2.6 The Saddle-Node Bifurcation

We can now use the same model to illustrate another important bifurcation, the saddle-node bifurcation. Suppose we have $k = 1$, as in (2.41)-(2.42), so that we deal with an attractive limit cycle. But we add a few further terms preceded by a parameter for which we use the same symbol as before:

$$\dot{x} = \left(1 - \sqrt{x^2 + y^2}\right)x + y - 2kxy^2 \tag{2.66}$$

$$\dot{y} = \left(1 - \sqrt{x^2 + y^2}\right)y - x + 2kx^2y \tag{2.67}$$

Again it is better to revert to the polar coordinates defined in (2.43)-(2.44). Then we have:

$$\dot{\rho} = (1 - \rho)\rho \tag{2.68}$$

$$\dot{\theta} = -1 + k\rho^2 \sin 2\theta \tag{2.69}$$

It is obvious that (2.68) is identical with (2.45), and that it therefore has the closed form solution (2.51). As for (2.69) it is now much more complicated than (2.46).

Let us first find the fixed points of the system (2.66)-(2.67). From (2.68) we find that either $\rho = 0$ or $\rho = 1$. In the first case, from (2.43)-(2.44) we have $x = y = 0$, which is the known fixed point at the origin. Substituting $\rho = 1$ in (2.66)-(2.67) we obtain:

$$2kxy = 1 \tag{2.70}$$

We also have:

$$x^2 + y^2 = 1 \tag{2.71}$$

which is just another way of writing $\rho = 1$, so from (2.70)-(2.71) we get:

$$x = \pm\frac{1}{\sqrt{2}}\left(1 \pm \sqrt{1 - \frac{1}{k^2}}\right) \tag{2.72}$$

$$y = \pm\frac{1}{\sqrt{2}}\left(1 \mp \sqrt{1 - \frac{1}{k^2}}\right) \tag{2.73}$$

Because the signs are always reversed for x and y this in all yields four different points on the unit circle. That is: provided that $k > 1$, otherwise there are no more fixed points than the origin. The emergence of the new fixed points is a sign of a bifurcation. Linearising (2.66)-(2.67) and substituting from (2.72)-(2.73) we find that they are alternating stable nodes and saddle points. This is illustrated in Fig. 2.15.

As the four new fixed points are located on the previous limit cycle, we may suspect that it has evaporated. This is indeed so. What happens is a so called saddle/node bifurcation, more precisely a global saddle node bifurcation occurring at several points simultaneously. In the following chapters we will find examples of such global saddle/node bifurcations. What happens is that for certain parameter values there arise pairs of coincident nodes and saddles on the limit cycle, which separate with further parameter changes. The cycle is hence broken and the trajectories take their path to one or the other of the stable nodes.

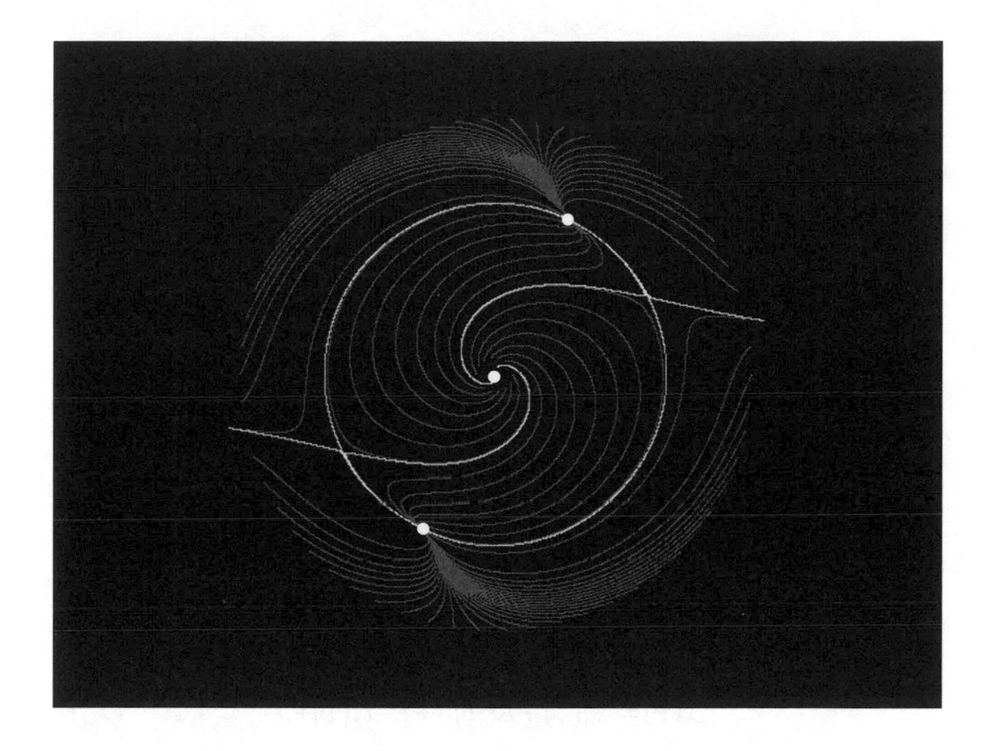

Fig. 2.15. Limit cycle undergiong global saddle/node bifurcation.

In a total scenario it may happen that a fixed point first loses stability in a Hopf Bifurcation and is replaced by a limit cycle, which then itself loses stability in a saddle/node bifurcation, and is replaced by a number of fixed points. Those stable fixed points, which may be any number in a global bifurcation, may themselves be further replaced by limit cycles in further Hopf Bifurcations, and so on.

2.7 Perturbation Methods: Poincaré-Lindstedt

Before the advent of computers and simulations the perturbation methods provided the most powerful tools for getting information about differential equations that could not be solved in closed form. These methods still are most useful for understanding nonlinear differential equations. We already encountered the concept of perturbation in the context of structural stability, but now the use is slightly different.

The idea is to regard a nonlinear equation as a slight modification of a linear equation whose solution we know. Suppose we deal with the equation for the pendulum $\ddot{x} + \sin x = 0$. We take the two first terms of the Taylor series for the sine: $\sin x \approx x - x^3/6$, and rescale the variable x by multiplication with $\sqrt{6}$. In this way we get:

$$\ddot{x} + x = x^3 \tag{2.74}$$

The left hand side is the simple harmonic oscillator whose solution we know, so suppose we regard the nonlinearity in the right hand side as small and write:

$$\ddot{x} + x = \varepsilon x^3 \tag{2.75}$$

Let us try a power series solution in the form:

$$x(\tau) = x_0(\tau) + \varepsilon x_1(\tau) + \varepsilon^2 x_2(\tau) + \ldots \tag{2.76}$$

where

$$\tau = \left(1 + \varepsilon\omega_1 + \varepsilon^2\omega_2 + \ldots\right)t \tag{2.77}$$

is a new time scale, the correcting terms of which are to be determined in the solution process.

We substitute the attempted solution and the new time variable into the differential equation, which itself becomes a power series in ε. From (2.76) and (2.77), by differentiation with respect to t, and using the chain rule:

$$\ddot{x} = \left(x_0'' + \varepsilon x_1'' + \varepsilon^2 x_2'' + \ldots\right)\left(1 + \varepsilon\omega_1 + \varepsilon^2\omega_2 + \ldots\right)^2 \tag{2.78}$$

To make things quite clear we denote differentiation with respect to real time t by dots, with respect to the new time variable τ with dashes. In order to see the power series character let us expand (2.78) in powers of ε, just keeping the first three terms. Thus:

$$\ddot{x} = x_0'' + \varepsilon\left(2\omega_1 x_0'' + x_1''\right) + \varepsilon^2\left(\left(\omega_1^2 + 2\omega_2\right)x_0'' + 2\omega_1 x_1'' + x_2''\right) \tag{2.79}$$

Likewise we deal with the third power of x, which we write as a series, though keeping in mind that the right hand side of (2.75) already contains one extra power of ε, so we just keep the first two terms:

$$\varepsilon x^3 = \varepsilon x_0^3 + \varepsilon^2 3x_0^2 x_1 \tag{2.80}$$

We now substitute from (2.76), (2.79) and (2.80) into the differential equation (2.75). The strategy is to require that the differential equation holds for all the powers separately. Never mind why, if it works then the power series is a solution. In this way we arrive at the following three equations:

$$x_0'' + x_0 = 0 \tag{2.81}$$

$$x_1'' + x_1 = -2\omega_1 x_0'' + x_0^3 \tag{2.82}$$

$$x_2'' + x_2 = -\left(\omega_1^2 + 2\omega_2\right)x_0'' - 2\omega_1 x_1'' + 3x_0^2 x_1 \tag{2.83}$$

We could have added any number of equations for higher powers. The interesting thing is that the equations form a sequence of linear differential equations which may be solved by standard methods. Noteworthy is that the right hand sides of the non-homogenous equations depend entirely on solutions to the earlier equations in the sequence.

Having obtained the differential equations in this form we next have to deal with the initial conditions. The formulation of the initial conditions for a series solution are a matter of convenience, so let us just put:

$$x_0(0) = A \tag{2.84}$$

$$x_i(0) = 0 \tag{2.85}$$

for $i > 0$, and

$$\dot{x}_i(0) = 0 \tag{2.86}$$

for all i. Given the initial conditions, we can now start solving the equations. First, we get from (2.81):

$$x_0 = A\cos\tau \tag{2.87}$$

because the zero derivative according to initial conditions (2.86) makes the sine term drop out. From the solution (2.87) we moreover have:

$$x_0'' = -A\cos\tau \tag{2.88}$$

and

$$x_0^3 = A^3\cos^3\tau = \frac{3}{4}A^3\cos\tau + \frac{1}{4}A^3\cos 3\tau \tag{2.89}$$

Substituting from (2.88)-(2.89) in the right hand side of (2.82), the next differential equation in the chain, we get:

$$x_1'' + x_1 = A\left(\frac{3}{4}A^2 + 2\omega_1\right)\cos\tau + \frac{1}{4}A^3\cos 3\tau \tag{2.90}$$

The core of this Poincaré-Lindstedt method is to realize that, if the periodic solution is to work, then we cannot have terms such as $\cos\tau$ in the right hand side, because when solved they would lead to terms of the form $\tau\sin\tau$, which would grow beyond limit. Those "secular terms" were eliminated by assuming that the coefficients had to be zero. To this end:

$$\omega_1 = -\frac{3}{8}A^2 \tag{2.91}$$

would have to hold. What remains of equation (2.82) is then:

$$x_1'' + x_1 = \frac{1}{4}A^3\cos 3\tau \tag{2.92}$$

This non-homogenous equation has a solution that is a sum of the solutions of any particular solution and the solution to the non-homogenous equation. A particular solution is:

$$x_1 = -\frac{1}{32} A^3 \cos 3\tau \tag{2.93}$$

The homogeneous solution is in terms of $\cos\tau$, but in order that the sum of the solutions match the initial conditions (2.85) we must have:

$$x_1 = \frac{1}{32} A^3 (\cos\tau - \cos 3\tau) \tag{2.94}$$

Now the following steps are similar to the previous, just progressively more messy. Using the first two solutions (2.87) and (2.94) in the right hand side of (2.83) we get:

$$x_2'' + x_2 = A\left(\frac{21}{128} A^4 + 2\omega_2\right)\cos\tau \tag{2.95}$$

$$+\frac{24}{128} A^5 \cos 3\tau - \frac{3}{128} A^5 \cos 5\tau$$

Again we cannot have the first secular term, so the parenthesis must be zero, which gives us the next correction for the period:

$$\omega_2 = -\frac{21}{256} A^4 \tag{2.96}$$

Given this the differential equation (2.83) is simplified to:

$$x_2'' + x_2 = \frac{24}{128} A^5 \cos 3\tau - \frac{3}{128} A^5 \cos 5\tau \tag{2.97}$$

and the particular solution is:

$$x_2 = -\frac{3}{128} A^5 \cos 3\tau + \frac{1}{1024} A^5 \cos 5\tau \tag{2.98}$$

The full solution again includes the basic frequency, but in view of the zero initial condition (2.85) for all higher powers we must have:

$$x_2 = \frac{23}{1024} A^5 \cos \tau - \frac{3}{128} A^5 \cos 3\tau + \frac{1}{1024} A^5 \cos 5\tau \tag{2.99}$$

Thus each new solution x_i adds a new (odd) harmonic, but it also adds new correcting coefficients to the lower harmonics. We note that the coefficients become progressively smaller for each new solution, something which favours convergence. Moreover, from the attempted series solutions (2.76), we find that an $\varepsilon < 1$ promotes convergence as well, just because each new solution is added to the series preceded by an increasing power of epsilon. Though the perturbation methods are developed for really small epsilons, it is true that even for our original equation (2.74) with $\varepsilon = 1$ things need not be too bad thanks to the progressive decrease of the numerical coefficients.

On the other hand, an initial amplitude larger than unity, i.e. $A > 1$, can easily destroy convergence. So, small amplitude oscillations are a condition for the usefulness of the method.

Let us now collect results from (2.77), (2.91) and (2.96) and assume $\varepsilon = 1$, as in our original equation. Then we have to our last approximation:

$$\tau = \left(1 - \frac{3}{8} A^2 - \frac{21}{256} A^4\right) t \tag{2.100}$$

Accordingly, the frequency depends on amplitude, larger oscillations taking longer time. The parenthesis becomes zero for $A \approx 1.37$. This represents the critical slowing down of the process when we approach the homoclinic saddle trajectory shown in Fig. 2.1, which we indicated would take infinite time. For such a value of the amplitude the convergence would however be really bad.

In Fig. 2.16 we display the approximations to the solution phase portrait with the three first equations. As we see the inner curves are rather round, as the higher harmonics count little, whereas the outer ones in fact resemble the shape of homoclinic trajectories. As a matter of fact, the picture is not so

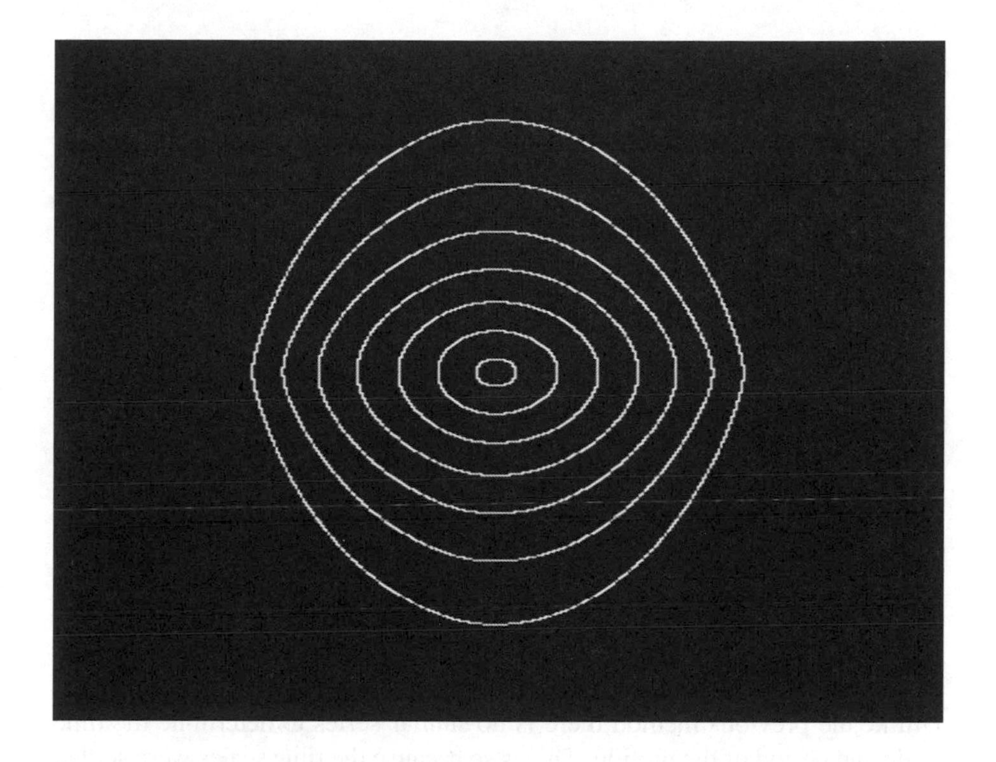

Fig. 2.16. Closed orbits obtained by the Poincaré-Lindstedt method.

unlike the central part of Fig. 2.1 above. More about the Poincaré-Lindstedt and other perturbation methods, of which we will present one more below, can be found in Stoker, Hayashi, and Jordan and Smith.

2.8 Perturbation Methods: Two-Timing

Suppose now that we intend to solve the differential equation:

$$\ddot{x} - \left(1 - x^2\right)\dot{x} + x = 0 \tag{2.101}$$

which is the much studied van der Pol's equation. We see that it is a harmonic oscillator which includes a friction term. But the friction term is now nonlinear, and such that there is damping for large amplitudes only, but the reverse, i.e., antidamping for small. Accordingly, close to the unstable singular point at the origin of phase space there is a region where the system is supplied with energy, whereas it loses energy further off in phase space. We can thus expect that the system has some kind of limit cycle.

We could try to prove its existence by the Poincaré-Bendixon Theorem, or locate it by the Poincaré-Lindstedt method just discussed, but we will do neither. Instead we will use the equation in order to present another perturbation method, the so called two-timing method.

To this end we first again pretend that the nonlinearity is small, and write:

$$\ddot{x}+x=\varepsilon\left(1-x^{2}\right)\dot{x} \tag{2.102}$$

Further we assume two different time scales, fast time t and slow time $T=\varepsilon t$, and we again try a power series solution:

$$x=x_{0}(t,T)+\varepsilon x_{1}(t,T)+\varepsilon^{2}x_{2}(t,T)+\ldots \tag{2.103}$$

Unlike the previous method there is no similar series to determine the time scale and period of the motion. This is so because the time scales were settled beforehand.

We, however, need to specify what the time derivative in (2.102) actually is in view of the multiple time scales. To this end we define the compound differential operator:

$$\frac{d}{dt}=\frac{\partial}{\partial t}+\varepsilon\frac{\partial}{\partial T} \tag{2.104}$$

The derivatives in (2.102) are understood as compounds of this whole operator, and whenever applied to the functions $x_i(t,T)$ it is understood that we have to take both partial derivatives, even the one with respect to slow time, preceded by the smallness coefficient. In the following we also write differentiation with respect to fast time alone by dashes, whereas all the other derivatives are fully written out.

Substituting the series (2.103) into the differential equation (2.102) we, of course, again get the differential equation as a power series in ε. This time, however, the complexity of the expressions progresses much faster than with the previous method, so we only write down the two first, and so at most keep the terms linear in ε. Accordingly, we obtain:

$$\ddot{x} = x_0'' + \varepsilon\left(2\frac{\partial^2 x_0}{\partial t \partial T} + x_1''\right) + \ldots \tag{2.105}$$

for the second derivative, and

$$\varepsilon\left(1 - x^2\right)\dot{x} = \varepsilon\left(1 - x_0^2\right)x_0' + \ldots \tag{2.106}$$

for the nonlinear friction term.

Substituting from (2.103), and (2.105)-(2.106) into (2.102), and again assuming that the coefficients of all the powers of ε vanish individually, we get the first two in the recursive sequence of differential equations:

$$x_0'' + x_0 = 0 \tag{2.107}$$

$$x_1'' + x_1 = \left(1 - x_0^2\right)x_0' - 2\frac{\partial^2 x_0}{\partial t \partial T} \tag{2.108}$$

Now, the solution to the first equation is obvious:

$$x_0 = A(T)\cos t + B(T)\sin t \tag{2.109}$$

The differential equation we have solved is actually a partial differential equation with respect to fast time, and though the arbitrary coefficients of the sine and cosine terms are constant with respect to fast time, they may depend on slow time, which has been taken in consideration in the solution. We will next use the second equation for determining these slowly varying coefficients, though we will not even attempt to solve that equation itself. It can be done, though, and for details the reader is referred to Kevorkian and Cole.

For convenience we note that from the solution (2.109):

$$\frac{\partial^2 x_0}{\partial t \partial T} = -\frac{dA}{dT}\sin t + \frac{dB}{dT}\cos t \tag{2.110}$$

whereas

$$x_0' = -A\sin t + B\cos t \tag{2.111}$$

We observe that the expression $x_0^2 x_0'$ upon substitution from (2.109) and (2.111) results in cubic expressions in the sines and cosines. To reduce such powers to the basic oscillation frequency and its harmonics we need the following trigonometric identities: $\cos^3 t = (3/4)\cos t + (1/4)\cos 3t$ and $\sin^3 t = (3/4)\sin t - (1/4)\sin 3t$. Further we also need corresponding expansions for the mixed cubics $\cos^2 t \sin t = (1/4)\sin t + (1/4)\sin 3t$ and $\cos t \sin^2 t = (1/4)\cos t - (1/4)\cos 3t$.

Using these identities along with (2.109) and (2.110)-(2.111) in (2.108), we now get:

$$x_1'' + x_1 = \tag{2.112}$$

$$-\left(2\frac{dB}{dT} - B + \frac{1}{4}(A+B)B\right)\cos t - \frac{1}{4}\left(B^2 - 3A^2\right)B\cos 3t$$

$$+\left(2\frac{dA}{dT} - A + \frac{1}{4}(A+B)A\right)\sin t - \frac{1}{4}\left(A^2 - 3B^2\right)A\sin 3t$$

The reasoning is as before. We cannot have the sines and cosines of the basic frequency in the right hand side, because this would result in secular terms and contradict the assumption that we have a periodic or even bounded solution. Therefore the coefficients of those terms must be put equal to zero.

This, however, now results in two differential equations, not in algebraic equations as before, thereby indicating that the coefficients of (2.109) indeed are no constants but vary with slow time. What we get is:

$$\frac{dA}{dT} = \frac{1}{2}A - \frac{1}{8}(A+B)A \tag{2.113}$$

$$\frac{dB}{dT} = \frac{1}{2}B - \frac{1}{8}(A+B)B \tag{2.114}$$

This system is very easy to deal with if we convert it into polar coordinates:

$$A = \mathrm{P}\cos\Theta \tag{2.115}$$

$$B = \mathrm{P}\sin\Theta \tag{2.116}$$

Then the system (2.113)-(2.114) becomes:

$$\frac{d\mathrm{P}}{dT} = \frac{1}{2}\mathrm{P} - \frac{1}{8}\mathrm{P}^3 \tag{2.117}$$

$$\frac{d\Theta}{dT} = 0 \tag{2.118}$$

The second equation (2.118) just results in a constant phase angle, whereas the first (2.117) can be handled by the separation of variables technique. We thus have:

$$dT = 8\frac{d\mathrm{P}}{\mathrm{P}(2-\mathrm{P})(2+\mathrm{P})} = 2\frac{d\mathrm{P}}{\mathrm{P}} - \frac{d\mathrm{P}}{\mathrm{P}-2} - \frac{d\mathrm{P}}{\mathrm{P}+2} \tag{2.119}$$

This is readily integrated to:

$$T = \ln K + 2\ln\mathrm{P} - \ln(\mathrm{P}-2) - \ln(P+2) \tag{2.120}$$

and exponentiation then yields:

$$\exp T = K\frac{\mathrm{P}^2}{\mathrm{P}^2-4} \tag{2.121}$$

Solving we get $\mathrm{P}^2 = 4/\left(1-K\exp(-T)\right)$, or, as only positive radii make any sense:

$$\mathrm{P} = 2\sqrt{\frac{1}{1-K\exp(-T)}} \tag{2.122}$$

Fig. 2.17. Approch to limit cycle obtained by two-timing.

We thus see that whereas the angle coordinate was constant, the radius coordinate for the slowly varying amplitudes goes asymptotically towards the value 2, no matter what its initial values was.

Though two-timing, due to progressive complexity, was used for one step only, so that we cannot expect much of convergence, we see that it has particular advantages pertaining to it. First, we automatically get an estimate for the amplitude of the limit cycle. Second, we get information about the transient process approaching the limit cycle. Third, by the last fact we get an automatic check on the stability of the limit cycle, which is valuable because there are, as we know, also unstable limit cycles.

Assembling all the pieces we can finally state the solution to the original van der Pol equation:

$$x = \frac{2\cos(t+\phi)}{\sqrt{1-K\exp(-\varepsilon t)}} \tag{2.123}$$

In Fig. 2.17 we display a number of different trajectories along with the limit cycle itself according to the approximate solution. As mentioned, we could do much better. We have not yet tried to solve the second differential equation (2.108) or (2.112) after the secular terms were eliminated, but the cost would be a lot more of computational work, so we stop here.

In our original system (2.101) we had $\varepsilon = 1$, i.e. there was actually no difference between fast time and slow time. This too might provide for poor convergence, but the qualitative results still hold.

The van der Pol oscillator also illustrates the Hopf Bifurcation. Suppose we put $\varepsilon = -1$, or equal to any negative value in equation (2.102). As the derivations in no way depend upon the sign of ε, we can just use the solution (2.123) as it stands.

In the intermediate case with $\varepsilon = 0$, we deal with the simple harmonic oscillator and a set of concentric closed orbits. The situations on either side of the zero value are, however, entirely different. We see from (2.102) that with a negative ε the system loses energy at small amplitudes, but gains energy at large, quite contrary to the case already discussed. Far out in the phase diagram the system explodes to infinity, whereas close to the origin it goes towards it. Looking at the final solution (2.123) we see that a negative ε reverses the direction of change, so that the amplitude deviation from the value 2 increases with time. Provided $K > 0$, we would sooner or later end up with having to take the square root of a negative number, but never mind, before that happens the process has gonc to infinity in finite time. The new situation is clear. The limit cycle still exists, but it has turned unstable. Any initial deviation from it will be enlarged in the course of time. So, the unstable limit cycle would act as a watershed between the attractors at the origin and at infinity. This is the Hopf Bifurcation as already encountered in Section 2.5.

2.9 Stability: Lyapunov's Method versus Linearisation

We should check that the origin indeed is an attractor while the limit cycle repels, and a repellor while the limit cycle attracts. To this end let us multiply through (2.102) by $\dot{x}$, thus obtaining:

$$\dot{x}\ddot{x} + x\dot{x} = \varepsilon\left(1 - x^2\right)\dot{x}^2 \tag{2.124}$$

The left hand side, however, is half the time derivative of the nonnegative sum of squares $\dot{x}^2 + x^2$, i.e.

$$\frac{1}{2}\frac{d}{dt}\left(\dot{x}^2 + x^2\right) = \varepsilon\left(1 - x^2\right)\dot{x}^2 \tag{2.125}$$

For an oscillator this nonnegative sum of squares represents total energy, kinetic plus potential. As, unlike the case of the undamped pendulum, the system is not conservative, energy will change during the course of time. Considering small amplitudes, $|x| < 1$, in the neighbourhood of the fixed point at the origin (in the stability in which we are at present interested) we see that the sign of the right hand side of (2.125) entirely depends on the sign of ε. If it is positive then the system gains energy, if it is negative, then the system loses energy.

Constant energy would be represented by closed orbits, to be exact, circles in phase space. When the system loses energy, then the spiralling trajectories always cross the circles of constant energy in an inward motion, and thus approach the origin asymptotically. This is to say that the origin is an attractor, a spiral sink when $\varepsilon < 0$. Likewise it is a repellor, a spiral source, when $\varepsilon > 0$.

What we have just done in equations (2.124)-(2.125) is to study the stability by Lyapunov's Direct Method. It is an efficient procedure, but it may be hard to apply in more difficult cases. It is about just as tricky to find the proper measure of "energy" as it was to find the proper annulus for the application of Poincaré-Bendixon's Theorem.

An alternative for studies of stability is linearisation. Though it gives no new interesting conclusions, it is instructive to use the method. We first write (2.102) as a pair of equations:

$$\dot{x} = y \tag{2.126}$$

$$\dot{y} = -x + \varepsilon\left(1 - x^2\right)y \tag{2.127}$$

Next substitute $x + \xi$ for x and $y + \eta$ for y in order to focus on small deviations, subtract the original system (2.126)-(2.127), and delete all powers and products of the small deviation variables ξ, η. In this way we arrive at:

$$\dot{\xi} = \eta \tag{2.128}$$

$$\dot{\eta} = -\xi + \varepsilon\left(-2xy\xi + \left(1 - x^2\right)\eta\right) \tag{2.129}$$

which is (2.126)-(2.127) linearised.

Suppose now that we want to study the neighbourhood of the origin. Accordingly put $x = y = 0$ in (2.128)-(2.129). Then the system simplifies to:

$$\dot{\xi} = \eta \tag{2.130}$$

$$\dot{\eta} = -\xi + \varepsilon\eta \tag{2.131}$$

which can be readily solved. Its characteristic equation is:

$$\lambda^2 - \varepsilon\lambda + 1 = 0 \tag{2.132}$$

and the eigenvalues are:

$$\lambda_1, \lambda_2 = \frac{\varepsilon}{2} \pm \frac{1}{2}\sqrt{\varepsilon^2 - 4} \tag{2.133}$$

Provided $|\varepsilon| < 2$ the discriminant is negative, and we have a spiral. This is actually the case we have been dealing with. What happens in the linearised system when the friction term becomes so large that the inequality $|\varepsilon| < 2$ is not fulfilled, is that the spiral is converted into a node. The interesting thing is the term $\varepsilon / 2$. If it is positive then the origin is unstable, if it is negative then the origin is stable. This corroborates what was concluded by Lyapunov's direct method, but yields no new results as anticipated.

Linearization is often done by replacing the given nonlinear system (2.126)-(2.127) by a linear system where the right hand side coefficients are the partial derivatives of the right hand sides of the nonlinear system. Thus by straightforward differentiation $\partial \dot{x} / \partial x = 0$ and $\partial \dot{x} / \partial y = 1$ from (2.126), whereas $\partial \dot{y} / \partial x = -1 - 2\varepsilon xy$ and $\partial \dot{y} / \partial y = \varepsilon\left(1 - x^2\right)$ from (2.127).

These provide precisely the right hand coefficients in equations (2.128)-(2.129).

Throughout the following text we will adhere to the first more direct procedure of introducing small deviation variables and deleting all powers and products of these small variables, whenever we need linearization. The methods are, however, equivalent and the choice a matter of taste.

2.10 Forced Oscillators, Transients, and Resonance

In what follows we are going to study forced oscillators. An oscillator is by its nature of second order, and forcing raises the order by one. Third order in continuous time processes is sufficient to produce chaos - provided the system is nonlinear. Indeed we are going to study two well known nonlinear cases: the van der Pol and Duffing models, but as we have not yet encountered forcing, we could as well introduce the concept in the most elementary of all cases: the simple harmonic oscillator.

We hence deal with

$$\ddot{x} + x + 2k\dot{x} = \Gamma \cos(\omega t) \tag{2.134}$$

where k denotes half the damping friction coefficient, Γ denotes the amplitude, and ω the frequency of the forcing term in the right hand side.

To see that the equation (2.134) is third order, note that, by defining the new variables $y = \dot{x}$ and $z = \omega t$, we can write it as:

$$\dot{x} = y \tag{2.135}$$

$$\dot{y} = -x - 2ky + \Gamma \cos z \tag{2.136}$$

$$\dot{z} = \omega \tag{2.137}$$

To start up the discussion, suppose the right hand side of (2.134) is zero, i.e. that there is no forcing at all. Then we have:

$$\ddot{x} + x + 2k\dot{x} = 0 \tag{2.138}$$

and, along the lines of Section 2.2, we try $x = A\exp(\lambda t)$ as a solution. Substituting for the attempted solution and its derivatives, and dividing through by the nonzero exponential, we obtain

$$\lambda^2 + 2k\lambda + 1 = 0 \tag{2.139}$$

which is the characteristic equation with eigenvalues: $\lambda_1, \lambda_2 = -k \pm \sqrt{k^2 - 1}$ or rather:

$$\lambda_1, \lambda_2 = -k \pm i\sqrt{1 - k^2} \tag{2.140}$$

whenever $k^2 < 1$ so that we indeed have an oscillatory solution. The complete solution then, put in trigonometric form, is:

$$x = C\exp(-kt)\cos(\overline{\omega}t + \phi) \tag{2.141}$$

There is exponential damping at the rate -*k* whenever the friction coefficient is positive. The cyclic motion has frequency $\overline{\omega} = \sqrt{1 - k^2}$, which is less than the unitary frequency of the undamped harmonic oscillator when $k = 0$. So, we see that friction, in addition to damping, also slows down the oscillations. If *k*, half the friction coefficient, approaches unity, then oscillations are slowed down critically, the cycle becomes infinitely long, and the trigonometric solution is no longer applicable. The amplitude *C*, and the phase lead ϕ are arbitrary constants whose values are fixed by the initial conditions. We recall that the choice of the cosine for the cyclic motion is arbitrary. As an alternative, expanding the solution with respect to ϕ produces sine and cosine terms, each with its arbitrary coefficient to accommodate the initial conditions.

This long recapitulation has relevance not only for the special case when forcing is not present, which is called the homogeneous case. It is part of the general solution to the original non-homogeneous equation (2.134) as well. To see this suppose we have found any solution $\overline{x}$ to (2.134). Such a solution is called a particular solution. In order to be a solution it fulfils:

$$\ddot{\overline{x}} + \overline{x} + 2k\dot{\overline{x}} = \Gamma\cos(\omega t) \tag{2.142}$$

Subtracting (2.142) from (2.134), defining $z=(x-\bar{x})$, and using the linearity property of the derivatives, we get $\ddot{z}+z+2k\dot{z}=0$, which is the homogeneous equation. Hence, as $x=\bar{x}+z$ according to assumption, we see that the solution to the homogeneous equation has to be added to any particular solution in order to arrive at the completely general solution of a non-homogeneous differential equation.

We already got the solution (2.141) to the homogeneous equation, sometimes called the complementary function. It now remains to find a particular solution. As we note that the complementary function vanishes when time goes to infinity, which is due to friction damping, we may guess that any persistent motion would be locked to the frequency of the driving force.

So let us try

$$\bar{x}=A\cos(\omega t)+B\sin(\omega t) \tag{2.143}$$

for a particular solution. Note that we have taken the frequency ω of the forcing term as given, but that we have introduced arbitrary amplitudes *A*, *B*. Differentiating the attempted solution we get

$$\dot{\bar{x}}=-\omega A\sin(\omega t)+\omega B\cos(\omega t) \tag{2.144}$$

$$\ddot{\bar{x}}=-\omega^2 A\cos(\omega t)-\omega^2 B\sin(\omega t) \tag{2.145}$$

Substituting these expressions (2.143)-(2.145) in (1.142) we have:

$$\begin{aligned}\left(A\left(1-\omega^2\right)+2kB\omega\right)\cos(\omega t)+\left(B\left(1-\omega^2\right)-2kA\omega\right)\sin(\omega t)\\ =\Gamma\cos(\omega t)\end{aligned} \tag{2.146}$$

In order that this be a solution the coefficients of the sine and cosine terms must vanish individually. This gives us two equations:

$$\left(1-\omega^2\right)A+2k\omega B=\Gamma \tag{2.147}$$

$$-2k\omega A+\left(1-\omega^2\right)B=0 \tag{2.148}$$

If we take squares of both sides and add we get the simple equation:

$$\left(A^2+B^2\right)\left(\left(1-\omega^2\right)^2+4k^2\omega^2\right)=\Gamma^2 \tag{2.149}$$

as the product terms cancel. Putting $A=K\cos\psi,\quad B=-K\sin\psi$, we get from (2.149)

$$K=\sqrt{A^2+B^2}=\frac{\Gamma}{\sqrt{\left(1-\omega^2\right)^2+4k^2\omega^2}} \tag{2.150}$$

and from (2.148)

$$\tan\psi=-\frac{B}{A}=\frac{2k\omega}{\omega^2-1} \tag{2.151}$$

and we can put the attempted solution (2.143) in the form:

$$\bar{x}=K\cos(\omega t+\psi) \tag{2.152}$$

where the amplitude and phase lead are determined by the dynamical system, through equations (2.150)-(2.151), not by the initial conditions as in the case of the complementary function (2.141). So, we actually found a particular solution.

Note from (2.150) that if we differentiate K with respect to ω, and solve, then we get the driving frequency $\omega=\sqrt{1-2k^2}$ at which the response amplitude is maximal. Substituting we find that this maximum value is $K=\Gamma/\left(2k\sqrt{1-k^2}\right)$, so it becomes a very sharp peak when friction approaches zero. Also note that this nonzero maximum resonance frequency only exists provided friction does not exceed a certain value, more precisely: $2k<\sqrt{2}\approx 1.4142$. Otherwise resonance amplitude K, according to (2.150) is a uniformly decreasing function of the driving frequency ω.

Adding the complementary function (2.141) and the particular solution (2.152), and using the expression (2.150) for the amplitude K, we finally get the general solution to the forced system:

$$x = C\exp(-kt)\cos(\overline{\omega}t + \phi) + \frac{\Gamma\cos(\omega t + \psi)}{\sqrt{(1-\omega^2)^2 + 4k^2\omega^2}} \tag{2.153}$$

We note the following interesting facts: 1) Only the complementary solution depends on the arbitrary initial conditions, which implies that those do not influence the long term behaviour of the system; 2) The complementary solution oscillates at the natural frequency, slightly slowed by friction, but it is damped out with time, and so it only contributes an initial so called transient; 3) In the long run only the motion of the particular solution remains, the system oscillating at the frequency of the forcing term; 4) The amplitude of the response to the forcing depends on how much the forcing frequency ω deviates from the unitary natural frequency of the undamped oscillator. 5) This is the phenomenon of resonance, the more the frequencies agree the larger is the response amplitude; 6) Only the presence of friction k saves the response from becoming infinite, and so blowing up the system at perfect resonance.

In general the addition of a forcing oscillation with another frequency than the natural frequency of the free oscillator itself gives rise to the phenomena of resonance, harmonic and also subharmonic (with any rational frequency to that of the driving force), used through ages in musical instruments. The vibrations in a violin string are not audible by themselves. It is only through the resonance in the bridge, the top and bottom plates, and the air cavity between them that the tone becomes powerful, and this makes all the difference between a million dollar Stradivarius and a plain instrument.

Resonance may also be a disadvantage. Military troop marching on a bridge do not march on the beat as usual, but in intentional disorder, because the ordered rhythm might by resonance make the whole structure collapse.

2.11 Forced Oscillators: van der Pol

In the following we will add forcing to the van der Pol model. Before continuing, we, however, have to stress the following fact about the van der Pol model. Suppose we define the new variable $\dot{y} = x$ and substitute in (2.101). Then we get:

$$\dddot{y} - \left(1 - \dot{y}^2\right)\ddot{y} + \dot{y} = 0 \tag{2.154}$$

which is easily integrated and yields:

$$\ddot{y} - \left(\dot{y} - \frac{1}{3}\dot{y}^3\right) + y = 0 \tag{2.155}$$

if we put the arbitrary integration constant zero as we may. This, in fact, is the same equation as the van der Pol and it behaves identically. Anyhow, it is interesting for us, as it is in this form that business cycle models arise.

It also has an even older historical provenance than the van der Pol equation. It is usually called Rayleigh's equation, and was proposed by Lord Rayleigh as a model for the bowed violin string in his classical studies on the theory of sound.

Returning now to the original van der Pol equation, including the perturbation parameter, we also introduce the forcing term and obtain:

$$\ddot{x} + x = \varepsilon\left(1 - x^2\right)\dot{x} + \Gamma\cos\omega t \tag{2.156}$$

Let us now consider equation (2.156) and try the solution:

$$x = A(t)\cos\omega t + B(t)\sin\omega t \tag{2.157}$$

We already encountered this type of solution in (2.109) when we introduced the two-timing method. The use is now somewhat different. We note that the frequency of the attempted solution is not the unit frequency of the unperturbed free oscillator, but that of the forcing term. This is so because, in the case of forced oscillators at resonance, the free oscillation only contributes an initial transient, so what remains in a longer perspective is the forcing frequency.

If we really have resonance, then A and B should turn out to be constants, if they vary, then the system is unable to settle at periodic motion, at least at the basic frequency. We may then still have subharmonic resonance, as we only check for entrainment at the basic frequency by (2.157). Or we could have quasiperiodic motion if the coefficients themselves display periodic motion of a frequency different from that of the forcing term. We may even have chaos.

We do not explicitly introduce any separate slow time variable, but the philosophy is the same as before. We assume A and B to be slowly varying, so second derivatives A'' and B'' are assumed to vanish, along with the products of the first derivatives with the smallness parameter, such as $\varepsilon A'$ and $\varepsilon B'$. From (2.157) we thus obtain:

$$\ddot{x} = \left(-\omega^2 A + 2\omega B'\right)\cos\omega t + \left(-\omega^2 B - 2\omega A'\right)\sin\omega t \quad (2.158)$$

where we skipped the second derivatives of the amplitudes. From (2.157) and (2.158) we now get for the left hand side of (2.156):

$$\ddot{x} + x = \left(\left(1-\omega^2\right)A + 2\omega B'\right)\cos\omega t + \left(\left(1-\omega^2\right)B - 2\omega A'\right)\sin\omega t \quad (2.159)$$

The right hand side of (2.156) is a little more laborious. First, note that, the nonlinear term being preceded by the small parameter ε, we can ignore the terms involving A' and B', and hence use the approximation:

$$\varepsilon\dot{x} = -\varepsilon\omega A\sin\omega t + \varepsilon\omega B\cos\omega t \quad (2.160)$$

The parenthesis $\left(1-x^2\right)$ in (2.156) is readily obtained from (2.157):

$$\left(1-x^2\right) = 1 - A^2\cos^2\omega t - B^2\sin^2\omega t - 2AB\cos\omega t\sin\omega t \quad (2.161)$$

Upon multiplying (2.160) and (2.161) together, we obtain cubic terms in the sines and cosines of ωt. Therefore we have to use the same trigonometric identities as before to transform them into the sines and cosines of ωt and $3\omega t$. We are, however, presently not interested in the latter, only in the fundamental harmonic, and so we obtain:

$$\begin{aligned}\varepsilon\left(1-x^2\right)\dot{x} &= \varepsilon\omega B\left(1-\frac{1}{4}\left(A^2+B^2\right)\right)\cos\omega t \\ &\quad - \varepsilon\omega A\left(1-\frac{1}{4}\left(A^2+B^2\right)\right)\sin\omega t\end{aligned} \quad (2.162)$$

We are now finally ready to use (2.159) and (2.162) in (2.156). Note that all expressions involve terms in $\cos\omega t$ and $\sin\omega t$. If the solution (2.157) is to work, then the coefficients of those terms have to equal zero individually, so we get a pair of equations:

$$2A' = \varepsilon\left(1 - \frac{1}{4}\left(A^2 + B^2\right)\right)A + \frac{1-\omega^2}{\omega}B \tag{2.163}$$

$$2B' = \varepsilon\left(1 - \frac{1}{4}\left(A^2 + B^2\right)\right)B - \frac{1-\omega^2}{\omega}A + \frac{\Gamma}{\omega} \tag{2.164}$$

Note that we have divided through by ω, and that we have arranged the terms so as to isolate the derivatives of the slowly varying amplitudes in the left hand sides. The result is a pair of differential equations, just as in the two-timing case. The last equation also contains the amplitude from the forcing term in (2.156), because it arbitrarily was introduced as a pure cosine.

Let us define the detuning coefficient:

$$\sigma = \frac{1-\omega^2}{\varepsilon\omega} \tag{2.165}$$

and a re-scaled forcing amplitude:

$$\gamma = \frac{\Gamma}{\varepsilon\omega} \tag{2.166}$$

Moreover define the sum of squares of the amplitudes:

$$\rho = A^2 + B^2 \tag{2.167}$$

With the definitions (2.165)-(2.167) the system (2.163)-(2.164) can be written:

$$2A' = \varepsilon\left(1 - \frac{1}{4}\rho\right)A + \varepsilon\sigma B \qquad (2.168)$$

$$2B' = \varepsilon\left(1 - \frac{1}{4}\rho\right)B - \varepsilon\sigma A + \varepsilon\gamma \qquad (2.169)$$

Now we are able to consider the problem of frequency entrainment, more specifically resonance at the basic frequency of the forcing term. Such entrainment implies that the system (2.168)-(2.169), or (2.163)-(2.164) has a fixed point.

A fixed point is defined by the system of algebraic equations:

$$\left(1 - \frac{1}{4}\rho\right)A + \sigma B = 0 \qquad (2.170)$$

$$\left(1 - \frac{1}{4}\rho\right)B - \sigma A = -\gamma \qquad (2.171)$$

Note that the parameter ε at this stage drops out as it is present in all the terms. Next take squares of (2.170)-(2.171) and add. Using (2.167) we obtain:

$$\left(\left(1 - \frac{1}{4}\rho\right)^2 + \sigma^2\right)\rho = \gamma^2 \qquad (2.172)$$

This is a relation between the squared amplitude variable ρ, the squared detuning coefficient σ^2 and the squared amplitude of the forcing term γ^2. The equation is cubic in the amplitude of the response. The detuning σ, depending on the frequency of the forcing term, and the amplitude γ of the forcing term can be taken as parameters describing the way the van der Pol equation is forced. Hence (2.172) gives various response amplitudes to each combination of such parameters. It is interesting to note that there may be up to three different responses to one single set of parameters. This is shown in Fig. 2.18. The detuning is on the horizontal axis whereas the response amplitude is on the vertical. The response curves are represented as a family

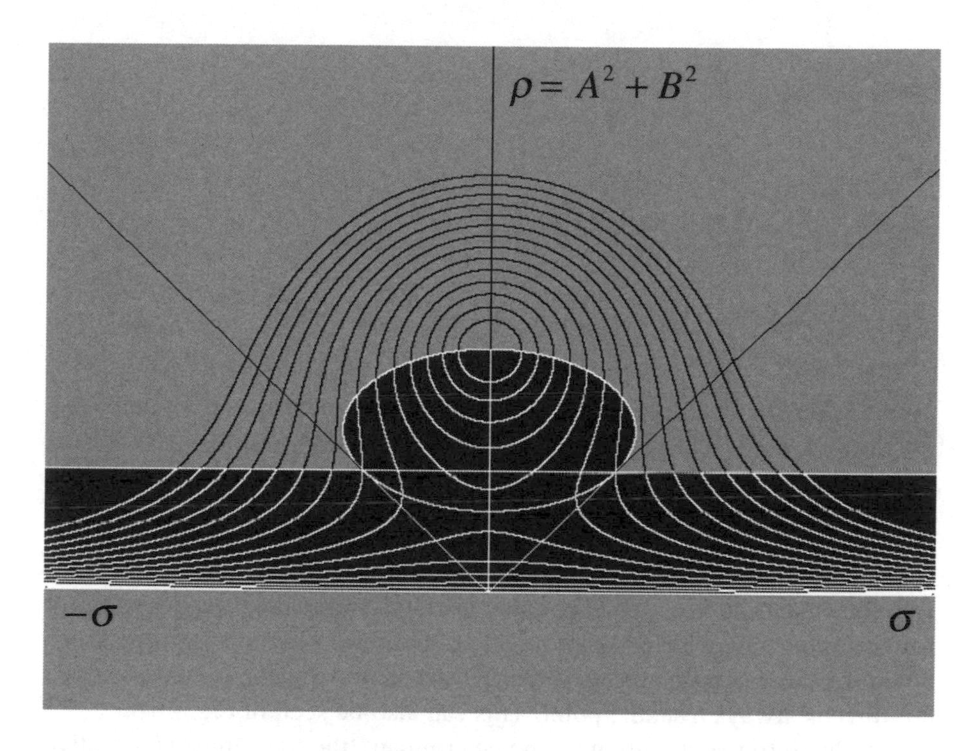

Fig. 2.18. Response amplitude diagram for various detuning.

for various forcing amplitudes. Some of them are backbending, or even disjoint, and so yield three response amplitudes.

In particular as we may deal with several response amplitudes it is of importance to find out the stability of the various fixed points of the system (2.168)-(2.169). We first linearise the system by substituting $A+\xi$ for A, $B+\eta$ for B, and subtracting the original system. As always at linearisation all powers and products of the small deviation variables are deleted. The linearised system reads:

$$\dot{\xi} = \frac{\varepsilon}{2}\left(1 - \frac{3}{4}A^2 - \frac{1}{4}B^2\right)\xi - \frac{\varepsilon}{2}\left(\frac{1}{2}AB + \sigma\right)\eta \tag{2.173}$$

$$\dot{\eta} = \frac{\varepsilon}{2}\left(1 - \frac{3}{4}B^2 - \frac{1}{4}A^2\right)\eta - \frac{\varepsilon}{2}\left(\frac{1}{2}AB - \sigma\right)\xi \tag{2.174}$$

The trace and determinant of this system can be easily computed using (2.167):

$$\mathrm{Tr} = \frac{\varepsilon}{2}(2-\rho) \tag{2.175}$$

$$\mathrm{Det} = \frac{\varepsilon^2}{64}\left((4-\rho)(4-3\rho)+16\sigma^2\right) \tag{2.176}$$

For stability the determinant must, as we have seen, be positive, otherwise we have a saddle. The determinant is zero on an ellipse in the ρ, σ -plane, and positive outside it. The centre of the ellipse is at $\rho = 8/3, \sigma = 0$, so the response amplitude and the detuning must deviate sufficiently to either side from these values in order that saddles be avoided. The interior of this ellipse is coloured dark in Fig. 2.18. We could note that, computing the discriminant for the cubic response equation (2.172), there are three distinct roots just whenever the determinant is negative. Therefore the middle root in case there are three is always a saddle point. This can also be seen in Fig. 2.18.

In addition to the condition on the determinant, the trace must be negative for stability, so we must also have $\rho > 2$ for the response amplitude. The area in the lower part of Fig. 2.18, where this does not hold, is also coloured dark.

It is also possible for the forced van der Pol oscillator to display quasiperiodic or even chaotic motion without any frequency entrainment at all. From our analysis it may appear that there always exist stable fixed points, but that analysis was based on the fact that the nonlinearity was small. If it is not, then the approximations used in the perturbation methods simply do not work. And then, of course, the analysis has no relevance.

The opposite case to that of a small perturbation of the linear oscillator has also been studied in the early literature, i.e., the case when the nonlinearity is so large that it dominates completely. The point is, however, easier to make if we deal with the Rayleigh equation (2.155) which as we saw was equivalent to the van der Pol. Making the nonlinearity large we have:

$$\ddot{y} + y = \frac{1}{\varepsilon}\left(\dot{y} - \frac{1}{3}\dot{y}^3\right) \tag{2.177}$$

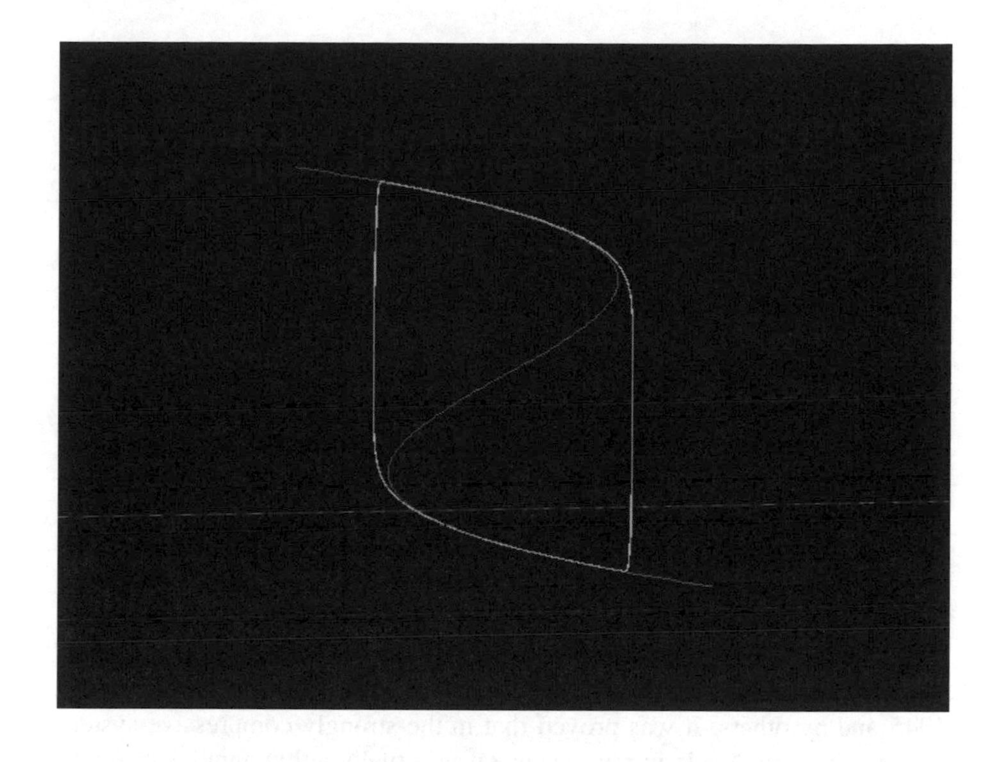

Fig. 2.19. The relaxation cycle of the Rayleigh oscillator.

where we for the moment ignore the forcing term and study the autonomous oscillator. Equation (2.177) can be written:

$$\dot{y} = z \tag{2.178}$$

$$\dot{z} = \frac{1}{\varepsilon}\left(z - \frac{1}{3}z^3\right) - y \tag{2.179}$$

Let us now divide (2.179) by (2.178). Thus we have the field directions:

$$\frac{dz}{dy} = \frac{1}{\varepsilon}\frac{\left(z - \frac{1}{3}z^3\right) - \varepsilon y}{z} \tag{2.180}$$

As ε is very small, the coefficient of the right hand side of (2.180) is very large. So, the field direction in y, z- space is vertical except when the numerator happens to be almost zero too. This happens on the so called characteristic curve:

$$\varepsilon y = z - \frac{1}{3} z^3 \tag{2.181}$$

which is shown in Fig. 2.19. From what has been said, the dynamic process follows the characteristic, as far as it can until it reaches any one of the two extremum points of the cubic. Then it has to leave the cubic and make a fast vertical transit through phase space, until it reaches the next branch of the cubic, and so it goes on. The resulting cycle is composed of sections of a cubic and of straight lines. The name relaxation cycle was coined as the system alternates between storing of energy during slow movement along the characteristic and sudden discharges of energy during the vertical transit.

The forced case of relaxation cycles for the van der Pol-Rayleigh oscillator have been studied early in the literature, by Cartwright and Littlewood in 1945 and by others. It was proved that in the strongly compressive system the seeming limit cycle in phase space was actually a thin annulus of thickness $1/\varepsilon^2$, and that within this thin annulus two very long and very different limit cycles did coexist. In 1981 Levi proved that the system was in fact what in the meantime had been baptized as chaotic, with an expected error in prediction of half a cycle. Guckenheimer and Holmes include an up to date and authoritative discussion of this topic.

In the sequel we will deal with the model extensively, as it in a natural way arises in classical business cycle modelling. It is interesting to note that chaos was detected in the systems that the multiplier-accelerator models produced already by the time they were formulated, and that this was long before the term chaos had been introduced.

Unfortunately, not much of intriguing fractal shapes are present in this type of chaos, and hence they are no favourites in the abundant literature on chaos. This is due to the very strongly compressive nature of the system. In phase space we need enormous magnification to see anything extraordinary at all, and so we might be led to think that the effects are slight, not only in the phase diagram but in the time series produced as well. Nothing could be more wrong! The expected error is, as already mentioned, half a cycle.

2.12 Forced Oscillators: Duffing

Fractal structures are better illustrated in a different type of oscillator once studied by Duffing, who like Rayleigh, but unlike van der Pol, studied a mechanical system. We encountered it in equation (2.74) when we presented the Poincaré-Lindstedt perturbation method. Recall that the model arouse as a Taylor expansion of the sine function in the mathematical pendulum model.

We now add forcing, and later on also friction. Suppose we have the original truncated Taylor series for the sine function, without the rescaling we used to arrive at (2.74), and add a forcing term:

$$\ddot{x} + x - \frac{1}{6}\varepsilon x^3 = \Gamma \cos \omega t \tag{2.182}$$

We now adopt the same solution strategy as in the previous case. Try entrainment at the basic frequency:

$$x = A(t)\cos \omega t + B(t)\sin \omega t \tag{2.183}$$

As in (2.158)-(2.159) we have:

$$\ddot{x} = \left(-\omega^2 A + 2\omega B'\right)\cos \omega t + \left(-\omega^2 B - 2\omega A'\right)\sin \omega t \tag{2.184}$$

and therefore

$$\ddot{x} + x = \left(\left(1-\omega^2\right)A + 2\omega B'\right)\cos \omega t + \left(\left(1-\omega^2\right)B - 2\omega A'\right)\sin \omega t \tag{2.185}$$

As for the cubic term, by exactly the same procedure as before, and again ignoring the third harmonics, we have:

$$x^3 = \frac{3}{4}A\left(A^2 + B^2\right)\cos \omega t + \frac{3}{4}B\left(A^2 + B^2\right)\sin \omega t \tag{2.186}$$

Substituting from (2.185)-(2.186) into (2.182) we again get terms involving the sines and cosines of the basic harmonic motion, and in order that (2.183) be a solution their coefficients have to vanish individually. By this strategy we again arrive at two differential equations:

$$2A' = -\left(\frac{1-\omega^2}{\omega} + \frac{1}{8}\frac{\varepsilon}{\omega}\left(A^2 + B^2\right)\right)B \tag{2.187}$$

$$2B' = \left(\frac{1-\omega^2}{\omega} + \frac{1}{8}\frac{\varepsilon}{\omega}\left(A^2 + B^2\right)\right)A + \varepsilon\frac{\Gamma}{\varepsilon\omega} \tag{2.188}$$

These are the equivalents of equations (2.163)-(2.164) above. By using the definitions (2.165)-(2.167), which we do not repeat here, we can write:

$$2A' = -\left(\sigma + \frac{1}{8}\frac{\varepsilon}{\omega}\rho\right)B \tag{2.189}$$

$$2B' = \left(\sigma + \frac{1}{8}\frac{\varepsilon}{\omega}\rho\right)A + \varepsilon\gamma \tag{2.190}$$

The fixed points of this system are defined by:

$$\left(\sigma + \frac{1}{8}\frac{\varepsilon}{\omega}\rho\right)B = 0 \tag{2.191}$$

$$\left(\sigma + \frac{1}{8}\frac{\varepsilon}{\omega}\rho\right)A = -\varepsilon\gamma \tag{2.192}$$

The situation is easier to analyse than the van der Pol case. We note that (2.191) is fulfilled if either the parenthesis is zero or if $B = 0$. If the parenthesis is zero, then the left hand side of (2.192) is, however, zero too, and then the forcing term would have to be zero. If it is not, which is a reasonable assumption as we are studying forced oscillations, then we must have $B = 0$. According to (2.167) then $\rho = A^2$, and (2.192) may be written:

Fig. 2.20. Phase plane for the forced Duffiing oscillator without friction.

$$\left(\sigma + \frac{1}{8}\frac{\varepsilon}{\omega}A^2\right)A + \varepsilon\gamma = 0 \tag{2.193}$$

This is of the third degree, and may so have three different response amplitudes.

The stability analysis can now proceed along the same lines as in the case of the van der Pol oscillator, by linearising, and computing the trace and determinant of the linearised equation system. We do not carry these details out here, because unlike the van der Pol model we will not use the Duffing model in the applications. Instead we display a phase portrait of the system (2.187)-(2.188) in Fig. 2.20.

It is easy to see that there are two centres, one to the left and one in the middle, and a saddle to the right. We recall that centres are structurally unstable. In addition there is a homoclinic saddle trajectory in the picture which separates the two centre regions. A homoclinic trajectory too is, as we know, highly unstable. Therefore the forced Duffing oscillator without friction is not a robust model, and it is essentially changed once we include friction.

Suppose we change (2.182) to:

$$\ddot{x}+x-\frac{1}{6}\varepsilon x^{3}+k\dot{x}=\Gamma\cos\omega t \tag{2.194}$$

To make a long story short, the derivation of (2.189)-(2.190) can be repeated step by step, and these equations are simply changed to:

$$2A'=-\left(\sigma+\frac{1}{8}\frac{\varepsilon}{\omega}\rho\right)B-kA \tag{2.195}$$

$$2B'=\left(\sigma+\frac{1}{8}\frac{\varepsilon}{\omega}\rho\right)A-kB+\varepsilon\gamma \tag{2.196}$$

The conditions for stationarity read:

$$\left(\sigma+\frac{1}{8}\frac{\varepsilon}{\omega}\rho\right)B+kA=0 \tag{2.197}$$

$$\left(\sigma+\frac{1}{8}\frac{\varepsilon}{\omega}\rho\right)A-kB=-\varepsilon\gamma \tag{2.198}$$

Taking squares, adding, and using (2.167) we get:

$$\left(\left(\sigma+\frac{1}{8}\frac{\varepsilon}{\omega}\rho\right)^{2}+k^{2}\right)\rho=\varepsilon^{2}\gamma^{2} \tag{2.199}$$

Equation (2.199) can have three real roots for the squared amplitude ρ. These can be substituted back into (2.197)-(2.198), which, as a pair of linear equations, give solutions for the stationary points A, B. The latter amplitude no longer needs be zero, as we see from (2.197). To check their character we can again linearise (2.197)-(2.198), and compute the trace and determinant.

Fig. 2.21. The forced Duffing oscillator with friction.

It is obvious from what was said above that the centres are converted into spirals. The saddle keeps its character, but the homoclinic trajectory is broken, and the stable saddle trajectory provides the basin boundary between the two stable spirals. A positive friction turns the centres into stable spirals. All this is seen in Fig. 2.21.

The closed orbits in Fig. 2.20 mean that the coefficients in (2.186) themselves undergo periodic motion. As there is no need whatever for the period of this motion to be in any rational relationship with the frequency of the forcing term, the model as a whole thus produces quasiperiodic motion.

In Fig. 2.21 these quasiperiodic orbits have disappeared, and only the periodic solutions, the two spiral foci, are the ultimate fate of the model, though the approach may be slow. With a low friction coefficient, the final approach may take many whirls, and if the trajectory passes close to the saddle point, then the motion may be slowed down for a very long period.

The change from Fig. 2.20 to Fig. 2.21 is a result of the structural instability of centres and homoclinic trajectories. It is a well known fact that certain quasiperiodic trajectories are structurally unstable, and are turned into periodic motion by the slightest change to the model. In reality, there are, however, lots of frequency locking at subharmonic resonance with very complicated, though rational, frequency ratios, and therefore indefinitely long periods. Such subharmonic resonance is, however, not caught by the present analysis which focuses the basic frequency alone.

Moreover, the reader should not conclude that quasiperiodic motion necessarily is a consequence of structural instability. Had we stabilized the Duffing model by replacing the centres with limit cycles instead, which is perfectly possible, then the resulting quasiperiodic trajectories would not have been structurally unstable.

Perturbation studies are one way of obtaining information about dynamical systems such as the van der Pol, Duffing, or Rayleigh oscillators. Another is, as mentioned, simulation. For popular items, such as the van der Pol and Duffing model, there exist ready made packages for their study, some of them, such as the Nusse and Yorke package, of high sophistication. Otherwise it is easy to make one's own simulation programs in any standard language following the recipes in Parker and Chua, and using simple algorithms such as the four point Runge-Kutta method, or even the Euler method. It should be kept in mind that simulation is no inferior method of doing scientific research with mathematical models. In case they are not solvable in closed form, simulations, in whatever application, becomes the only means of obtaining information, and the scientific procedure becomes itself empirical/experimental. The procedure is hence as respectable as any experimental one in sciences where we are by historical experience accustomed to these facts.

By simulation studies many more interesting things can be found out about the forced Duffing model. Fig. 2.22, for instance, shows four periodic attractors in phase space. The interesting thing is that they are all obtained for one and the same parameter combination. They even have different periodicities!

We also note that they are pairwise mirror images of each other. The Duffing system is symmetric, which means that, if we reverse the signs of all the variables, then the system is not changed. Such symmetry implies that attractors are either symmetric themselves, or, if they are not, they come in pairs so as to make up for symmetry together. These are very basic facts about the breaking of symmetry.

As the attractors coexist in phase space, they represent the asymptotic futures for different initial points in phase space. This coexistence may seem a bit confusing. If we put all the four orbits from Fig. 2.22 in the same picture, then they seem to intersect. This they, of course, do not do as our forced system is three-dimensional, including time, and the phase space we actually

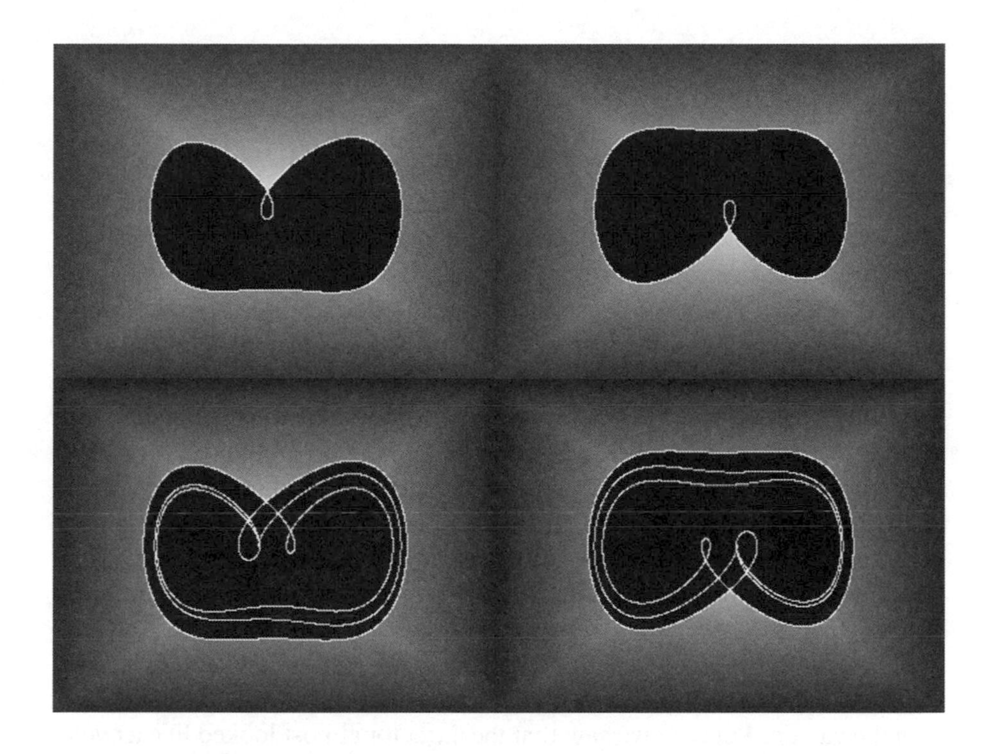

Fig. 2.22. Four coexistent periodic solutions to the forced Duffing model.

look at is a two-dimensional projection of the complete phase space. As we know, solid objects that definitely do not overlap, such as two people, may well cast overlapping shadows.

We can nevertheless study, for each initial point, which one of the four coexistent attractors is the final state of the process, and hence subdivide phase space in attraction basins. In view of what was just said we must then agree upon an implicit convention that the forcing term always starts at a given phase. Such basins are in general very complex, and their boundaries may be fractal objects. Fractals generally occur as attractors when a system becomes chaotic, but they may show up as basin boundaries even when a system is not yet chaotic. Examples of fractal basin boundaries are given in the next chapter and in the application work later in this book.

2.13 Chaos

In Fig. 2.23 we show a more complex tangle of a trajectory. This one is chaotic. It is not possible to say when looking at a picture of an attractor in phase space whether the motion is quasiperiodic or chaotic. In both cases the trajectory would seem to cover an area in phase space of nonzero measure. This, of course, is deceptive as it is the result of the finite resolution of a computer screen. We simply cannot see real one-dimensional curves.

At a live simulation we could see that quasiperiodic trajectories move more smoothly, whereas chaotic ones seem to be wilder and more capricious. But, neither this is a test on chaos. The real test is the separation of very close trajectories, that first cannot be seen as distinct objects, but later diverge so as to end up in completely different parts of the attractor. It is relatively easy to prepare software, where two different processes, with very close initial conditions, are simulated together, and the trajectories are drawn in different colour. Then the critical separation can be seen visually.

This separation can be quantified and precisely measured by the Lyapunov exponents. These will be defined in the next chapter.

In the van der Pol case we saw that the attractor almost looked like a cycle, even if it was not, so there was no structure to be seen. In general there is, however, an intriguing geometrical structure of the attractors. Often they are fractals, i.e. objects with fractional dimension, neither points, nor curves, nor surfaces, nor solids. Even though fractal structures often are the imprints of chaotic processes there is no perfect correlation. There may be fractals without chaos, and chaos without fractals.

It is very difficult to see any fractal structure in a trajectory such as that of Fig. 2.23. To see more we first consider presenting the information in another form. The first step would be to consider the fact that the real phase space is three-dimensional. However, time, one of the dimensions, influences the process through a periodic function contributing the same sequence of impacts over and over with a certain periodicity. In order to take advantage of this we take the phase diagram of Fig. 2.23, which at a certain moment only contributes a point, and fix it radially to a horizontal turning wheel. The wheel is turned at the velocity of the periodic function in the driving force in the Duffing equation. While it turns, the point on the phase plane, of course, moves too, and the result is a winding trajectory in space. It seems to be wound around a torus, which is what is shown in Fig. 2.24. This strategy is a good way of visualization with all forced oscillators. Abraham and Shaw

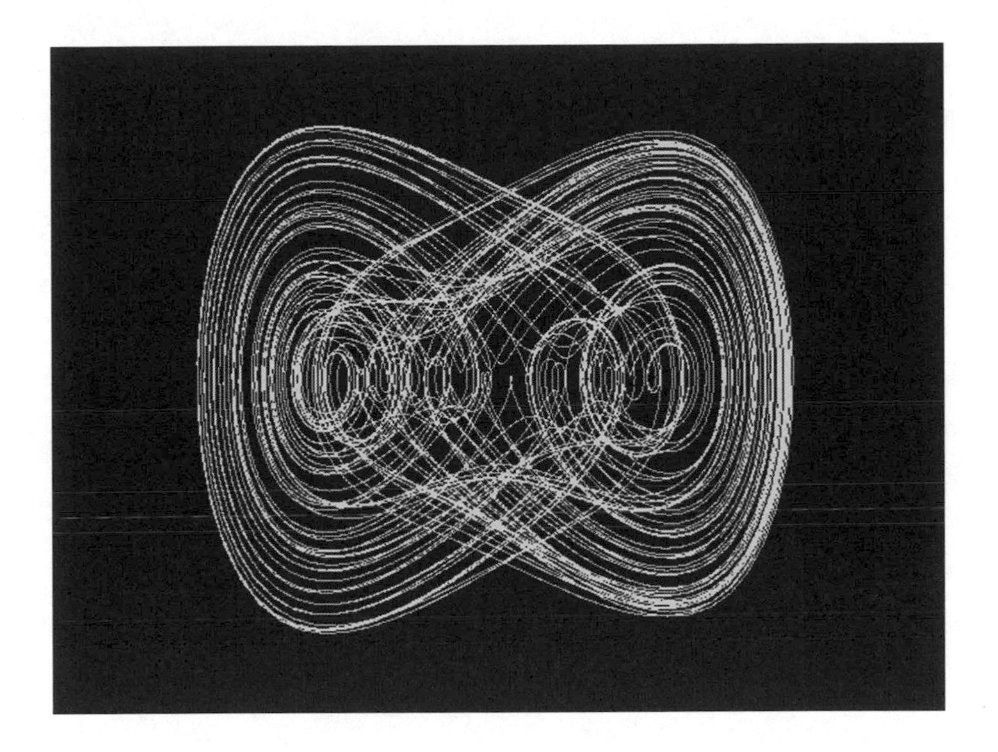

Fig. 2.23. Chaotic attractor of the forced Duffing model.

make wonders with general considerations of the geometry of such tori in their dynamics picture book.

Now, this is not yet the end of the visualization strategy. The next move is to cut the torus through with a plane more or less perpendicular to its length direction. To put it vulgarly, we cut a thin slice of the sausage and see what it looks like. This slice is called the Poincaré section, and it is fundamental for the study of all forced dynamical systems. Fig. 2.25 displays the Poincaré section of the torus of Fig. 2.24, which was itself the trajectory of Fig. 2.23 wrapped around a torus.

This procedure, first wrapping the trajectory around a sausage, and then cutting a thin slice, is indeed efficient. There is no chance that the picture of Fig. 2.25 could be seen in Fig. 2.23. The picture in Fig. 2.25 is a fractal. It seems to be an object consisting of a finite number of curves, but it is not. At every level of magnification, we would see any curve dissolve into an intriguing tangle of thinner curves, and so on ad infinitum. For comparison we

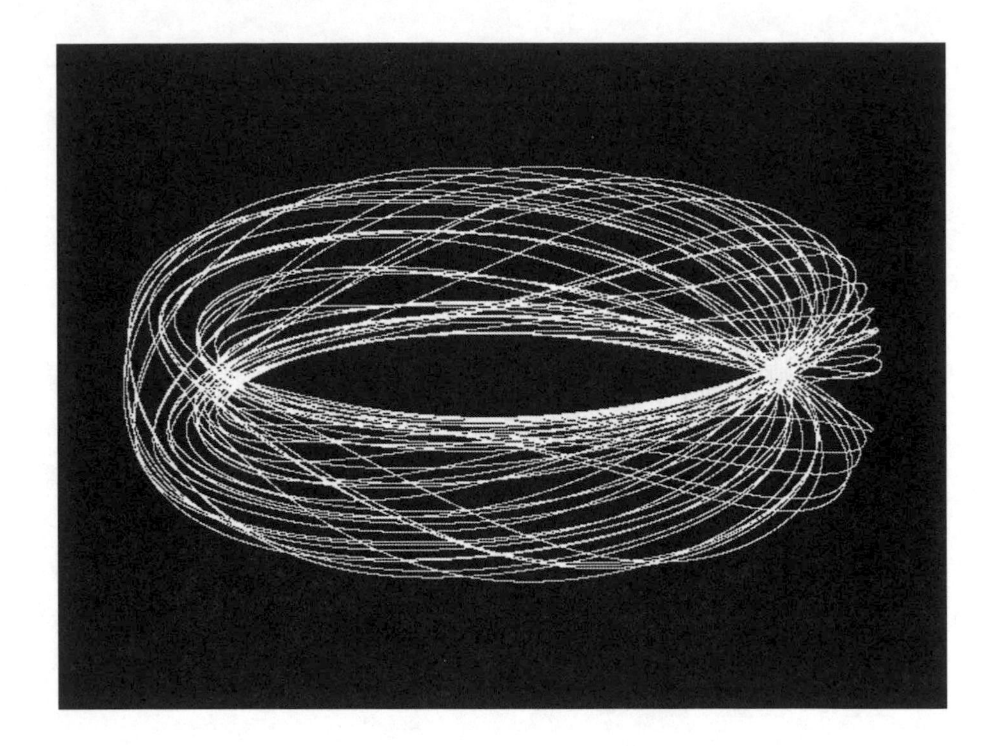

Fig. 2.24. Chaotic trajectory wrapped on a torus.

should say that a quasiperiodic trajectory would normally have a Poincaré section that is a closed curve, and a periodic solution would be a collection of points. This discussion of the Duffing model has rather closely followed the elaborate study by Ueda.

On the Poincaré section we can study such things as the approach of a dynamical process to a limit cycle, and thus study the stability of limit cycles. The formal tool to this end is to define the first return map for any point on the Poincaré section. We can take any point of the section, take the trajectory that passes the point and follow it a full round until it hits the section again. This is another point, and we can thus, by repeating the procedure for any starting point find a mapping of the Poincaré section onto itself. This first return map is an important tool, because it is used in defining the Lyapunov exponents, and provides a general link to maps, which are the basis for all the formal tools used in the study of chaotic systems. Maps, or, to use a familiar word for economists, difference equations, are studied in the Chapter 4.

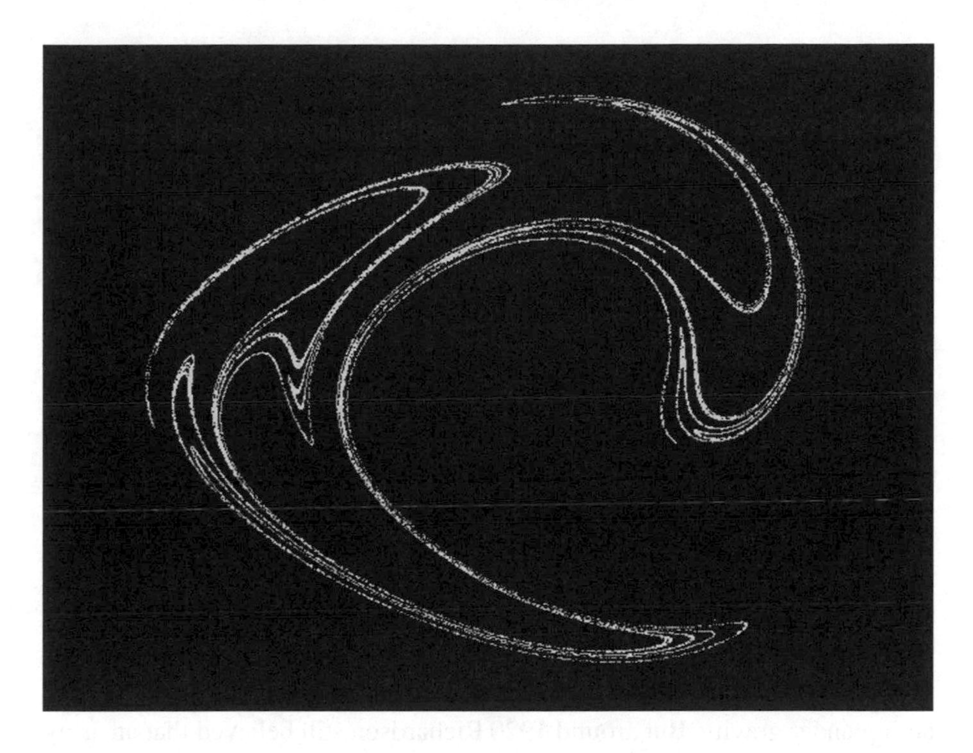

Fig. 2.25. Poincaré section of the forced Duffing equation.

2.14 A Short History of Chaos

Though the formal tools for the analysis of chaos apply to maps, the development of chaos theory is strongly linked up with nonlinear differential equations, so a short history of chaos has its proper place here. The study of differential equations started in the 17th Century with the work by Sir Isaac Newton. By his newly formulated infinitesimal calculus he was not only able to formulate the laws of motion for any number of particles interacting under gravity. He was also able to completely solve the system for two interacting bodies, and to actually derive the Keplerian Laws.

As the solar system essentially dissolves into two body problems whenever a subsystem of interactions between for instance the earth and the sun, or the earth and the moon are considered, all the masses of the other bodies of our planetary system being ignorable due to minor size or large distance, the solution proved very accurate for any prediction. Events such as eclipses could be prognosticised thousands of years ahead with great accuracy. As the laws worked equally well backwards, it was even possible to use the information by historians who wanted to date an event which coincided with an eclipse according to contemporary chronicles.

Now, the laws looked as simple for three, or any number of interacting bodies, and thousands of scientists tried their hands on finding the closed form solutions for those problems. It was of course assumed that the solutions would be more complicated than for the two body problem, but there were no doubts that the proper closed form solutions did exist.

Around 1800 Laplace still expressed his belief that a sufficiently powerful mind would be able to calculate the complete future and past of the universe if he only had information about the positions and momenta of all the particles making up this universe.

A slightly different type of problem was that of the diffusion of heat and convectional motion in a medium such as a fluid or the atmosphere. The laws of motion, the so called Navier-Stokes equations, fully described the problem. They were partial differential equations, hence infinitely more complex than the ordinary differential equations for a finite set of particles interacting under gravity. But around 1920 Richardson still believed that an army of scientists working with slide rulers on telegraphically delivered data on air pressures, wind velocities and temperatures all over the world, would be able to forecast the development of the weather for any perspective. He only despaired at the estimated number of scientists needed for the task, 64,000 only for predicting weather at the rate at which it actually developed, and of course more to make a real forecast. It would seem that the prospects would be incomparably improved today when the slide rulers have been replaced by supercomputers and the army of human observers by weather satellites.

Statistics was needed for problems which were not as well understood as the ones mentioned, where many unknown factors of influence on a problem had to be ignored, but otherwise there was a firm belief in the scientific community that deterministic well understood processes, such as the motion of a collection of particles, or the convection of the atmosphere, would as a matter of course also be predictable.

Chaos theory demonstrated that determinism and predictability are two completely different things. As mentioned, thousands of treatises were written over more than a century on the three-body problem, as a first stepping stone on the road in the direction of Laplace's grandiose perspective. It was considered wise to even make the problem simpler, by considering a particle moving in the double well of two equivalent gravitational potentials, such as those of two big planets. The particle itself, a speck of dust, would, of course, have an influence back on the planets, but it would be so small that it could be neglected, and the double well potential taken as given.

Even so the problem was evasive and defied any attempt on closed form solution. The scientist who gave the answer why was Henri Poincaré in a prize winning treatise around 1900. The contest was announced by the Swedish King Oscar II, who originally challenged the scientists to show whether the solar system was stable or unstable. It is no wonder why so many methods touched upon bear Poincaré's signum, for instance, the Poincaré-Bendixon Theorem, the Poincaré Sections, and the Poincaré-Lindstedt perturbation method. Poincaré gave up closed form solutions, and developed a geometrical approach focusing the geometry of the phase space. This is the secret why Poincaré alone had made any progress on this rather old problem.

Poincaré showed that the movement of the speck could not be predicted, that its movement was more like that of a random sequence than of any closed form solution to a differential equation. Poincaré's original method of geometrical reasoning has been taken up recently by Abraham and Shaw, who, in an unbelievable number of intricate diagrams, dissect the anatomy of dynamical systems.

Though crowned by Royal Prize, Poincaré's contribution made no immediate impact on the direction of research. The next milestone was around 1960, when Edward Lorenz blew up Richardson's dream, as Poincaré had blown up Laplace's. Like the case of Poincaré, Lorenz made no immediate success. The reason in this case was obvious. Lorenz used mathematics of a sophistication that no other meteorologist of that time was able to cope with, and the path breaking article was literally buried for a decade in a specialized journal on atmospheric sciences.

Quite as in the case of Poincaré, lots of simplifications were needed for the Navier-Stokes equations. First it was assumed that atmospheric change only took place in the vertical and one geographical direction. In the other geographical direction things were assumed invariant, so that convection for instance would be in terms of infinite rolls. The next step was to convert the system of partial differential equations to one of ordinary differential equations. These were set up in terms of various characteristics of the convective rolls, and finally Lorenz ended up with a system of three coupled ordinary differential equations in three variables. Though there was only mild nonlinearity in terms of products in two of the otherwise linear differential equations, the result was amazing.

Lorenz detected critical sensitivity by accident. Having interrupted a long simulation he later made it anew but with a different number of significant digits. To his astonishment the slight difference in initial conditions was progressively magnified, until the trajectories were in entirely different parts of the now well known Lorenz attractor, the so called "mask". He then concentrated on the reasons for this and presented a highly sophisticated mathematical analysis.

It is surprising that chaos actually is something as trivial as the magnification of computation errors. Once we set up a deterministic dynamical system it is a matter of course that given initial conditions give rise to a unique trajectory, which would hence seem to be fully predictable. This, however, is only true if we specify the initial conditions exactly to the last decimal in an infinite sequence. In practice this can never be done. Computation errors were, of course, not unknown, but it was believed that they would be kept in bounds during the evolution of the dynamic process. Statistics was used to provide these bounds.

The problem with chaotic systems, however, is that computation errors are progressively increased without bounds. As the process is generally compressed to a small portion of phase space, the attractor, the uncertainty eventually becomes as large as the whole extension of this attractor. And this happens in an uncomfortably short while. Remedies relying on increased precision in stating the initial conditions are not practical, because the rate of error propagation is exponential. As there is a considerable cost attached to the acquisition of improved data one can easily incur astronomical cost just to obtain a marginal extension of the possible prediction period.

Chaos can thus not be wiped away from dynamical systems if we do not like this uncomfortable state of things. Meteorologists have taken the consequences and only make five day forecasts. In economics, on the contrary, long term forecasting still is an important means of earning for many scientists advising banks, governments, the media and the general public. Economists are accustomed to believe that things are rational, manageable, and predictable, and, though they cannot define chaos away, they can believe that the well regulated self-governing economic systems are more like the well behaved sun-planet system, than the double well dust speck system. This, however, is no more than a pious hope. The practical difficulties of forecasting might in fact be an indication of the contrary.

In this brief history, we concentrated on the most important milestones in the development of chaos theory. There was, of course, much more to it. The forced Duffing and van der Pol oscillators were invented around 1920, and, as we saw, the extraordinary tangled annulus of the latter was analysed in 1945.

The whole field of chaos became a real success once the personal computers were there, and anybody could make simulations. Meanwhile, Benoit Mandelbrot, around 1970 had launched the concept of fractals, which were intimately linked to chaotic processes. Those object fascinated many both because they were the first break with Euclid's smooth geometrical objects in 2000 years, and because of their aesthetic appeal, and capability of resembling complex natural objects, such as leaves, trees, clouds, smoke, mountains, or entire landscapes.

During the recent decades many more chaotic systems were detected, such as the frictionless double pendulum, i.e., a pendulum hanging from another pendulum. This is an example of a conservative dynamical system producing chaos. A system similar to Poincaré's double well system is produced as a toy, a pendulum, an iron bob hanging in a string above a bottom plate to which several magnets are fixed. And there are all the trivial everyday phenomena of dripping faucets, and turbulent flow, which fascinated the inquisitive minds from Leonardo on.

There exist lots of literature on differential equations and chaos. The books by Strogatz and by Thompson and Stewart can be recommended as particularly readable and enjoyable.

3 Differential Equations: Partial

3.1 Vibrations and Waves

We have been studying nonlinear oscillatory processes, starting with the simplest autonomous case, and later incorporating forcing. The modelling was in terms of ordinary differential equations (ODE), and the most important tool used was the phase diagram.

In physics an oscillatory process applies to a zero dimensional "mass point", as in the case of the mathematical pendulum. This, of course, is an abstraction, the material bob on a material rod of a pendulum clock being replaced by a point (the centre of gravity of the bob) constrained to move along a circular path by the ideally rigid rod. The spatial extension of these three-dimensional objects and any spatial pattern, such as undulations in imperfectly rigid rods, are abstracted away. The system can hence be completely described by the angular deflection and velocity of the idealized rod from its equilibrium position in the abstract state space whose picture is the phase diagram.

There are, however, physical systems where the focus is the temporal change of spatially extended patterns, such as transverse oscillations of a string under tension. Such oscillations involve both space and time, they are *waves*, not just vibrations. Each particle in the continuum of the string vibrates as an individual oscillator, but they are all coupled, and at each frozen moment of time there is a wavy shape to be observed as a spatial pattern.

The simplest case is where the shape studied is a one-dimensional continuum, such as the idealized string. The waves on the surface of the sea, like those put up in the soundboard of a musical instrument are two-dimensional, and most wave phenomena in physics are, of course, three-dimensional.

The study of waves, be it in one, two, or three dimensions, requires the use of partial differential equations (PDE), called so because they involve partial derivatives, with respect to time *and* with respect to space coordinates (one, two, or three). In an oscillator the temporal process is of the second order, i.e. the highest time derivative involved is second order. The *Wave Equation* is a typical example. One can, of course, also consider first order (temporal) processes. In physics such processes are referred to as diffusion, the *Heat Diffusion Equation*, describing the dispersion of heat in continuous matter is the classical example.

It is easy to find candidates for both types of models in economics: The propagation of business cycles in geographical space due to interregional trade for second order processes, the spatial diffusion of a migrating population for first order processes, just to mention one case of each.

The natural spatial setting for economics is geographical space, which is two-dimensional, quite as the normal setting of classical nonrelativistic physics is three-dimensional. In this sense things are easier in economics as the dimension of the space we really want to consider is one lower than in the case of physics. Like in physics, one can also introduce ideas by the easier case of one dimension. This, however, should be regarded as a pedagogical and heuristic strategy only. There has been a relative overproduction of models of "the long narrow city" and the like, in both economics and regional science, with no attempt at all to use the case as a stepping stone to the real two-dimensional cases. This is more of a limitation than one might at first think, because a region in one dimension is just an interval, and so any communication between endpoints just *must* pass the centre. In two dimensions communication lines can for instance avoid a congested centre. Conclusions from one dimension may therefore be completely misleading for important issues in two dimensions.

3. 2 Time and Space

Above we distinguished between first order diffusion and second order wave processes. The difference was in terms of the order of the highest *time* derivative. It may be a surprise that the lowest space derivatives involved are *always* second order.

This is due to the essential difference between time and space - even in the case of one dimensional space! The reason for this difference is that there is an inherent forward direction in time. A positive first derivative of any variable with respect to time says that the variable increases in the future. If we interchange forward and backward directions, the derivative reverses sign, and says that the variable decreases if we go back in history.

In the case of space there is no forward and no backward, right and left are equivalent. So, if we want to compare the value of a variable at a point with those at the points next to it - to the right *and* left - then we have one first order derivative to the right and another to the left. A measure comparing conditions at a point with conditions at the surrounding points has to take account of both right *and* left.

For comparison with the point to the right, the first derivative says how things change when we move *away* from the point, whereas for comparison with the point to the left, the first derivative says how things change when we move *towards* the point. It is therefore not the sum but the *difference* of the first derivatives that matters.

In the limit this difference goes to the second derivative. The second derivative also is the lowest which does not change under interchanging left and right directions. This is the reason why we always find second order space derivatives in wave and diffusion processes alike, no matter what the temporal order is. It may be a bit surprising at first that it is nothing more than the antisymmetry implied by the forward direction inherent in time, but lacking in space, which makes all the difference concerning which order derivatives to use in modclling.

There is no particular advantage in beginning with the study of processes of the lower order diffusion type, it is as easy to jump directly to the wave case, because, as we have seen, second order in the space coordinates is involved anyhow.

3.3 Travelling Waves in 1D: d'Alambert's Solution

There is, however, a point in starting with the wave equation in one dimension, and then generalize to two dimensions, because the latter involves some simple concepts from vector analysis which we first have to introduce. The one dimensional wave equation reads:

$$\frac{\partial^2 y}{\partial x^2} = \frac{1}{c^2}\frac{\partial^2 y}{\partial t^2} \tag{3.1}$$

Here we study the variable y in its dependence on the location in space x and in its variation over time t. The solution to (3.1) we seek is of the form $y(x, t)$. Equation (3.1) applies for instance to the much studied case of transverse vibrations in a tensioned string, where y is the transverse deflection of the string at point x from its equilibrium location, and where the constant c represents the square root of the ratio of tension to weight per unit length of the string. The same equation also appears in models of longitudinal sound waves in narrow pipes, and in current and voltage waves in electrical transmission lines. As a rule c represents the speed of wave propagation in space.

The most classical solution to (3.1) is due to d'Alambert, one of the famous authors of the 18th Century French *Encyclopédie*: Replacing the two separate variables x, t by their compound $x - ct$, or $x + ct$, we find that *any* function of the alternative single argument satisfies (3.1). The arbitrary function defines a wave shape, and, considering the visually simple example of the string, the solution then says that this shape without change of form travels along the string to the right or left, depending on the sign before c. This is a *travelling wave*.

Thus, any function of the form:

$$y = f\left(x \mp ct\right) \tag{3.2}$$

is a solution to (3.1) no matter what the form of the function.

The arbitrariness of the function also makes it possible to accommodate the solution to any *initial conditions* prescribing an initial wave profile $y = \phi(x)$ at $t = 0$, obtained for instance by pinching the string at a point and letting it go. To the initial conditions also belongs an initial velocity profile. If we just pinch the string and let it go, then every point on it starts from rest, and we just have $\partial y / \partial t = 0$ at $t = 0$. This is what we will use in exemplifying, but it is of course also possible to prescribe any initial nonzero velocity profile.

Note that, quite like the case of second order processes referring to mass points, we can impose conditions both for initial position and initial velocity, but, as we now deal with the evolution of a continuous shape over time, the conditions take the form of functions rather than single numerical values.

It is easy to check that (3.2) fulfils (3.1). Differentiating (3.2) twice with respect to x we obtain:

$$\frac{\partial^2 y}{\partial x^2} = f'' \tag{3.3}$$

likewise differentiating twice with respect to t we get:

$$\frac{\partial^2 y}{\partial t^2} = (\mp c)^2 f'' = c^2 f'' \tag{3.4}$$

so eliminating f'' between (3.3) and (3.4) we obtain (3.1).

Next, we note that, (3.1) being linear, it is true that if $y = f(x \mp ct)$ and $y = g(x \mp ct)$ both are solutions, then so is their sum, or the sum of any number of such solutions. However we never need more than two to be completely general. As the solution functions are completely arbitrary, any number of shapes can be combined into one, *given that they travel in the same direction*. However, waves travelling in opposite directions cannot be combined to a given shape, so therefore we need two, one eastbound and one westbound.

The most general d'Alambert solution therefore is the combination:

$$y = f(x - ct) + g(x + ct) \tag{3.5}$$

where again f and g are any arbitrary functions. Note that we always take the minus sign first because it represents a wave travelling in the positive direction, i.e. to the right.

3.4 Initial Conditions

Let us now see how the initial conditions can be used. Substituting $y = \phi(x)$ and $t = 0$ in (3.5) we find:

$$f(x) + g(x) = \phi(x) \tag{3.6}$$

Further, differentiating (3.5) with respect to t, and using $\partial y / \partial t = 0$ at $t=0$, we find:

$$-cf'(x) + cg'(x) = 0 \tag{3.7}$$

Dividing (3.7) by c, which is permitted as it is nonzero, and integrating (3.7), we can easily solve (3.6)-(3.7) for

$$f(x) = \frac{1}{2}\phi(x) \qquad g(x) = \frac{1}{2}\phi(x) \tag{3.8}$$

Now we can substitute x - ct for x in the first equation of the pair (3.8), and x + ct for x in the second, and write (3.5) as:

$$y = \frac{1}{2}\phi(x - ct) + \frac{1}{2}\phi(x + ct) \tag{3.9}$$

Accordingly, the initial conditions imply that half the initial displacement travels right and half of it travels left. Initially they combine to produce the specified profile. This, of course, only holds if the process starts from a displacement at rest. A more complicated initial velocity condition produces other decompositions of the initial profile into travelling waves.

These waves travel to left and right, getting ever more apart. If the initial disturbance was just one single hump, of finite spatial extension, then the components disappear in the directions of infinity, the space region of the initial disturbance coming to rest again. Such solitary waves are a hot subject in modern nonlinear dynamics research and are called *solitons*. As a rule they involve third order space derivatives, which implies spatial asymmetry, quite as in the case of time. In physics solitons appear in models of shallow water waves in narrow ducts and the like.

In the classical wave equation, which concerns us at present, the solution functions considered are periodic functions of the sine type rather than consisting of merely one hump. Then the disturbance never dies out in the location where it initially occurred. Even with old waves disappearing from any observed region new ones enter it incessantly.

3.5 Boundary Conditions

Besides initial conditions it is also possible to impose so called boundary conditions on solutions to partial differential equations. These are conditions that must be fulfilled at given locations in space at all times. The simplest are those that state that the solution take on some given value in some point at all times, but one can also think of periodic boundary conditions which constrain the solution in a point to oscillate at some given frequency.

Such conditions are called boundary conditions because they apply to the boundary of a spatial region. A region in one dimension is just an interval, and its boundary consists of the two endpoints. Accordingly, conditions to hold at all times imposed at two points in space actually *define* such a region. If there is also a condition imposed at an intermediate point, then the region simply splits in two which can be analysed individually.

Hence the most general region we need to study in 1 D is the interval, its boundary consists of the two endpoints, and the boundary conditions apply to these endpoints. Note that in 2 D things become much more complex. In two dimensions regions have not only size but also shape, they are enclosed areas and their boundaries are the closed boundary curves. Boundary conditions then prescribe conditions for all points located on these curves to hold at all times.

Suppose for illustration that we want (3.5), the general solution to the one dimensional wave equation (3.1), to fulfil the condition $y = 0$ at $x = 0$ for all t. In view of the physical model of transverse waves in a string this means that the string is clamped at the point of origin and so constrained to remain at rest. Substituting in (3.5) we obtain:

$$f(-ct) + g(ct) = 0 \tag{3.10}$$

and substituting $z = ct$:

$$g(z) = -f(-z) \tag{3.11}$$

which *defines* the function g from the function f. Geometrically f is both reflected in the x-axis and reversed in the right/left directions. Only in this way do the travelling waves combine to make the spatial origin be a point at rest at all times.

Using (3.11) in (3.5) we get:

$$y = f(x - ct) - f(-x - ct) \tag{3.12}$$

Thus we no longer have two arbitrary solution functions but only one, a situation which is due to the boundary constraint. Note that, (3.11) just being a definition of the shape of the g function, there is nothing to prevent us from using $z = x - ct$ in the original f-function but $z = x + ct$ in the derived g-function, as indeed prescribed by (3.5). Also note that the two wave components now seem to move in the same direction.

We can also see what happens if we combine initial and boundary conditions. Applying (3.11) to (3.8) we find: $\phi(x) = -\phi(-x)$. This just means that the initial disturbance profile is an odd function, like a sine or a power series with only odd numbered powers. This may seem to constrain the initial condition, which would be absurd. However, the origin is a boundary point of the region studied, so only one half of the odd function matters, the other half is not operative. Accordingly, the only constraint is that the initial profile must intersect the origin to fulfil the boundary condition. This is just a matter of consistency in assumptions and shows that initial and boundary conditions can be easily combined.

Further, suppose we have boundary conditions prescribed at both endpoints, the origin $x = 0$ as before and the other $x = L$, and assume that $y = 0$ at both points. Then (3.12) still applies, though we now also have:

$$f(L - ct) - f(-L - ct) = 0 \tag{3.13}$$

by substitution of $x = L$ in (3.12). This must hold for all time points, so putting $-z = (L+ct)$ and rearranging we have:

$$f(z) = f(z + 2L) \tag{3.14}$$

which requires that f be a periodic function with period $2L$. The combined facts that the solution is periodic, in time as well as in space, implying that the motion never takes an end, and that the endpoints of the region are points of rest introduce the idea of a *standing wave*.

3.6 Standing Waves: Variable Separation

As an example of a periodic function satisfying (3.14) let us take the sine function which is periodic in 2π, so for convenience put $L = \pi$. Then (3.12) becomes:

$$y = \sin(x - ct) - \sin(-x - ct) \tag{3.15}$$

Expanding we have $\sin(x - ct) = \sin(x)\cos(ct) - \cos(x)\sin(ct)$, where we have taken in account that $\cos(-A) = \cos(A)$ is even and $\sin(-A) = -\sin(A)$ is odd. Similarly, $\sin(-x - ct) = -\sin(x)\cos(ct) - \cos(x)\sin(ct)$. Accordingly (3.15) may be written:

$$y = 2\cos(ct)\sin(x) \tag{3.16}$$

We see how the travelling wave (3.15) is converted to the standing wave (3.16). The sum of two terms has been converted to a product, one depending on time alone and one depending on space alone. The space function determines a fixed profile, with the zeros at $x = 0$ and $x = \pi$ as a guarantee that the boundary conditions are fulfilled. This fixed spatial profile is periodically scaled up and down between the values ± 1 by the factor $\cos(ct)$. The factoring in (3.16) is called *separation of variables* and it is a powerful method of solution for partial differential equations.

So, let us directly attempt a solution of the type:

$$y = T(t)S(x) \tag{3.17}$$

for (3.1), with one factor depending on time only and another depending on space only. As we have $\partial^2 y / \partial x^2 = T(t)S''(x)$ and $\partial^2 y / \partial t^2 = T''(t)S(x)$ substitution of (3.17) into (3.1) yields:

$$T(t)S''(x) = \frac{1}{c^2} T''(t)S(x) \tag{3.18}$$

or, dividing through by $T(t)S(x)$,

$$\frac{S''(x)}{S(x)} = \frac{1}{c^2}\frac{T''(t)}{T(t)} \tag{3.19}$$

Now comes the really important consideration. Equation (3.19) states that the left hand side, which only depends on the space coordinate, always equals the right hand side, which only depends on the time coordinate. Hence, both must be constant. To see that this is so we differentiate (3.19) with respect to time. As there is no time dependence in the left hand side, the derivative becomes zero, which from the equation also holds for the right hand side. The derivative being zero is the same as the right hand side being a constant. Quite similarly we can differentiate with respect to the space coordinate, and show that the left hand side is a constant. These constants, of course, will be the same, and for convenience we denote them $-i^2$.

Restating (3.19) as

$$\frac{S''(x)}{S(x)} = \frac{1}{c^2}\frac{T''(t)}{T(t)} = -i^2 \tag{3.20}$$

we obtain two differential equations:

$$S''(x) + i^2 S(x) = 0 \tag{3.21}$$

and

$$T''(t) + i^2 c^2 T(t) = 0 \tag{3.22}$$

Note the important fact that (3.21)-(3.22) are ordinary differential equations (ODE), no longer partial (PDE), and hence much easier to handle.

The normal procedure is to deal with the spatial equation (3.21) first. It is a classical eigenvalue problem and serves to determine the eigenvalue i which is then substituted in (3.22) before proceeding with the temporal equation. The possible eigenvalues depend on the size of the region and on the boundary conditions.

As the region is the interval $(0, \pi)$, and the boundary conditions prescribe that the system be at rest at both endpoints at all times, i.e. $S(0) = S(\pi) = 0$, the obvious solution candidates for (3.21) are of the type:

$$S(x) = \sin(ix) \tag{3.23}$$

where $i = 1, 2, 3, \ldots$ is any positive integer. In (3.16) we already encountered the case $i = 1$, but now we find an infinity of new possibilities. Any integral i still makes the sine function take on zero values at the endpoints of the interval $(0, \pi)$. They also, given $i > 1$, produce new points of rest, so called *nodes* between the endpoints. This is so because $\sin(x)$ on the interval $(0, \pi)$ is just the positive half of the complete sine cycle, whereas $\sin(2x)$ on the same interval squeezes in a complete cycle, with an additional node in the middle, and $\sin(3x)$ one and a half cycles, with two additional nodes, equally spaced by one third of the whole interval, and so on. The different $\sin(ix)$ are called *eigenfunctions*. They are companions to the eigenvalues i, and in a sense correspond to the eigenvectors in ordinary differential equations.

Returning to the example of the string it is known that it can vibrate as a whole, in halves, in thirds and so on. In applications to sound these represent the natural harmonics, the basic pitch, the octave, the octave plus fifth etc. Speaking in these terms we are ahead of the discussion. The octave and fifth refer to *pitches* and have to do with *temporal frequency*, not with *spatial mode*. But, as we will soon see, the temporal frequency is linked to the spatial mode by the value of i which, as chosen in (3.23), feeds back into the temporal equation (3.22).

From (3.16) we find that, for $t = 0$, $\cos(ct) = 1$ and accordingly $y = 2\sin(x)$. The question is: How can this fit an arbitrary initial condition? It seems that choosing a higher harmonic does not help - it just squeezes a larger range of the sine function with its regular peaks and valleys. However, we can form a weighted sum of all the sine solution curves for the different modes, and it can be shown that such a sum of solutions still fits the differential equation (3.1). Due to a very powerful theorem by Fourier it can be fitted to any profiles we might wish to prescribe as initial conditions.

Let us now return to the formal discussion. Once we have solved (3.21) and chosen i, we substitute this back into (3.22) which is now readily solved:

$$T(t) = K\cos(ict - \phi) \tag{3.24}$$

K is an arbitrary amplitude and ϕ an arbitrary phase lead. As S and T are multiplicative we did not care to introduce any amplitude in the solution (3.23). Defining $A = K\cos\phi$ and $B = K\sin\phi$, we may expand the cosine and write:

$$T(t) = A\cos(ict) + B\sin(ict) \tag{3.25}$$

This shows how the temporal frequency depends on the eigenvalue i. The frequency being $f = ic/(2\pi)$, we see that it is proportional to the eigenvalue. Thus the string vibrating in halves indeed produces octaves of the basic pitch and so forth, quite as conjectured.

We can now piece together the complete solution (3.17) from (3.23) and (3.25):

$$y = \left(A\cos(ict) + B\sin(ict)\right)\sin(ix) \tag{3.26}$$

applicable to the i:th vibration mode. Again we note that the shorter the spatial wave the faster it oscillates.

3.7 The General Solution and Fourier's Theorem

We can now proceed to the general solution for the standing wave by adding together all different mode solutions (3.26), only noting that we have to index the coefficients of the terms. Hence:

$$y = \sum_{i=1}^{\infty}\left(A_i\cos(ict) + B_i\sin(ict)\right)\sin(ix) \tag{3.27}$$

We note that for $t = 0$ the sines of the time argument in (3.27) vanish and only the cosines remain as unitary constants. Denoting the initial displacement profile by $\bar{y}$ we accordingly have:

$$\bar{y} = \sum_{i=1}^{\infty} A_i \sin(ix) \tag{3.28}$$

Likewise we can first differentiate (3.27) with respect to time, whereby again at $t = 0$ only the cosines of time remain, now the second terms, because sines become cosines and cosines become sines by differentiation. We thus obtain the initial velocity profile which we denote $\bar{y}'$:

$$\bar{y}' = \sum_{i=1}^{\infty} icB_i \sin(ix) \tag{3.29}$$

Actually, we have the left hand sides of (3.28)-(3.29) as initial conditions, whereas the coefficients A_i, B_i are the unknowns.

In order to determine these unknown coefficients we multiply (3.28) and (3.29) through by sin(jx), where j is any integer equal to or different from i. Finally we integrate the expressions with respect to x over the whole space interval (0, π). Thus from (3.28):

$$\int_0^{\pi} \bar{y} \sin(jx) dx = \sum_{i=1}^{\infty} A_i \int_0^{\pi} \sin(ix) \sin(jx) dx \tag{3.30}$$

and from (3.29):

$$\int_0^{\pi} \bar{y}' \sin(jx) dx = \sum_{i=1}^{\infty} icB_i \int_0^{\pi} \sin(ix) \sin(jx) dx \tag{3.31}$$

Now, the right hand side integrals are all zero except when $i = j$, when they take the value $\pi/2$. Accordingly:

$$A_j = \frac{2}{\pi} \int_0^{\pi} \bar{y} \sin(jx) dx \qquad B_j = \frac{2}{\pi} \int_0^{\pi} \frac{\bar{y}'}{jc} \sin(jx) dx \tag{3.32}$$

This is the contents of Fourier's Theorem and shows how we can calculate the coefficients from any initial displacement and velocity profile.

It may be appropriate to illustrate how this works. Suppose that initially the string is at rest, i.e. $\bar{y}' \equiv 0$. As a consequence of this all $B_j = 0$. Moreover, suppose it was pinched at the middle and released from the shape of an isosceles triangle, i.e. $\bar{y} = x$ for $x < \pi/2$ and $\bar{y} = \pi - x$ for $x > \pi/2$.

Then

$$A_i = \frac{2}{\pi}\int_0^{\pi/2} x\sin(ix)dx + \frac{2}{\pi}\int_{\pi/2}^{\pi}(\pi - x)\sin(ix)dx \tag{3.33}$$

Though messy, this expression can be evaluated in closed form and we get:

$$A_i = \frac{4}{\pi i^2}\sin\left(\frac{\pi i}{2}\right) \tag{3.34}$$

Every even numbered coefficient is zero, as we see, and the squared eigenvalue in the denominator makes the coefficients decrease very fast for higher modes. To see this we can approximate the first nonzero (odd numbered) coefficients $A_1 \approx 1.27, A_3 \approx -0.14, A_5 \approx 0.05$. Given the coefficients from (3.34) we substitute in the general solution (3.27) and obtain for our example case:

$$y = \sum_{i=1}^{\infty}\frac{4}{\pi i^2}\sin\left(\frac{\pi i}{2}\right)\cos(ict)\sin(ix) \tag{3.35}$$

Taking only eight nonzero terms as shown in Fig. 3.1 already makes it difficult to distinguish the series from the actual triangular shape. We find that, as time proceeds, the wave proceeds as a truncated triangle. A middle section begins to move and becomes ever wider as time goes on, while the points of the region close to the endpoints remain at rest until the flattened moving section has reached them. This takes longer the closer to the endpoints they are located. The endpoints themselves remain at rest all the time, just as prescribed by the boundary conditions. We will meet this triangular type of wave again in the nonlinear context.

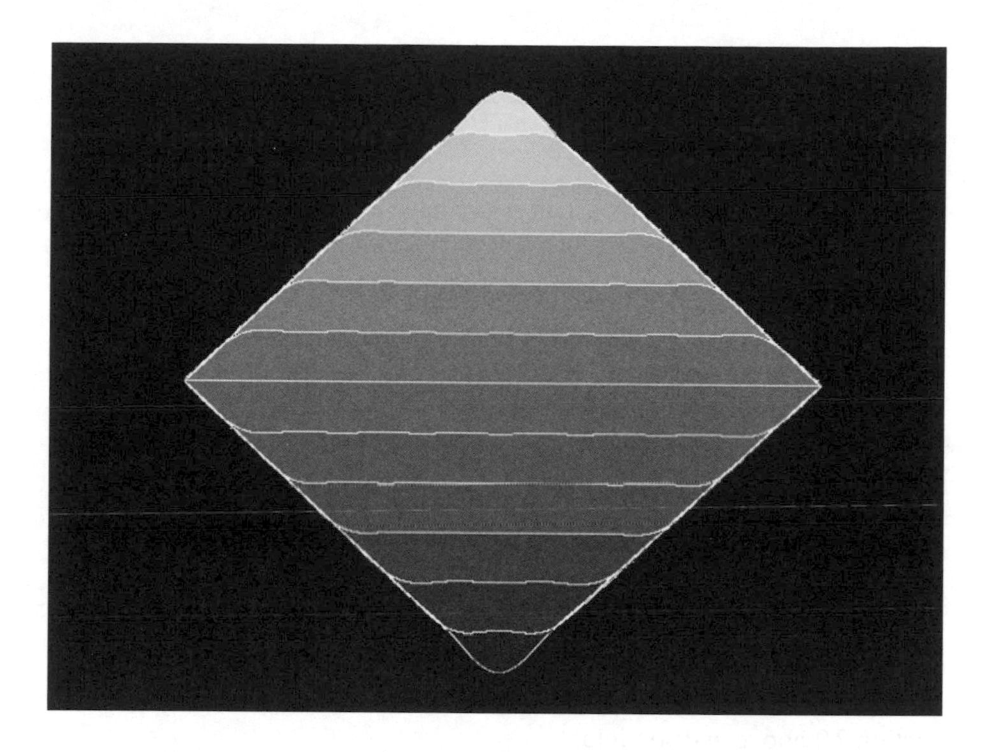

Fig. 3.1. Triangular wave at various moments approximated by Fourier Series.

3.8 Friction in the Wave Equation

The differential equation (3.1) had no friction term. Therefore the oscillating waves never come to an end, but continue for ever with undamped amplitudes which are determined by the arbitrary initial conditions. In the case of ordinary differential equations, such as the simple harmonic oscillator, we characterised such models as structurally unstable. They of course are so even now, and equally unrealistic as any vibrating string loses energy due to air viscosity and torsional heat losses. The natural thing to do is to include friction and to restate (3.1) as:

$$c^2 \frac{\partial^2 y}{\partial x^2} = \frac{\partial^2 y}{\partial t^2} + 2k \frac{\partial y}{\partial t} \tag{3.36}$$

Attempting the solution (3.17) $y = T(t)S(x)$ we obtain upon substitution, differentiation, and division by $T(t)S(x)$:

$$c^2 \frac{S''(x)}{S(x)} = \frac{T''(t)}{T(t)} + 2k \frac{T'(t)}{T(t)} \tag{3.37}$$

Again, both sides must be constant so we have as before:

$$S''(x) + i^2 S(x) = 0 \tag{3.38}$$

identical to (3.21), and

$$T''(t) + 2kT'(t) + i^2 c^2 T(t) = 0 \tag{3.39}$$

which is a little different from (3.22). The solutions to (3.38) are

$$S(x) = \sin(ix) \tag{3.40}$$

as in (3.23) above, whereas (3.39) can be solved along the lines indicated in section 2.2 above and so yield:

$$T(t) = \exp(-kt)\big(A\cos(\omega t) + B\sin(\omega t)\big) \tag{3.41}$$

where

$$\omega = \sqrt{i^2 c^2 - k^2} \tag{3.42}$$

It is worth noting that according to (3.42) the frequency depends on the eigenvalue i, and also that the temporal frequencies are a little slowed down by friction from their free values ic. For this reason frequency ratios for ascending modes are not rational numbers, and the overtones produced by a string not quite harmonic. It is therefore important in a musical instrument that friction is kept as small as possible.

From (3.41) we infer that friction also leads to damping at the exponential rate $-k$, which, unlike the slowing of oscillations, is uniform for all modes. We can now piece together the general solution as a product of (3.40) and (3.41) summed over all the different modes:

$$y = \exp(-kt)\sum_{i=1}^{\infty}\left(A_i \cos(\omega_i t) + B_i \sin(\omega_i t)\right)\sin(ix) \tag{3.43}$$

This solution corresponds to (3.27) above, except for the frequencies $\omega_i = \sqrt{i^2c^2 - k^2}$, which replace *ic* above, and the exponential damping factor, which, being the same for all modes, can be placed outside the series.

As for the initial distribution (3.28) still applies because exp(-*kt*) = 1 for $t = 0$, but we restate it here for convenience:

$$\bar{y} = \sum_{i=1}^{\infty} A_i \sin(ix) \tag{3.44}$$

On the other hand, differentiation with respect to time now yields two terms as we also differentiate the exponential. Hence:

$$\bar{y}' = \sum_{i=1}^{\infty}(\omega_i B_i - kA_i)\sin(ix) \tag{3.45}$$

which differs from (3.29). By the same procedure as in the derivations (3.30)-(3.31), skipping details, we obtain:

$$A_j = \frac{2}{\pi}\int_0^{\pi} \bar{y}\sin(jx)dx \quad B_j = \frac{2}{\pi}\int_0^{\pi}\left(\frac{\bar{y}'}{\omega_j} - k\bar{y}\right)\sin(jx)dx \tag{3.46}$$

Hence again it is possible to obtain the Fourier coefficients from the initial displacement and velocity profiles.

3.9 Nonlinear Waves

Above we indicated that a nonlinear PDE could produce a triangular wave. There is a nice example of this from a model of wind induced waves in overhead lines, analysed in great detail by Myerscough. There is friction, though nonlinear, providing damping at large oscillations but antidamping at small. The latter means that the system is constantly fed with energy (from the wind) and thus never comes to rest. This energy feeding is modelled quite as Lord Rayleigh modelled the motion of a bowed violin string. The string sticks to the resin on the bow as long as it can, then suddenly slips and makes one free oscillation, but is caught again by the bow the moment its velocity agrees with that of the bow. The wind acts similarly on the overhead line, though it operates at every point of it. We already met this nonlinear type of friction term repeatedly.

So let us rephrase equation (3.1) like this:

$$\frac{\partial^2 y}{\partial x^2} = \frac{1}{c^2}\frac{\partial^2 y}{\partial t^2} - \left(\frac{\partial y}{\partial t} - \left(\frac{\partial y}{\partial t}\right)^3\right) \tag{3.47}$$

We will deal with this equation in a very simple heuristic manner. The full analysis, as can be seen from Myerscough, is very complicated.

So, suppose we have a triangular wave solution. Its sides are flat at all moments, as we see from Figure 3.1, so the first space derivative is constant and the second is hence zero (everywhere except at the junction points, but never mind about those). Hence $\partial^2 y / \partial x^2 = 0$. Moreover, suppose the first time derivative is constant everywhere, either zero as in the side segments which do not move, or else moving at a constant speed up or down, which refers to the horizontal middle segments. Then the second time derivative would vanish too (again except the moments a point at rest is caught into movement, or a moving point shifts to rest). Hence $\partial^2 y / \partial t^2 = 0$ as well.

Substituting these zero second derivatives in (3.47) we get:

$$\frac{\partial y}{\partial t} - \left(\frac{\partial y}{\partial t}\right)^3 = 0 \tag{3.48}$$

which is satisfied for

$$\frac{\partial y}{\partial t} = 0 \quad \text{and for} \quad \frac{\partial y}{\partial t} = \pm 1 \tag{3.49}$$

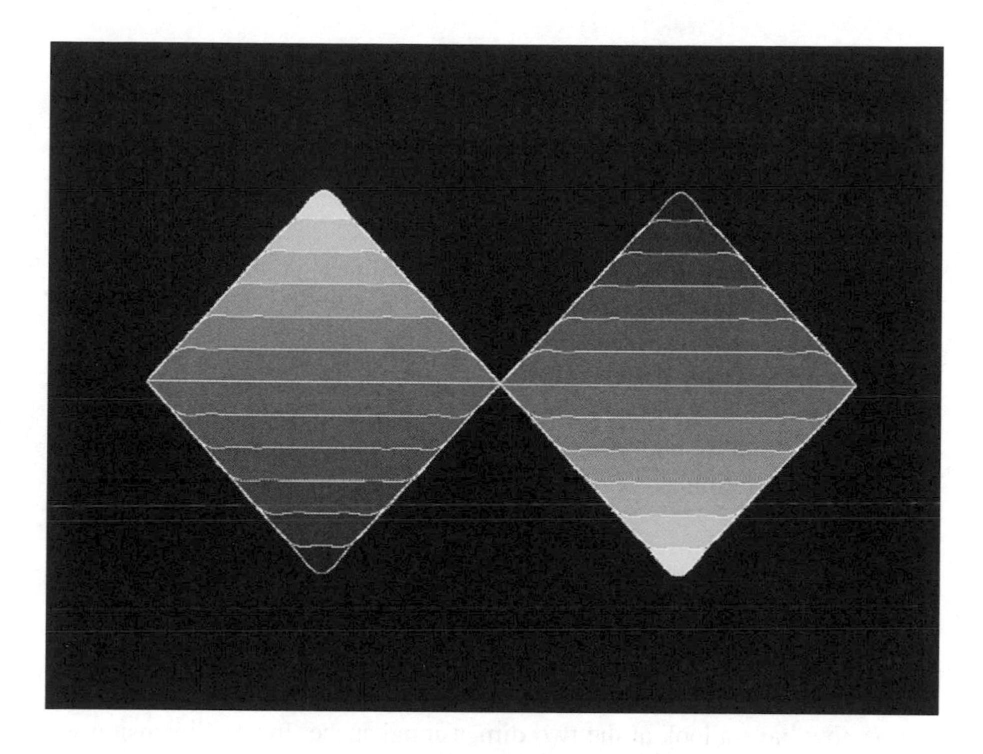

Fig. 3.2. Octave of triangular wave.

Accordingly, if the constant speeds are either zero or plus/minus unity, then the differential equation (3.47) is indeed satisfied. Note that this is quite like the relaxation cases we dealt with earlier, as the nonlinearity dominates so completely in the solution.

However, the triangular wave illustrated in Figure 3.1 is not the only possibility, given (3.49), $\partial^2 y / \partial t^2 = 0$, and $\partial^2 y / \partial x^2 = 0$. We can easily arrange the triangular octave. Suppose we initially have $\bar{y} = x$ for $0 < x < \pi/3$, $\bar{y} = \pi / 2 - x$ for $\pi/3 < x < 2\pi/3$, and $\bar{y} = x - \pi$ for $2\pi/3 < x < \pi$. This squeezes a complete triangularised sine wave into the interval $(0, \pi)$ with an additional node in the midpoint. The wave can then proceed exactly as before by moving horizontal segments.

In the same way we can arrange any of the triangular harmonics. It is possible to simulate the equation (3.47) by feeding in a regular sine wave displacement of any number of peaks and valleys. What one finds is that the system always transforms the smooth sine wave into the sharp triangular shape, but it keeps the number of peaks and valleys of the original profile. Accordingly, the system seems to have many attractors, actually an infinity, and is capable to oscillate in any of those. As they coexist the space of initial conditions (a space of functions) is divided into an infinity of attraction basins. The stability of the different attractors was dealt with by Myercough and is a complicated issue.

However, there is something new to the nonlinear system. The different modes can no longer be combined! It is also noteworthy that if we try the separation of time and space variables technique with our nonlinear equation (3.47) we get into difficulties. The cubic term spoils the possibility of splitting the equation into one space-dependent and one time-dependent side. So, though we in a sense undoubtedly have to do with a standing wave we cannot write the solution as a product $T(t)S(x)$.

3.10 Vector Fields in 2 D: Gradient and Divergence

Let us now have a look at the two dimensional space. In one dimension we deal with numbers, constants and variables, in two we have *pairs* of numbers, i.e. *vectors*. To emphasize the difference, the single numbers are called *scalars* in contrast to the vectors. We know that geometrically scalars are representable as points on a line, and, likewise, that vectors are representable as points in the plane.

In one dimensional analysis we also deal with functions of one variable $y = f(x)$, and with ordinary derivatives, dy/dx, and integrals, $\int f(x)dx$, of such functions. Geometrically the functions represent curves, the derivatives slopes, and the integrals areas under curves.

In two dimensions functions $y = f(x_1, x_2)$ represent surfaces, and we deal with *partial* derivatives $\partial y / \partial x_i$, representing slopes on tangent planes, and *multiple* integrals $\iint f(x_1, x_2)dx_1dx_2$, representing volumes under the surfaces.

But there are also particular differential operators, above all the divergence, which operate at once on both functions in a pair, a so called vector field.

There are also special integrals, line or curve integrals which we also have to introduce.

So, let us start with the concept of the *vector field*

$$\phi(x) = (\phi_1(x_1,x_2), \phi_2(x_1,x_2)) \tag{3.50}$$

which is a vector with components $\phi = (\phi_1, \phi_2)$, associated with each point $x = (x_1, x_2)$ in the plane. The vector ϕ has length, or magnitude, given by the Euclidean norm

$$|\phi| = \sqrt{\phi_1^2 + \phi_2^2} \tag{3.51}$$

and direction

$$\frac{\phi}{|\phi|} = \left(\frac{\phi_1}{\sqrt{\phi_1^2 + \phi_2^2}}, \frac{\phi_2}{\sqrt{\phi_1^2 + \phi_2^2}} \right) = (\cos\theta, \sin\theta) \tag{3.52}$$

obtained dividing the vector (3.50) by its norm (3.51) and so obtaining a vector of unit length. We can hence geometrically consider the vector ϕ as an arrow pointing in the direction θ, having length $|\phi|$, and being placed at a point $x = (x_1, x_2)$. It is essential to understand that the vector ϕ as a rule changes, in length *and* direction, as the basing point x varies.

We will supply three examples of such vector fields, the two first of them purely mathematical, the third substantial in terms of physics, and also with an economics interpretation.

i) A good example of a vector field, with which we are already familiar, is the direction field defined by the right hand sides of a pair of ordinary differential equations:

$$\frac{dx_1}{dt} = f(x_1, x_2) \tag{3.53}$$

$$\frac{dx_2}{dt} = g(x_1, x_2) \tag{3.54}$$

We already considered such tangent vectors to first integral curves in the flow portrait, and displayed them in Figures 2.1 and 2.3 by the tiny gray segments.

ii) Another important example is the gradient field to some function $y = f(x_1, x_2)$:

$$\nabla y = \left(\frac{\partial y}{\partial x_1}, \frac{\partial y}{\partial x_2} \right) \tag{3.55}$$

The gradient consists of the partial derivatives of the given function. The vectors point in the direction of steepest ascent as projected to the x_1, x_2-plane and have the length of the norm:

$$|\nabla y| = \sqrt{(\partial y / \partial x_1)^2 + (\partial y / \partial x_2)^2} \tag{3.56}$$

representing the increase in height of the surface in the direction of this steepest ascent.

iii) Finally, a physical example is the planar flow of some incompressible liquid in a shallow layer, where the direction of the vector field is the actual direction of flow and the norm is the volume of flow at any point in space. An economics parallel is the commodity flow in interregional trade in the ingenious models proposed by Beckmann in the early 50es.

Given an arbitrary vector field $\phi = (\phi_1, \phi_2)$ we now define the following important composite differential operator:

$$\nabla \cdot \phi = \frac{\partial \phi_1}{\partial x_1} + \frac{\partial \phi_2}{\partial x_2} \tag{3.57}$$

which is called the *divergence* of the vector field. Note the difference in notation between the gradient (without a dot) and the divergence (with a dot) after the same "nabla" symbol. Despite the tiny difference in notation the operators are very different: The gradient makes a vector field from a scalar field, the divergence makes the reverse, a scalar field out of a vector field. Later we will see how these operators can be chained.

It may seem arbitrary to take the partial derivative of the first component function in a vector with respect to the first space coordinate and that of the second component with respect to the second coordinate and add. What about cross derivatives, and why just add?

We will give two different reasons for this combination, one heuristic right now and a formal in the following section.

Consider the case of fluid flow in two dimensions. It is a composite of two components, one horizontal ϕ_1 and one vertical ϕ_2. So, imagine a small box placed in this flow, where the component ϕ_1 enters the box from the left side and leaves through the right, whereas the component ϕ_2 enters from the lower side and leaves through the upper. How much does the flow change in volume while passing through the box? In the limit, when the size of the box vanishes, differences become derivatives, and the horizontal component obviously changes by the $\partial\phi_1 / \partial x_1$, while the vertical changes by $\partial\phi_2 / \partial x_2$.

Considering both components, it seems that the volume as a whole changes by the sum $\partial\phi_1 / \partial x_1 + \partial\phi_2 / \partial x_2 = \nabla \cdot \phi$. If the flow is not compressible, then any change of volume must be due to a local source whenever the divergence is positive, and to a local sink whenever it is negative. In economics terms we would speak of local excess supply and excess demand, and equating the divergence of the trade flow to excess supply, or the negative of excess demand, provides a condition for interregional trade equilibrium. In physics one speaks of a conservation equation that guarantees that flow components are not just created or disappear unless there is a physical source or sink. Likewise, in economics, at equilibrium any excess supply must be entered in the trade flow, any excess demand be withdrawn from it.

We will now see what happens if we apply the divergence (3.57) to the gradient (3.55):

$$\nabla^2 y = \nabla \cdot \nabla y = \nabla \cdot \left(\frac{\partial y}{\partial x_1}, \frac{\partial y}{\partial x_2} \right) = \frac{\partial^2 y}{\partial x_1^2} + \frac{\partial^2 y}{\partial x_2^2} \tag{3.58}$$

Thus we first convert the scalar $y = f(x_1, x_2)$ to a vector, and then back to a scalar again. This sum of the two direct second derivatives of a function is called the Laplacian, and it has important applications in the two-dimensional wave equation, and other partial differential equations in mathematical physics.

3.11 Line Integrals and Gauss's Integral Theorem

Let us now introduce the concept of a "line", or rather, curve or path integral. Its conventional form is:

$$\oint_C P(x,y)dx + Q(x,y)dy \tag{3.59}$$

and has the following interpretation: The curve C, a closed loop, is located in two dimensional x, y -space, and is parameterized by

$$x = x(t) \qquad y = y(t) \tag{3.60}$$

The reader will forgive us for using the same letters for function form and function value in (3.60), and also for shifting for a while to denoting the space coordinates by x and y instead of x_1 and x_2. The reason for this shift is that we need the subscripts for other purposes and wish to avoid the mess of multiple subscripts. The small increments in (3.59) are interpreted as

$$dx = x'(t)dt \qquad dy = y'(t)dt \tag{3.61}$$

Upon substitution from (3.60) and (3.61), (3.59) becomes a normal integral of a function of t with respect to that variable itself, and there is no longer anything mysterious about it. The sense of this style of writing is that one does not emphasize any particular parameterisation of the curve C, we just refer to it as a geometrical object.

Now, suppose C is a simple closed curve, delimiting some area R in the two-dimensional plane. Moreover, suppose the area enclosed is convex, such as illustrated in Fig. 3.3. The curve C is given an orientation, indicated by the arrows. The positive sense is chosen so that the interior of the enclosed region is to the left when passing along the boundary in the positive direction.

Given the area is convex, we can decompose the boundary C in two ways, using the natural coordinates as parameters: Either by the pair of functions $y_2(x) \geq y_1(x)$, with $a \leq x \leq b$, using the first coordinate, or by the alternative pair of functions $x_2(y) \geq x_1(y)$ with $c \leq y \leq d$, using the second, as illustrated in the upper and lower parts of Figure 3.3.

Suppose now that we want to evaluate the following double integral on the region R:

$$\iint_R \frac{\partial P}{\partial y} dx\,dy \tag{3.62}$$

where $P(x,y)$ is some arbitrary function defined over the region we study. We can be more precise than that, as we have two parameterisations of the boundary curve. Choosing the first alternative, illustrated on the top part of Fig. 3.3 we get:

$$\int_a^b \int_{y_1(x)}^{y_2(x)} \frac{\partial P}{\partial y} dy\,dx = \int_a^b (P(x,y_2(x)) - P(x,y_1(x)))dx \tag{3.63}$$

Note that integrating with respect to y undoes the differentiation with respect to that variable. In the antiderivative we just substitute the values of y corresponding to the values of x as obtained from the integration limits for the inner integral.

The right hand side of equation (3.63) has a nice interpretation: It is in fact a line integral of the function $P(x,y)$ taken along the boundary curve C.

We only have to consider that integrating with respect to x along $y_2(x)$ is in the reverse direction of the orientation of the boundary - just compare the arrows on the horizontal axis and on the boundary. Taking an integral in a reverse direction reverses the sign. The integration on $y_1(x)$ is in the right direction, but it is preceded by a minus sign. In all, the right hand side of (3.63) equals the negative of the line integral, i.e.

$$\iint_R \frac{\partial P}{\partial y} dx\,dy = -\oint_C P dx \tag{3.64}$$

Analogously consider

$$\iint_R \frac{\partial Q}{\partial x} dx\,dy \tag{3.65}$$

where $Q(x,y)$ is another arbitrary function defined over the same region as the previous function. Now, using the second parameterisation for a change, (3.65) can be specified as:

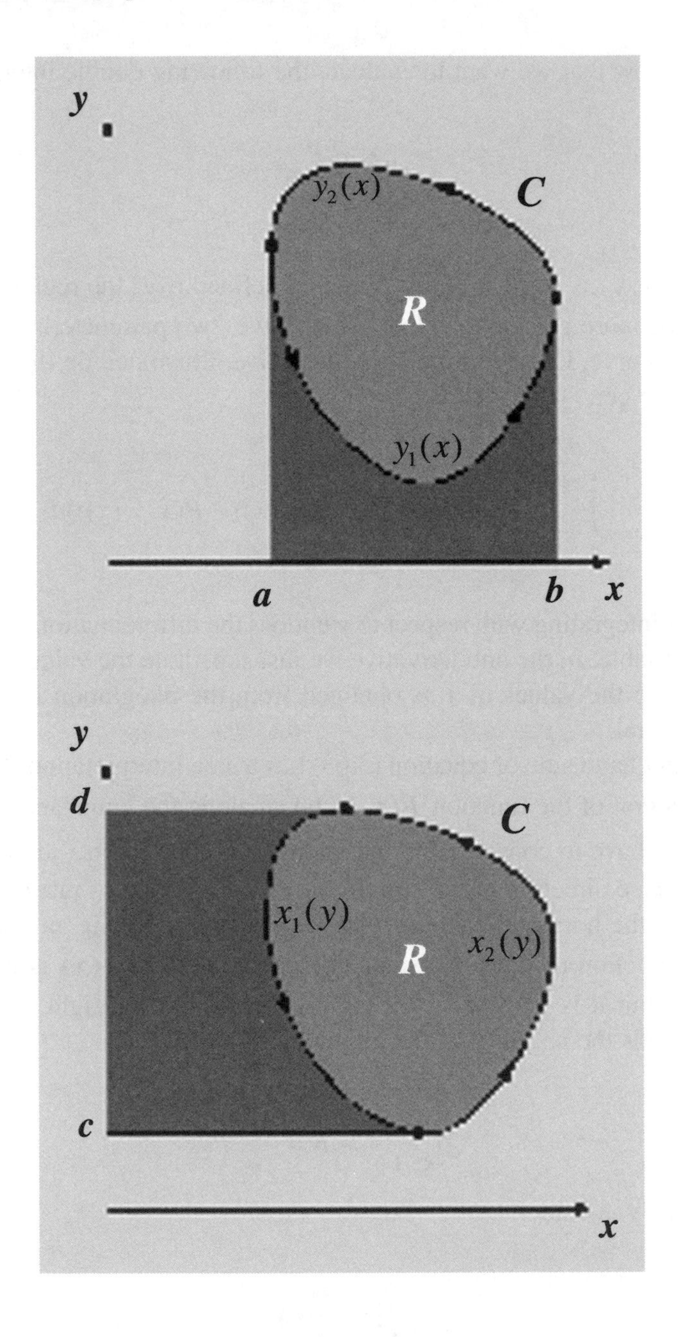

Fig. 3.3. Line integral and Green's Theorem.

$$\int_{c}^{d}\int_{x_1(y)}^{x_2(y)} \frac{\partial Q}{\partial x} dx\,dy = \int_{c}^{d} (Q(x_2(y), y) - Q(x_1(y), y))dy \tag{3.66}$$

The right hand side of (3.66) is again a line integral of Q, taken along C. But, as we see from Figure 3.3, it is now the negative term that is taken in a direction opposite to the orientation of the boundary, and hence has to have the sign reversed. We thus get:

$$\iint_R \frac{\partial Q}{\partial x} dx\,dy = \oint_C Q\,dy \tag{3.67}$$

without any preceding negative sign.

Collecting results, subtract (3.64) from (3.67):

$$\iint_R \left(\frac{\partial Q}{\partial x} - \frac{\partial P}{\partial y} \right) dxdy = \oint_C P dx + Q dy \tag{3.68}$$

This formula is called Green's Theorem in the Plane, and is one of the most basic results in vector analysis. It relates the double integral of derivatives of some pair of functions over an area to a line integral of the functions along the boundary of that area.

Green's Theorem was derived for a simply connected convex region, but the result is very general. Any nonconvex area can be partitioned into a set of convex areas, and the theorem applied to each of them. The inner boundaries in such subdivisions always belong to two adjacent regions, and, when applying Green's Theorem to the entire subdivision, they are taken twice, once in a positive, once in a negative sense.

Thus, the inner boundaries make no net contribution, and the theorem applies to nonconvex regions as well. Similarly, we can deal with regions that are not simply connected, i.e. such that have holes. Then, we only need to note that the boundaries of the holes have to be taken in consideration, always orienting them in the right sense, with the interior to the left.

The use to which we put the theorem requires a form of it called Gauss's Integral Theorem. The best way of introducing it is by applying Green's Theorem to the vector field (3.50) $\phi(x) = (\phi_1(x_1, x_2), \phi_2(x_1, x_2))$. So, let us define $P(x_1, x_2) = -\phi_2(x_1, x_2)$ and $Q(x_1, x_2) = \phi_1(x_1, x_2)$. Note that we reintroduced the original notation for space coordinates. Substituting into (3.68) we get:

$$\iint_R \left[\frac{\partial \phi_1}{\partial x_1} + \frac{\partial \phi_2}{\partial x_2} \right] dx\,dy = \oint_C \phi_1 dx_2 - \phi_2 dx_1 \tag{3.69}$$

We now need an interpretation of the right hand side (3.69). Consider the boundary curve as parameterized by $x_1(t), x_2(t)$. The tangent vector to the boundary curve is then $\mathbf{t} = (x_1'(t), x_2'(t))$ and the outward normal vector is accordingly $\mathbf{n} = (x_2'(t), -x_1'(t))$. The right hand side of (3.69) is thus the dot product of two vectors (multiplied componentwise and added), i.e. the flow vector (3.50), and the outward normal to the boundary curve. Geometrically, it is the projection of the flow field vector in the boundary normal direction.

As for the left hand side we just need the piece of notation, the divergence, as introduced in (3.57):

$$\nabla \cdot \phi = \frac{\partial \phi_1}{\partial x_1} + \frac{\partial \phi_2}{\partial x_2} \tag{3.70}$$

Collecting what we have, (3.69) can be written in the following concise form:

$$\iint_R \nabla \cdot \phi \, dx_1 \, dx_2 = \oint_C \phi \cdot \mathbf{n} \, dt \tag{3.71}$$

This is Gauss's Integral Theorem. It states that the double integral of the divergence of a vector field over any bounded region equals the line integral of the outward normal component of that vector field taken along the boundary.

The right hand side of (3.71) is the net outflow through the boundary according to the flow vector. The normal component indeed is a measure of outflow, because a flow that is tangential to the boundary does not actually leave the region. On the other hand the flow maximally penetrates through the boundary if it is normal, and in all the intermediate cases we get the correct estimate of the outflow by taking the projection on the normal. If the flow is inward, the projection will give a negative number. The normal projection of the flow vanishes if, as mentioned, the flow is tangential, and, of course, also if the flow stagnates.

Going around the entire boundary and integrating as prescribed in (3.71), we get net outflow from the region. Hence the right hand side is net outflow through the boundary, and the left hand side is the integral of the divergence over the area.

Shrinking together the region around a point we are in the limit left with just the divergence at the point, whereas the right hand side says that it equals net outflow, all directions taken in consideration. This shows that the heuristic reasoning around the divergence in terms of the little box indeed is relevant, because Gauss's Theorem is independent of the size and shape of the area. The box can be deformed to an amoeba and then shrunk to a point.

We can now apply the theorem to a flow field which is a gradient, such as (3.58). Then we have:

$$\iint_R \nabla^2 y dx_1 dx_2 = \iint_R \nabla \cdot (\nabla y) dx_1 dx_2 = \oint_C \nabla y \cdot \mathbf{n} dt \qquad (3.72)$$

As the right hand side integrand is the projection of the gradient of the variable y on the direction normal to the boundary, the right hand side integral represents the net change of the variable $y(x_1, x_2)$ as we leave the region, all points of departure taken into consideration. Again shrinking the region around any given point, we in the limit just retain the Laplacian at that point, $\nabla^2 y = \partial^2 y / \partial x_1^2 + \partial^2 y / \partial x_2^2$, on the left, and the net change of that variable as we leave the point, all directions of departure taken in consideration, on the right.

Hence the Laplacian measures the spatial difference of a variable in a point and all its surrounding points in the two dimensional plane. It is therefore the obvious generalisation of the second order space derivative as we consider two dimensions, and enters partial differential equations such as the wave equation, dealt with above.

The simplest PDE in this connection is Laplace's differential equation:

$$\nabla^2 y = 0 \qquad (3.73)$$

or, with the Laplacian operator written out in partial derivatives:

$$\frac{\partial^2 y}{\partial x_1^2} + \frac{\partial^2 y}{\partial x_2^2} = 0 \qquad (3.74)$$

The Laplacian operator obtained its name from equation (3.74) which seeks a function $y = f(x_1, x_2)$ whose gradient is a vector field with zero divergence. Such functions are called potential functions, their study establishes a branch of mathematics called potential theory, and their field of application is the distribution of electrical charges in space. Of course the context in physics is normally three dimensions, but the study of charge distributions on the surface of a capacitor is for instance truly two dimensional. An example of a function for which (3.74) is fulfilled is the saddle $y = x_1^2 - x_2^2$, which the reader can easily check.

3.12 Wave Equation in Two Dimensions: Eigenfunctions

The generalisation of the wave equation to two dimensions is now quite straightforward. We just replace the left hand side of (3.1) by the Laplacian, i.e. by the sum of both second order space derivatives:

$$\frac{\partial^2 y}{\partial x_1^2} + \frac{\partial^2 y}{\partial x_2^2} = \frac{1}{c^2}\frac{\partial^2 y}{\partial t^2} \tag{3.75}$$

or in the more compact style using the symbol for the Laplacian:

$$\nabla^2 y = \frac{1}{c^2}\frac{\partial^2 y}{\partial t^2} \tag{3.76}$$

Dealing with (3.76) we can again try the separation of time and space variables technique, in terms of the attempted solution:

$$y = T(t)S(x_1, x_2) \tag{3.77}$$

Note that the spatial Laplacian operator only works on the spatial function, and hence $\nabla^2(TS) = T\nabla^2 S$, just as $\partial^2 y / \partial t^2 = ST''$ only operates on the temporal function. Hence, substituting from (3.77) into (3.76) and dividing through by the product TS we obtain:

$$\frac{\nabla^2 S}{S} = \frac{1}{c^2}\frac{T''}{T} \tag{3.78}$$

This time we skipped the arguments for the functions. Quite as in the onedimensional case we have a left hand side which only depends on the space coordinates and a right hand side which only depends on the time coordinate. Accordingly, by the same argument as in Section 3.6, we can state that both sides be constant. That constant we now denote $-\lambda^2$.

Accordingly, (3.78) again splits in two equations, a spatial eigenvalue problem:

$$\nabla^2 S + \lambda^2 S = 0 \tag{3.79}$$

and a temporal equation:

$$T'' + \lambda^2 c^2 T = 0 \tag{3.80}$$

The latter, (3.80), presents nothing new. As before the solution is:

$$T(t) = A\cos(\lambda ct) + B\sin(\lambda ct) \tag{3.81}$$

with the frequency depending on the solution to the eigenvalue problem. The former, (3.79), however, is different as it is still a partial differential equation, somewhat in the style of Laplace's equation. Written out in derivatives we have (3.79) as:

$$\frac{\partial^2 S}{\partial x_1^2} + \frac{\partial^2 S}{\partial x_2^2} + \lambda^2 S = 0 \tag{3.82}$$

The solution to such eigenvalue problems highly depends on the shape of the region and on the boundary conditions. We will deal with three stylised cases, the square, the circular disk, and the surface of a sphere. Before that we have to settle a matter which provides a nice use for the concepts from vector analysis.

We have again introduced the separation constant as the negative square $-\lambda^2$, but this only makes sense if $\lambda^2 > 0$, i.e. if λ itself is a real number. To prove this multiply (3.79) through by S and integrate over the region of interest:

$$\iint_R S\nabla^2 S dx_1 dx_2 + \lambda^2 \iint_R S^2 dx_1 dx_2 = 0 \qquad (3.83)$$

Now, we can use the chain rule for differentiation in vector analysis too. Hence considering the vector: $S\nabla S$ and applying the divergence, we get $\nabla \cdot (S\nabla S) = \nabla S \cdot \nabla S + S\nabla^2 S$. The first term on the right is the scalar product of the gradient vector with itself $(\nabla S)^2 = (\partial S / \partial x_1)^2 + (\partial S / \partial x_2)^2$, and so nonnegative, whereas the second term is the first of the integrands in equation (3.83). Now apply Gauss's Integral theorem to $\nabla \cdot (S\nabla S) = (\nabla S)^2 + S\nabla^2 S$:

$$\iint_R \nabla \cdot (S\nabla S) dx_1 dx_2 = \oint_C S\nabla S \cdot \mathbf{n} dt = 0 \qquad (3.84)$$

The zero comes from a boundary condition stating that the system be at rest along the boundary C at all times, i.e. that the integrand be zero. Now substitute for $\nabla \cdot (S\nabla S) = (\nabla S)^2 + S\nabla^2 S$ in (3.84) and we have:

$$\iint_R (\nabla S)^2 dx_1 dx_2 + \iint_R S\nabla^2 S dx_1 dx_2 = 0 \qquad (3.85)$$

which gives a substitution for the first term in equation (3.83), so that we get:

$$-\iint_R (\nabla S)^2 dx_1 dx_2 + \lambda^2 \iint_R (S)^2 dx_1 dx_2 = 0 \qquad (3.86)$$

As both integrals are nonnegative, λ^2 is indeed a nonnegative number. It is zero in one case, when the gradient is identically zero on the entire region, otherwise it is positive. Note that this proof of the nonnegativity of λ^2 rests on the fact that we choose the region so that the system always is at rest on the boundary, $S = 0$ on C, i.e. all the boundary points are nodes. Though it is possible to choose more complicated boundary conditions, it is fairly nonrestrictive to choose the region of analysis such that the entire boundary becomes a node line. Observe that in two dimensions the node points are replaced by node lines or curves. There may, of course be additional interior node lines in two-dimensional space, in addition to the boundary, defining more or less refined subdivisions of the region.

3.13 The Square

The easiest type of region to deal with is the square, for which we can again take a side length of π. Then we can continue the process of variable separation by trying the solution:

$$S(x_1,x_2)=X_1(x_1)X_2(x_2) \tag{3.87}$$

Obviously we have $\partial^2 S/\partial x_1^2 = X_1''X_2$ and $\partial^2 S/\partial x_2^2 = X_1 X_2''$. Substituting in (3.82) and dividing through by $S(x_1,x_2)=X_1(x_1)X_2(x_2)$ we get:

$$\frac{X_1''}{X_1}+\frac{X_2''}{X_2}+\lambda^2=0 \tag{3.88}$$

Accordingly, we have two different terms X_1''/X_1 and X_2''/X_2, which have to be constant. Let us denote them $-i^2$ and $-j^2$ respectively. Then (3.88) splits in the two equations:

$$X_1''+i^2X_1=0 \qquad X_2''+j^2X_2=0 \tag{3.89}$$

which have the obvious solutions:

$$X_1=\sin(ix_1) \qquad X_2=\sin(jx_2) \tag{3.90}$$

Of course, i and j can be any integers in view of $X_1(0)=X_1(\pi)=0$ and $X_2(0)=X_2(\pi)=0$, which is what the boundary conditions boil down to once the variables are separated.

Even if the choice of i and j is free, we see that if we solve for X_1''/X_1 and X_2''/X_2 from equations (3.89) and substitute in (3.88) we get:

$$\lambda^2 = i^2 + j^2 \tag{3.91}$$

so the choice determines the value of λ^2 to be substituted back in the temporal equation (3.80) and its solution (3.81).

Assembling the pieces, we have the solution to the eigenvalue problem:

$$S(x_1, x_2) = \sin(ix_1)\sin(jx_2) \tag{3.92}$$

and from (3.77), (3.81), and (3.92) the solution

$$y = \left(A\cos(\lambda ct) + B\sin(\lambda ct)\right)\sin(ix_1)\sin(jx_2) \tag{3.93}$$

where the integers i and j are free but λ must fulfil the constraint (3.91).

Finally, the modes can be combined, and we have the general solution:

$$y = \sum_{i=1}^{\infty}\sum_{j=1}^{\infty}\left(A_{ij}\cos(\lambda_{ij}ct) + B_{ij}\sin(\lambda_{ij}ct)\right)\sin(ix_1)\sin(jx_2) \tag{3.94}$$

where $\lambda_{ij} = \sqrt{i^2 + j^2}$ according to (3.91). We of course index both this frequency component and the coefficients. Quite as in Section 3.7 we can set $t = 0$ in (3.94) and so obtain the initial distribution, which now is a surface over x_1, x_2-space:

$$\bar{y} = \sum_{i=1}^{\infty}\sum_{j=1}^{\infty} A_{ij}\sin(ix_1)\sin(jx_2) \tag{3.95}$$

In the same way, by first differentiating with respect to time and then putting $t = 0$, we obtain the initial velocity distribution:

$$\bar{y}' = \sum_{i=1}^{\infty}\sum_{j=1}^{\infty} \lambda_{ij} c B_{ij} \sin(ix_1)\sin(jx_2) \tag{3.96}$$

Again we multiply these expressions through by $\sin(hx_1)\sin(kx_2)$ and integrate over the region, i.e. with respect to x_1 and x_2 over the interval $(0,\pi)$. Each term in the series, for which either $h \neq i$ or $k \neq j$, becomes zero, so only those for $h = i$ and $k = j$ remain. There remains just one in the series, for which each of the integrals on the right again gets the value $\pi/2$ or, as we have products of two, $\pi^2/4$. So, solving for the Fourier coefficients we have:

$$A_{hk} = \frac{4}{\pi^2}\int_0^{\pi}\int_0^{\pi} \bar{y}\sin(hx_1)\sin(kx_2)dx_1dx_2 \tag{3.97}$$

and

$$B_{hk} = \frac{4}{\pi^2}\int_0^{\pi}\int_0^{\pi} \frac{\bar{y}'}{\lambda_{hk}c}\sin(hx_1)\sin(kx_2)dx_1dx_2 \tag{3.98}$$

As we see from (3.94) the general solution is a combination of all the different vibration modes, and from (3.97)-(3.98) we see how the Fourier coefficients can be fitted to suit any initial displacement and velocity distributions on the square. This is quite as it was in the one dimensional case. There are, however a few features which are specific for two dimensions.

First, despite the lack of friction, the frequencies of the modes are no longer in rational proportions, i.e. they are not harmonic. In the context of sound this means that whereas an ideal string can only produce notes of a specific pitch, though coloured to different "timbres" by different mixes of the harmonic overtones, a soundboard can produce mere noise. This is due to the square root $\lambda_{ij} = \sqrt{i^2 + j^2}$ which results in nonintegral numbers. Pythagoras, who believed that the entire universe was built upon harmonic proportions, and who let drown a former student who insisted that $\sqrt{2}$ was irrational, would have hated this conclusion.

The irrationality of eigenvalues for the square also implies that for higher values of i and j the frequencies come quite densely packed in the intervals between integers, which means that a two dimensional membrane easily comes in near resonance to almost any note and therefore is an excellent amplifier.

As an example, three octaves, $2^3 = 8$, above the basic pitch of the square we obtain six nonintegral frequencies, approximately 8.06, 8.24, 8.48, 8.54, 8.60, 8.94 before the next harmonic note.

Second, the vibration modes defined by the node lines corresponding to a given frequency are no longer unique. As an example, suppose we deal with the series (3.94), and that we focus on the spatial aspects of the solution, by considering the initial distribution (3.95). Further we concentrate on the terms leading to the single eigenvalue $\sqrt{10} = \sqrt{1^2 + 3^2}$:

$$A_{13}\sin(x_1)\sin(3x_2) + A_{31}\sin(3x_1)\sin(x_2) \tag{3.99}$$

Note that the eigenvalue, and therefore the temporal frequency, corresponding to these modes, is not affected by which values we give the constants. Using the formula for expansion of the sine of thrice an argument, we see that (3.99) factorises into:

$$\sin(x_1)\sin(x_2)\left(4A_{13}\cos^2(x_1) + 4A_{31}\cos^2(x_2) - A_{13} - A_{31}\right) = 0 \tag{3.100}$$

We put the expression equal to zero in order to find the node lines, i.e. the lines of rest. The factors before the parenthesis take on zero value if any of the arguments takes on the values 0 or π, and they therefore just define the edges of the square studied.

To find the internal node lines we have to put the parenthesis equal to zero, i.e.

$$4A_{13}\cos^2(x_1) + 4A_{31}\cos^2(x_2) = A_{13} + A_{31} \tag{3.101}$$

There are now several possibilities. Suppose first that $A_{31} = 0$. Then we get $\cos^2(x_1) = 1/4$ which has the solutions $x_1 = \pi/3, 2\pi/3$, which means that there are two node lines dividing the square vertically in three equal strips.

This is quite like the case in one dimension, the square oscillates in three equal pieces, though, unlike the case of the string, it is not the third harmonic we get, but rather the irrational $\sqrt{10}$. We can, however, imagine that, for the same irrational eigenvalue, the square could as well oscillate in horizontal strips. This indeed is so, which we find out by putting $A_{13}=0$, in which case (3.101) reduces to $\cos^2(x_2)=1/4$. Then $x_2=\pi/3, 2\pi/3$ which corroborates the conjecture.

But the square can also be split by the diagonals. We see this by putting $A_{13}+A_{31}=0$ in (3.101), which yields $\cos^2(x_1)=\cos^2(x_2)$ with the solutions $x_1=x_2$ *and* $x_1+x_2=\pi$. Accordingly, we get *both* diagonals at once, and the mode of oscillation is in four triangles instead of in three strips. Not even the number of oscillating pieces hence need be the same for one given eigenvalue.

The square can also oscillate in two pieces only, and yet yield the same eigenvalue. To see this, put $A_{13}=A_{31}$. Then we get $\cos^2(x_1)+\cos^2(x_2)=1/2$ from (3.101), which defines a closed curve, almost like a circle, splitting the square in one central part and a periphery, oscillating in opposite phase, and separated by the nodal curve where the points are always at rest.

By these four cases we have exhausted all the different pure modes of oscillation associated with the eigenvalue $\sqrt{10}$. There are: three vertical strips, three horizontal strips, four triangles split by the diagonals, and two parts separated by a closed curve. Note that this was the result of just interchanging the two values 1 and 3 for *i* and *j* and giving the coefficients various values. In Figure 3.4 we display all these various vibration modes, and also some intermediate cases.

But this is not yet the end of the story. For higher eigenvalues a given sum of squares can even be obtained in different ways. Let us for instance take $\sqrt{325}=\sqrt{1^2+18^2}=\sqrt{6^2+17^2}=\sqrt{10^2+15^2}$. It can be obtained for three different pairs of integers, which can then be interchanged as before. Accordingly, there will be six different terms in the solution (3.94) which are associated with one eigenvalue, and they produce very complex spatial modes.

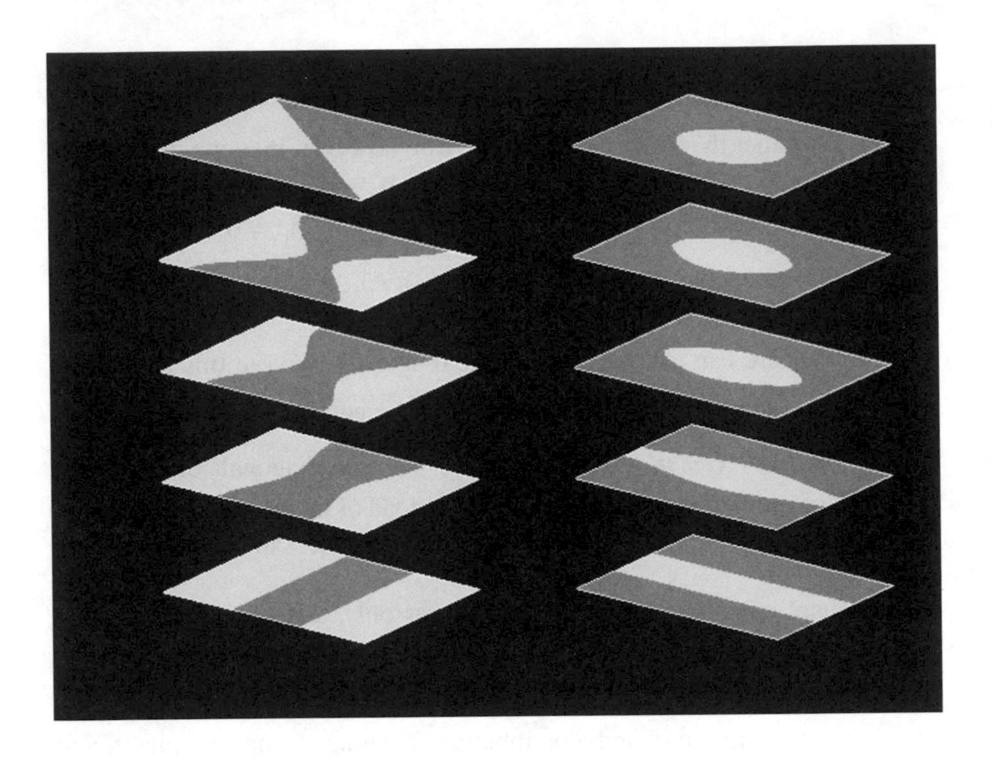

Fig. 3.4. Vibration modes for the square associated with eigenvalue $\sqrt{10}$.

3.14 The Circular Disk

Suppose next that the region is a disk, and that the boundary conditions state that the system be at rest along its circular boundary. The way of solution we are going to present is applicable whenever the boundary conditions prescribe values for points only with reference to the length and the angle of radius vector, but we can as well stick to this simplest type prescribing rest on the entire boundary.

It then seems that we should introduce polar coordinates:

$$x_1 = \rho\cos\theta \tag{3.102}$$

$$x_2 = \rho\sin\theta \tag{3.103}$$

We now need to work out the Laplacian in these polar coordinates for the function

$$S(x_1, x_2) = S(\rho\cos\theta, \rho\sin\theta) \tag{3.104}$$

for substitution into equation (3.82). To this end we differentiate (3.104) with respect to the polar coordinates to obtain:

$$\frac{\partial S}{\partial \rho} = \cos\theta \frac{\partial S}{\partial x_1} + \sin\theta \frac{\partial S}{\partial x_2} \tag{3.105}$$

$$\frac{\partial S}{\partial \theta} = -\rho\sin\theta \frac{\partial S}{\partial x_1} + \rho\cos\theta \frac{\partial S}{\partial x_2} \tag{3.106}$$

Though more messy, the calculation of the second derivatives is quite straightforward too:

$$\frac{\partial^2 S}{\partial \rho^2} = \cos^2\theta \frac{\partial^2 S}{\partial x_1^2} + 2\cos\theta\sin\theta \frac{\partial^2 S}{\partial x_1 \partial x_2} + \sin^2\theta \frac{\partial^2 S}{\partial x_2^2} \tag{3.107}$$

$$\begin{aligned}\frac{\partial^2 S}{\partial \theta^2} = {} & \rho^2\sin^2\theta \frac{\partial^2 S}{\partial x_1^2} - 2\rho^2\cos\theta\sin\theta \frac{\partial^2 S}{\partial x_1 \partial x_2} + \rho^2\cos^2\theta \frac{\partial^2 S}{\partial x_2^2} \\ & - \rho\cos\theta \frac{\partial S}{\partial x_1} - \rho\sin\theta \frac{\partial S}{\partial x_2}\end{aligned} \tag{3.108}$$

We can now divide (3.108) through by ρ^2 and add to (3.107). In this way we obtain, recalling that as usual $\cos^2\theta + \sin^2\theta = 1$,

$$\frac{\partial^2 S}{\partial \rho^2} + \frac{1}{\rho^2}\frac{\partial^2 S}{\partial \theta^2} = \frac{\partial^2 S}{\partial x_1^2} + \frac{\partial^2 S}{\partial x_2^2} - \frac{1}{\rho}\left(\cos\theta \frac{\partial S}{\partial x_1} + \sin\theta \frac{\partial S}{\partial x_2}\right) \tag{3.109}$$

But the parenthesis equals the right hand side of (3.105), so substituting and rearranging we finally have:

$$\frac{\partial^2 S}{\partial \rho^2}+\frac{1}{\rho}\frac{\partial S}{\partial \rho}+\frac{1}{\rho^2}\frac{\partial^2 S}{\partial \theta^2}=\frac{\partial^2 S}{\partial x_1^2}+\frac{\partial^2 S}{\partial x_2^2} \tag{3.110}$$

which is the Laplacian in polar coordinates. Substituting this in (3.82) and multiplying by ρ^2 we get the wave equation in polar coordinates:

$$\rho^2\frac{\partial^2 S}{\partial \rho^2}+\rho\frac{\partial S}{\partial \rho}+\frac{\partial^2 S}{\partial \theta^2}+\lambda^2\rho^2 S=0 \tag{3.111}$$

The stage is now again set for a further separation of coordinates in terms of:

$$S=\mathrm{P}(\rho)\Theta(\theta) \tag{3.112}$$

Substituting from (3.112) in (3.111), and dividing through by S we get:

$$\rho^2\frac{\mathrm{P}''(\rho)}{\mathrm{P}(\rho)}+\rho\frac{\mathrm{P}'(\rho)}{\mathrm{P}(\rho)}+\frac{\Theta''(\theta)}{\Theta(\theta)}+\lambda^2\rho^2=0 \tag{3.113}$$

Again we have: three terms that depend on ρ only, and one term that depends on θ only. If we now put the quotient Θ''/Θ equal to the constant $-i^2$, then (3.113) splits into:

$$\Theta''(\theta)+i^2\Theta(\theta)=0 \tag{3.114}$$

and

$$\rho^2\mathrm{P}''(\rho)+\rho\mathrm{P}'(\rho)+\left(\lambda^2\rho^2-i^2\right)\mathrm{P}(\rho)=0 \tag{3.115}$$

We again have two different ordinary differential equations. As a solution for (3.114) we can choose:

$$\Theta=\cos(i\theta) \tag{3.116}$$

Sine or cosine makes no difference, neither does any phase lead or lag, because (3.116) just defines $2i$ equally spaced radials as node lines which divide the entire disk into $2i$ sectors oscillating in opposite phase.

Equation (3.115), which is nonlinear, is well known and is called Bessel's differential equation of order i. Its solutions are called Bessel Functions of the first and second kind (conventionally denoted by the symbols J and K). Only the former make sense for the case of a disk including the centre, because the other take on infinite values at the origin. The solution functions to (3.115) are:

$$\mathrm{P} = J_i(\lambda\rho) \tag{3.117}$$

Though not as well known as sine or cosine functions, the Bessel functions $J_i(z)$ are well defined and tabulated in most standard, printed or electronic, handbooks. Like the sine they oscillate and define a number of zeros as nodes. We hence see that the solution to the Bessel equation (3.115) depends on the number of radials we have chosen in the solution for the radial equation (3.116). The solution (3.117) itself defines a number concentric rings as node lines.

The combination of radial and sectorial modes

$$S = \mathrm{P}(\rho)\Theta(\theta) = J_i(\lambda\rho)\cos(i\theta) \tag{3.118}$$

defines a net of internal node lines, consisting of both radials and concentric circles. We have the boundary condition that the oscillations be always at rest on the boundary circle. So, assume that the region has unit radius. Then from (3.118) we see that we must have:

$$J_i(\lambda) = 0 \tag{3.119}$$

which determines the value of λ to be used in the temporal equation (3.81). The complete solution for a given mode is hence:

$$y = \left(A\cos(\lambda t) + B\sin(\lambda t)\right)J_i(\lambda\rho)\cos(i\theta) \tag{3.120}$$

where λ is any of the zeros as determined by (3.119). Again we can superpose the different modes and determine the Fourier coefficients so as to fit any given initial distributions, but this time we contend ourselves with pointing out the fact that it can be done.

3.15 The Sphere

The final example of a spatial region is the two-dimensional surface of a sphere. While the square and disk can be idealized examples of urban regions, such as the classical Roman camp or a medieval city, surrounded by the shortest possible and hence most easily defended circular walls, it is the entire globe that comes into mind in the case of the sphere. This region is special, as it is curved. It is also one, which, taken as a whole, has no boundary, and hence no boundary conditions.

Points on the sphere are most conveniently identified by spherical coordinates, the colatitude θ and the longitude ϕ, just as by navigators at sea.

Embedding the sphere in three-dimensional Euclidean space with the centre at the origin, we can identify the locations in threespace by the coordinate transformation:

$$x_1 = \sin\theta\cos\phi \tag{3.121}$$

$$x_2 = \sin\theta\sin\phi \tag{3.122}$$

$$x_3 = \cos\theta \tag{3.123}$$

provided we for simplicity normalise the radius of the sphere to unity. If we let θ vary in the interval $(0,\pi)$ and ϕ in the interval $(0,2\pi)$, then we can locate the entire surface of the sphere. Fixing θ we get parallel circles, whereas fixing ϕ we get great circles through the poles according to (3.121)-(3.123).

In view of the three coordinates of the embedding space (x_1, x_2, x_3) we should, of course, generalise the Laplacian to the sum of the three direct second order derivatives $\partial^2 S / \partial x_1^2 + \partial^2 S / \partial x_2^2 + \partial^2 S / \partial x_3^2$. In threespace we should then also precede each of the coordinate function in (3.121)-(3.123) by a radius coordinate ρ. Then the Laplacian in the spherical coordinates could be worked out along the same lines as in the case of the disk, though the derivatives with respect to the radius coordinate would drop out as we consider things on the surface of the sphere and so keep radius constant at unity. Skipping the details the Laplacian in the coordinates (3.121)-(3.123) becomes:

$$\nabla^2 S = \frac{\partial^2 S}{\partial \theta^2} + \cos\theta \frac{\partial S}{\partial \theta} + \frac{1}{\sin^2\theta} \frac{\partial^2 S}{\partial \phi^2} \tag{3.124}$$

The eigenvalue problem (3.79) accordingly becomes:

$$\frac{\partial^2 S}{\partial \theta^2} + \cos\theta \frac{\partial S}{\partial \theta} + \frac{1}{\sin^2\theta} \frac{\partial^2 S}{\partial \phi^2} + \lambda^2 S = 0 \tag{3.125}$$

We can now again attempt the separation of variables:

$$S = \Theta(\theta)\Phi(\phi) \tag{3.126}$$

The calculation of derivatives being obvious, substitution in (3.125) and division by S as usual yields:

$$\frac{\Theta''(\theta)}{\Theta(\theta)} + \cos\theta \frac{\Theta'(\theta)}{\Theta(\theta)} + \frac{1}{\sin^2\theta} \frac{\Phi''(\phi)}{\Phi(\phi)} + \lambda^2 = 0 \tag{3.127}$$

Suppose we put

$$\Phi''(\phi) + i^2 \Phi(\phi) = 0 \tag{3.128}$$

Then substituting for $\Phi''(\phi) / \Phi(\phi) = -i^2$ in (3.127) and slightly rearranging we get:

$$\Theta''(\theta) + \cos\theta\,\Theta'(\theta) + \left(\lambda^2 - \frac{i^2}{\sin^2\theta}\right)\Theta(\theta) = 0 \tag{3.129}$$

Provided we put:

$$\lambda^2 = j(j+1) \tag{3.130}$$

we can write (3.129) as

$$\Theta''(\theta) + \cos\theta\,\Theta'(\theta) + \left(j(j+1) - \frac{i^2}{\sin^2\theta}\right)\Theta(\theta) = 0 \tag{3.131}$$

which again is a standard equation, called Legendre's differential equation, and has solutions in terms of Legendre Polynomials:

$$\Theta = P_j^i(\cos\theta) \tag{3.132}$$

whereas (3.128) is again solved by

$$\Phi = \cos(i\phi) \tag{3.133}$$

The complete solution to the eigenvalue problem then is:

$$S = P_j^i(\cos\theta)\cos(i\phi) \tag{3.134}$$

The Legendre polynomials being at least as unfamiliar as the Bessel functions, we have to give a simple definition of $P_j^i(\cos\theta)$. The functions are defined for $i \le j$, and the most handy definition is:

$$P_j^i(\cos\theta) = \frac{\sin^j\theta}{2^j j!}\frac{d^{i+j}(\sin^{2j}\theta)}{d(\cos\theta)^{i+j}} \tag{3.135}$$

Fig. 3.5. Mixed vibration mode for the sphere.

The first function, for which (3.135) hardly works, $P_0^0(\cos\theta)$ is defined to equal unity. We easily see that if we put $i = j = 0$ in (3.131) then $\Theta = 1$ indeed works, all its derivatives becoming zero.

With a little more labour, but still using the elementary rules for differentiation, we get the spatial modes as shown in the Table 3.1 below.

Observe that there are two pure cases for the modes: for $i = 0$, and for $i = j$.

As we see from the table, the former, for $i = 0$, only involve terms in θ. They hence just define parallel circles. For $j = 1$ we just get the equator from $\cos\theta = 0$, for $j = 2$ we get two parallel circles from $\cos^2\theta = 1/3$, and for $j = 3$ we get the equator *and* two other parallel circles from $\cos\theta = 0$ and $\cos^2\theta = 3/5$ respectively. These oscillation modes with parallel circles as node lines are called *zonal*.

On the other hand, the latter modes, for $i = j$, only involve the coordinate ϕ, and so define great circles through the poles as node lines, one, two, or three according to the factors: $\cos\phi$ for $j = 1$, $\cos 2\phi$ for $j = 2$, and $\cos 3\phi$ for

$j = 3$. It is true that we also find powers of $\sin\theta$, but they only define degenerate parallel circles of zero radius, i.e. just points, at the poles, and those points we already have anyhow, because the great circles pass them. The oscillation modes with polar great circles as node lines are called *sectorial*.

The rest of the modes for $i < j$ are called mixed. In Figure 3.5 we display a more complicated mixed mode.

Table 3.1 Spatial modes for the sphere.

j	i	$P_j^i(\cos\theta)\cos i\phi$
0	0	1
1	0	$\cos\theta$
1	1	$\sin\theta\cos\phi$
2	0	$3\cos^2\theta - 1$
2	1	$\cos\theta\sin\theta\sin\phi$
2	2	$\sin^2\theta\cos 2\phi$
3	0	$5\cos^3\theta - 3\cos\theta$
3	1	$\sin\theta(5\cos^2\theta - 1)\cos\phi$
3	2	$\cos\theta\sin^2\theta\cos 2\phi$
3	3	$\sin^3\theta\cos 3\phi$

Note again that we do not have any boundary conditions for the sphere as there is no boundary. On the other hand the initial conditions can be imposed quite as usual, and there is again a way of computing the Fourier coefficients by multiplying initial distributions with the Legendre functions and integrating. The reason this always works is that the different eigenfunctions are what is called orthogonal. Multiplying equal functions and integrating yields a nonzero integral, multiplying different ones results in zero integrals. This sense of orthoganality is a generalisation of orthogonality for the eigenvectors in ordinary differential equations. The reader can find a lot more information in the classics, Duff and Naylor, and Courant and Hilbert.

3.16 Nonlinearity Revisited

Consider again the nonlinear case discussed in Section 3.9, though now in two-dimensional space. Equation (3.47) then reads:

$$\frac{\partial^2 y}{\partial x_1^2}+\frac{\partial^2 y}{\partial x_2^2}=\frac{1}{c^2}\frac{\partial^2 y}{\partial t^2}-\left(\frac{\partial y}{\partial t}-\left(\frac{\partial y}{\partial t}\right)^3\right) \tag{3.136}$$

It is obvious that we can consider the same heuristic solution as in one dimension: Linear spatial wave shapes, i.e. now plane facets, and movement at any of the three constant speeds, defined by (3.49), making the right hand sides of (3.136) and (3.47) zero. As the facets are plane, all first order space derivatives become constant, and hence all second order ones zero. Likewise, temporal first derivatives being constant (at any of the three possible values), the second time derivative becomes zero too. So, with all second derivatives zero, and the first temporal derivative equal to any of the three roots (3.49), equation (3.136) is satisfied.

Again there are continuity problems at the junctions between the plane facets, as there are at the moments in time when a point shifts from being at rest to moving in either direction, but this is as it was in one dimension. Such shifting plane patterns can still be attractive, in two dimensions as well as in one.

There is one obvious candidate for a solution which we have to consider now: The spatially homogeneous equilibrium solution $y(x_1,x_2,t)\equiv 0$, which obviously satisfies (3.136). It is, however, unstable. To show this gives a nice opportunity to use some of the concepts from vector analysis introduced above.

Let us first rephrase (3.136) in a more compact style, by writing $\dot{y}=\partial y/\partial t$, and $\ddot{y}=\partial^2 y/\partial t^2$, in spite of the fact that they are partial derivatives, and the Laplacian $\nabla^2 y=\partial^2 y/\partial x_1^2+\partial^2 y/\partial x_2^2$ for the left hand side. Also, put the constant propagation speed $c = 1$, to make the formulas shorter. This actually makes no difference. Hence we write:

$$\nabla^2 y = \ddot{y} - \left(\dot{y} - \dot{y}^3\right) \tag{3.137}$$

Next, multiply through (3.137) by $\dot{y}$, and rearrange:

$$\dot{y}\ddot{y} - \dot{y}\nabla^2 y = \dot{y}^2 - \dot{y}^4 \tag{3.138}$$

Note also that:

$$\dot{y}\ddot{y} = \frac{1}{2}\frac{\partial}{\partial t}\left(\dot{y}^2\right) \tag{3.139}$$

Integrating (3.138) over the region of interest and using (3.139) we get:

$$\frac{1}{2}\frac{\partial}{\partial t}\iint_R (\dot{y})^2 dx_1 dx_2 - \iint_R \dot{y}\nabla^2 y dx_1 dx_2 = \iint_R \dot{y}^2 dx_1 dx_2 - \iint_R \dot{y}^4 dx_1 dx_2 \tag{3.140}$$

Further, consider the divergence:

$$\nabla \cdot \left(\dot{y}\nabla y\right) = \nabla \dot{y} \cdot \nabla y + \dot{y}\nabla^2 y \tag{3.141}$$

and integrate:

$$\iint_R \nabla \cdot \left(\dot{y}\nabla y\right) dx_1 dx_2 = \iint_R \nabla \dot{y} \cdot \nabla y dx_1 dx_2 + \iint_R \dot{y}\nabla^2 y dx_1 dx_2 \tag{3.142}$$

Applying Gauss's Divergence Theorem to the left hand side of (3.142) we have:

$$\iint_R \nabla \cdot \left(\dot{y}\nabla y\right) dx_1 dx_2 = \oint_C \dot{y}\nabla y \cdot \mathbf{n} ds = 0 \tag{3.143}$$

The right hand side is zero, because we assume the boundary points to be in constant equilibrium, $\dot{y} \equiv 0$ on C. Next, applying (3.143) to (3.142) we obtain:

$$\iint_R \dot{y}\nabla^2 y dx_1 dx_2 = -\iint_R \nabla\dot{y}\cdot\nabla y dx_1 dx_2 \tag{3.144}$$

Note also that $(\partial/\partial t)(\nabla y)^2 = 2\nabla\dot{y}\cdot\nabla y$, so that (3.144) can be written:

$$\iint_R \dot{y}\nabla^2 y dx_1 dx_2 = -\frac{1}{2}\frac{\partial}{\partial t}\iint_R (\nabla y)^2 dx_1 dx_2 \tag{3.145}$$

which provides a handy substitution for (3.140). Hence, finally:

$$\frac{1}{2}\frac{\partial}{\partial t}\iint_R \left((\dot{y})^2 + (\nabla y)^2\right) dx_1 dx_2 = \iint_R \dot{y}^2 dx_1 dx_2 - \iint_R \dot{y}^4 dx_1 dx_2 \tag{3.146}$$

We see that on the left we integrate a nonnegative sum of squares, and that its integral is zero only if the system is in equilibrium $\dot{y} \equiv 0$, and the spatial pattern $\nabla y \equiv (0,0)$ is homogeneous everywhere. Otherwise it is strictly positive. The time derivative of this left hand side integral is positive, i.e. the system diverges from homogeneous equilibrium, whenever the right hand side of (3.146) is positive, it is negative, i.e. the system converges towards homogeneous equilibrium, whenever the right hand side is negative.

On the right we have the difference of the integrals of the time derivatives to the second power and to the fourth power. Accordingly, when the rates of change are low, i.e. we are close to equilibrium, then the second power dominates, and the system moves away from equilibrium, whereas for large rates of change the fourth power dominates, so the system approaches equilibrium.

The system (3.136) sets the stage for bounded motion for ever without exploding and without actually converging on homogeneous equilibrium.

3.17 Tessellations and the Euler-Poincaré Index

The similarity between one and two dimensions being pointed out, we might now imagine that there is nothing new to nonlinearities in two dimensions. This, however, is completely erroneous. Again we face the fact that in two dimensions regions have shape in addition to size. Considering shapes of waves made from plane facets, we conclude that the pattern of nodal curves must consist of straight lines forming some set of polygons.

Quite as in the onedimensional case, we are interested in *regular* patterns, now called *tessellations*, of symmetric polygons of equal size. As already known to Kepler around 1600, there exist only three regular tessellations of the plane: by triangles, by squares, and by hexagons.

In the triangular tessellation, six triangles come together in each corner. As each of the angles of an equilateral triangle is $60°$, six of them make a full round of $360°$. Likewise, in the case of squares, each angle is $90°$, so four of them in a corner again make $360°$. Finally, three hexagons can meet in each corner, so, each of the angles being $120°$, three make $360°$ as before. This is it.

Why is the octagon not a candidate? Its angles are $135°$ each, so two of them at a corner occupy as much as $270°$. This leaves over an angle of $(360°-270°) = 90°$ only, which can only accommodate a little square, but not another octagon. Accordingly this does not work. The same is true already for the heptagon, but, the computation involving fractions of degrees, we leave that case.

In the list we admitted polygons with 3, 4, and 6 corners, while 7 and 8 were too much, as would any higher number be, of course. But what about 5, i.e. pentagons? This is an interesting issue. The angles of a pentagon are $108°$. Three of them at a corner make $324°$, which leaves an excess of $(360°-324°) = 36°$ only. This does not even accommodate an equilateral triangle. However, suppose the surface was curved, a solid made of pentagonal patches, so that three patches meet at each corner. Then twelve such patches would in fact close a solid called the dodecahedron.

Even this was known to Kepler: That regular solids can be made with triangular, square, or pentagonal faces, but no other. There are three with triangular faces: the tetrahedron, the octahedron, and the icosahedron, with 4, 8, and 20 faces respectively. There is just one with 6 square faces, the cube, and one with 12 pentagonal ones, the decahedron.

The reason hexagons do not work on solids is exactly the same as that they do work in the plane, viz. that three of them at a junction fall flat in the plane. So, the surface cannot be curved. There is a deep theorem due to Euler and Poincaré which summarises all this. The number of *vertices*, plus the number of *faces*, minus the number of *edges* is a characteristic number for each type of surface.

This *Euler-Poincaré Index* must be 1 for the plane, but 2 for a simply connected solid without holes. As an example take the cube. It has 8 vertices, 6 faces, and 12 edges. The index is (8 + 6 -12) = 2, as it should. Or take the tetrahedron, with 4 vertices, 4 faces, and 6 edges. Again (4 + 4 - 6) = 2. This holds for all the other regular polyhedra, and also for any more complex solid, such as a football made of alternating pentagonal and hexagonal patches. On the other hand, take the plane, paved by just one hexagon. We have 6 vertices, 1 face, and 6 edges, so the index is (6 + 1 - 6) = 1. Now make them two. The vertices increase to 10 (two being shared), the faces to 2, and the edges to 11 (one being shared). As a result we get the index (10 + 2 - 11) = 1, exactly as before.

3.18 Nonlinear Waves on the Square

Given we can consider triangular, square, and hexagonal tessellations, the question arises whether they are equally good candidates for standing waves. One could suspect the hexagons to be particularly good as they turn up so often in pattern formation, both in theory and experiment/experience.

For the case of standing waves, the hexagonal nets of node lines would, however, not do. The reason is that the regions on either side of a node line always have to display alternating motion in opposite phase. In a tessellation of hexagons, three of them meet in a point. It is, however, impossible to orient a cycle of three regions (or any other odd number) so that they all are in opposite phase. We need an even number of regions, so both squares and triangles would do, but not hexagons.

It is simplest to analyse the case of squares, so we take that as an example. For a start let us return to the linear wave equation (3.76), which we restate here for convenience

$$\nabla^2 y = \frac{1}{c^2}\frac{\partial^2 y}{\partial t^2} \tag{3.147}$$

Suppose we deal with a region in the shape of a square of side π , as already discussed in Section 3.13. Consider a two dimensional wave with the initial shape of a pyramid on this square base of node lines.

The pyramid has its top located over the centre $(\pi/2, \pi/2)$ of its square base, and its sides aligned along the x_1, x_2 -axes.

Assume again that the initial velocity distribution is zero, as in previous examples, but that we have the described initial displacement distribution: $\bar{y} = \left\| x_1 + x_2 - \pi \right| + \left| x_1 - x_2 \right\|$. This is the succinct formula for a pyramid. Some software for symbolic mathematics, such as Derive, is able to evaluate the integrals (3.97) for A_{ij} directly from this expression for $\bar{y}$, but, in order to understand better what is going on, we use the more pedestrian way of braking the pyramid down into four triangular flat surfaces as follows:

$$\text{W)}\quad \bar{y} = x_1 \qquad \text{if } 0 \le x_1 < \pi/2 \quad \text{and} \quad x_1 \le x_2 \le \pi - x_1 \tag{3.148}$$

$$\text{E)}\quad \bar{y} = \pi - x_1 \qquad \text{if } \pi/2 \le x_1 \le \pi \quad \text{and} \quad \pi - x_1 \le x_2 \le x_1 \tag{3.149}$$

$$\text{S)}\quad \bar{y} = x_2 \qquad \text{if } 0 \le x_2 < \pi/2 \quad \text{and} \quad x_2 \le x_1 \le \pi - x_2 \tag{3.150}$$

$$\text{N)}\quad \bar{y} = \pi - x_2 \qquad \text{if } \pi/2 \le x_2 \le \pi \quad \text{and} \quad \pi - x_2 \le x_1 \le x_2 \tag{3.151}$$

The inequalities define the W(est), E(ast), S(outh), and N(orth) triangles into which the diagonals divide the square. As we see, the surface is defined by four different linear functions over the four triangles. Together they define a right pyramid standing on the square basis described above. In order to derive the Fourier coefficients for this initial displacement pattern, we need to evaluate the four different integrals for the four triangles. These integrals will be denoted by the same symbols as the regions above.

From (3.97) we now have:

$$A_{ij} = \frac{4}{\pi^2}(W + E + S + N) \tag{3.152}$$

with the following integrals to be substituted

Fig. 3.6. Standing wave on a square tessellation.

$$W = \int_{0}^{\pi/2} \int_{x_1}^{\pi - x_1} x_1 \sin(ix_1)\sin(jx_2)dx_2dx_1 \tag{3.153}$$

$$E = \int_{\pi/2}^{\pi} \int_{\pi - x_1}^{x_1} (\pi - x_1)\sin(ix_1)\sin(jx_2)dx_2dx_1 \tag{3.154}$$

$$S = \int_{0}^{\pi/2} \int_{x_2}^{\pi - x_2} x_2 \sin(ix_1)\sin(jx_2)dx_1dx_2 \tag{3.155}$$

$$N = \int_{\pi/2}^{\pi} \int_{\pi - x_2}^{x_2} (\pi - x_2)\sin(ix_1)\sin(jx_2)dx_1dx_2 \tag{3.156}$$

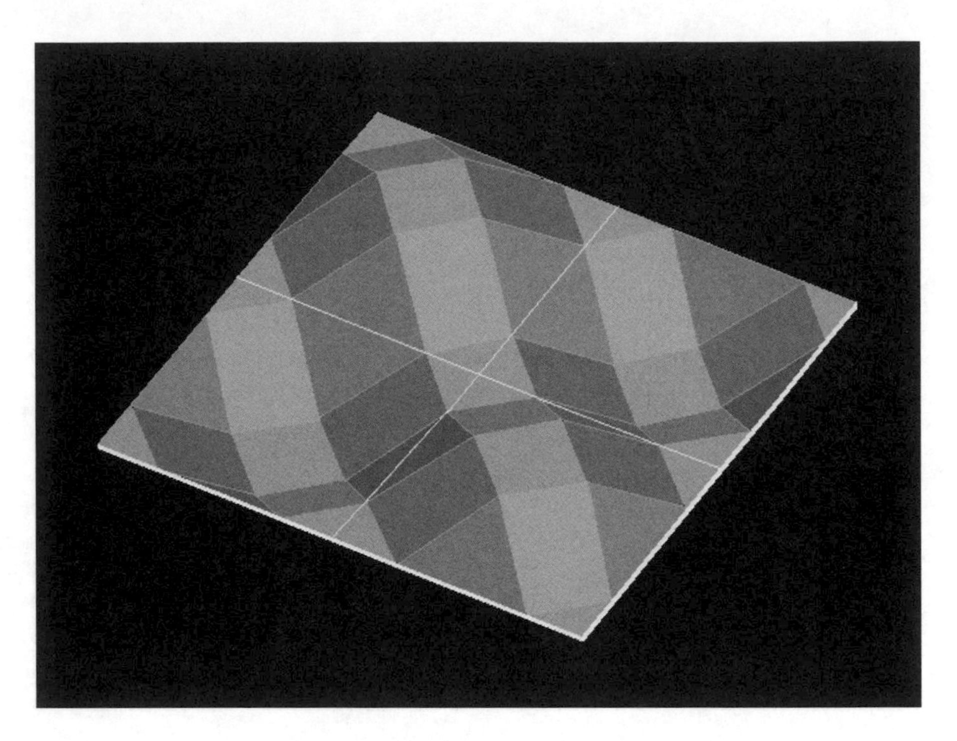

Fig. 3.7. Standing wave on a square tessellation in progress.

The evaluation of these integrals is messy in detail, though relatively elementary, and the final solution is simple:

All off diagonal elements in the matrix of coefficients A_{ij} are zero, i.e. $A_{ij} = 0$ for $i \neq j$. So, the only possibly nonzero elements remaining are:

$$A_{ii} = \frac{4}{\pi i^2} \sin\left(\frac{\pi i}{2}\right) \tag{3.157}$$

The associated spatial mode is accordingly:

$$S_i(x_1, x_2) = \frac{4}{\pi i^2} \sin\left(\frac{\pi i}{2}\right) \sin(i x_1) \sin(i x_2) \tag{3.158}$$

We display the initial displacement distribution in Figure 3.6, though not only for the basic pyramid in the lower corner, but for a larger part of the square tessellation where we see the initial picture of two pyramids and two inverted ones.

The coefficients A_{ii} are the same as in (3.34), and the $\sin(\pi i/2)$ function makes all even terms zero again. As in the one dimensional case, we also have $B_{ij}=0$ from $\bar{y}'=0$ and from (3.98).

So, we are ready to write down the complete solution according to equation (3.94), where we also substitute $\lambda_{ii}=\sqrt{2}i$ according to (3.91). Thus:

$$y=\sum_{i=1}^{\infty}\frac{4}{\pi i^2}\sin\left(\frac{\pi i}{2}\right)\cos\left(\sqrt{2}ict\right)\sin(ix_1)\sin(ix_2) \tag{3.159}$$

This expression is quite similar to (3.35). Despite the fact that we now have two space coordinates, we only need one single sum for the solution. This is so because all off diagonal A_{ij} are zero.

The solution proceeds, according to the temporal cosine factor, so that the pyramid becomes truncated as t increases from 0. It is completely erased level with the ground at the moment $\sqrt{2}ct=\pi/2$. Further it is excavated below earth level as the mirror image of the initial shape at $\sqrt{2}ct=\pi$, and then filled out again level with the ground at $\sqrt{2}ct=3\pi/2$. Finally the original pyramid is recreated anew at $\sqrt{2}ct=2\pi$, and so forth, over and over.

We see that the wave equation (3.147) is fulfilled in a trivial sense: All sides of the truncated pyramid being flat at any moment, the Laplacian is zero. Parts of the sloping sides of the pyramid stay at rest, but the truncated horizontal top and bottom surfaces move at constant speed, up or down depending on phase. As there are three possible speeds: constant up, constant down, and rest, the second time derivative is zero as well as the space derivatives. Accordingly both sides of the wave equation are zero.

We display the wave in progress in Figure 3.7. It is interesting to note how it proceeds. The flat horizontal squares, resulting from truncation, are tilted by $45°$, as compared to the orientation of the original squares. This is a result of the fact that the edges of the pyramid are shaved off, so it is not just truncated, and in the process of development the pyramids acquire an octagonal outline.

Despite this complex evolution of the wave, it still is true that all points move at one constant speed up or down, or else stay at rest.

This type of solution is of particular interest in cases with cubic nonlinearities of the van der Pol type, such as (3.136), when the system itself picks out solutions of the type now described, approaching them asymptotically even if a smooth sine wave is fed into the system as initial condition. As so often, nonlinearity makes attractors of some solutions which in the linear case are just consequences of accidental initial conditions.

We should also remember that, quite as in the one dimensional case, there are infinitely many such attractors of different wave length, i.e. we could have squeezed in just one, or nine, or sixteen, or twentyfive squares in the Figures 3.6-3.7, and the solution would work equally well, asymptotically picking the number of pyramids which best corresponds to the number of waves first fed into the system as initial conditions.

Note that we could equally well have carried out the analysis in this section for a triangular tessellation, but it would be computationally much more messy, so we chose the squares.

3.19 Perturbation Methods for Nonlinear Waves

The perturbation methods, Poincaré-Lindstedt as well as two-timing, discussed in Sections 2.7-2.8, and used throughout Sections 2.11-2.12, are applicable in the case of nonlinear spatial waves as well. Of course, the calculations become much more involved, so we can only give a very short outline here.

To this end we reformulate the wave equation by including a nonlinearity of the Rayleigh type as written in (3.137), with some slight modification:

$$\ddot{y} + y - \nabla^2 y = \varepsilon\left(\dot{y} - \dot{y}^3\right) \tag{3.160}$$

We just added y on the left, which produces what physicists know as the Klein-Gordon equation for a charged particle in an electromagnetic field, and we also included the perturbation parameter. Moreover the terms were slightly redistributed. This PDE comes very close to those that arise in business cycle modelling, and it has the advantage of having been intensively studied by perturbation methods in the applied mathematics literature around 1970. See for instance Millman and Keller, and Keller and Kogelman.

These authors use a different style for the Poincaré-Lindstedt method than the one we presented in Section 2.7, but there is no doubt that they are equivalent, so we adhere to the style we started with.

So suppose we can write the solution as a Taylor series in ε

$$y = y_0(\tau,x) + \varepsilon y_1(\tau,x) + \varepsilon^2 y_2(\tau,x) + \ldots \tag{3.161}$$

quite as in (2.76). The various solutions in the series $y_i(\tau,x)$ now depend on the time-like variable τ, but also on the space coordinates. We can leave it open for the moment if we consider x as a single space coordinate in 1 D, or as a vector $x = (x_1, x_2)$ in 2 D. The existing literature deals with one dimensional space, but there is nothing to prevent us from considering things in two dimensions.

Quite as in (2.77) we define the new time scale:

$$\tau = (\omega_0 + \varepsilon\omega_1 + \varepsilon^2\omega_2 + \ldots)t \tag{3.162}$$

There is one difference: The first term in the parenthesis, representing the basic frequency ω_0, was in (2.77) put equal to unity from the outset, while the successive corrections were determined in the solution process. The inclusion of a possibly nonunitary basic frequency is due to a substantial difference. Without the perturbation term, (2.75) was the simple harmonic oscillator, whose frequency we know is unitary. At present we also have the Laplacian in the left hand side of the unperturbed equation, and therefore we cannot be sure that the basic frequency is unitary. As a matter of fact it is not!

We next substitute the attempted solution (3.161) in (3.160). By the chain rule we get powers of the parenthesis in (3.162) in the process of differentiation, the first power for first derivatives, the second power for second derivatives. From the attempted solution and these powers we can get all the terms of (3.160) expressed as power series in ε. For simplicity we just retain the terms of zero and first order, the following ones becoming increasingly messy.

Hence:

$$\ddot{y} = \omega_0^2 y_0'' + \varepsilon(\omega_0^2 y_1'' + 2\omega_0\omega_1 y_0'') + \ldots \tag{3.163}$$

where, as in Section 2.7, we denote (partial) derivatives with respect to the new time variable τ by dashes, whereas those with respect to real time t were denoted by dots. Further:

$$\nabla^2 y = \nabla^2 y_0 + \varepsilon \nabla^2 y_1 + \ldots \tag{3.164}$$

where we again leave it open whether we have $\nabla^2 y_i = \partial^2 y_i(\tau, x) / \partial x^2$ in 1 D, or $\nabla^2 y_i = \partial^2 y_i(\tau, x_1, x_2) / \partial x_1^2 + \partial^2 y_i(\tau, x_1, x_2) / \partial x_2^2$ in 2 D.

We are now able to assemble the terms for the left hand side of (3.160) from (3.161), (3.163), and (3.164). As for the right hand side, the perturbation parameter is already present, so there are no zero order terms, and the first order terms just contain the zero order ones from the expansion for the first derivative. Thus:

$$\varepsilon\left(\dot{y} - \dot{y}^3\right) = \varepsilon\left(\omega_0 y_0' - \left(\omega_0 y_0'\right)^3\right) \tag{3.165}$$

We are now finally ready to substitute from (3.161), (3.163), (3.164), and (3.165) in (3.160). Again it is required that the terms for each power of the perturbation parameter ε match the differential equation (3.160) individually. In this way we get a series of differential equations of which we write down the first two:

$$\omega_0^2 y_0'' + y_0 - \nabla^2 y_0 = 0 \tag{3.166}$$

$$\omega_0^2 y_1'' + y_1 - \nabla^2 y_1 = \omega_0 y_0' - \left(\omega_0 y_0'\right)^3 - 2\omega_0 \omega_1 y_0'' \tag{3.167}$$

These are quite like (2.81) and (2.82) from Section 2.7, though there we stated one equation more in the chain.

The facts are also quite similar. Unlike our original equation (3.160), which was nonlinear, (3.166)-(3.167) are *linear*, and can hence be solved in closed form by standard methods. Also note that they can be *solved in sequence*, starting with (3.166), then substituting its derivatives in the right hand side of (3.167) as a nonhomogeneous term which is bound to lead to a particular solution. The following equation, not stated, would have a right hand side, now quite messy, which then depends on (3.166) and (3.167), and so forth to any desired degree of approximation in the Taylor series.

We will also find that amplitudes and frequency corrections will be determined by the elimination of secular terms quite as in the case of the Poincaré-Lindstedt method applied to ordinary differential equations.

There is just one big difference: Equations (3.166)-(3.167) are partial differential equations, PDE, whereas (2.81)-(2.82) were ordinary, ODE. This, of course, can be expected to result in some additional complications.

Equation (3.166) is not hard to deal with. We just try separation of variables:

$$y_0(\tau,x)=T_0(\tau)S_0(x) \tag{3.168}$$

Again leave it open whether x is one space coordinate or a vector of two, it does not matter yet. Substituting from (3.168) in (3.166) and dividing through by (3.168) we get:

$$\omega_0^2\frac{T_0''(\tau)}{T_0(\tau)}+1-\frac{\nabla^2 S_0(x)}{S_0(x)}=0 \tag{3.169}$$

As the different terms only depend on either time or space, we note that they must be constants. The last term results in a traditional eigenvalue problem:

$$\nabla^2 S_0+\lambda^2 S_0=0 \tag{3.170}$$

where the admissible eigenvalues λ depend on the dimension of the space, and on the shape and boundary conditions of the region in consideration.

If we deal with one dimensional space, say the line segment of length π, and if the boundary conditions prescribe rest at the boundary points, then we have $\lambda^2=i^2$ with i equal to any positive integer. If we have a square of side π, with rest prescribed for the edges, then all $\lambda^2=i^2+j^2$ work for i, j integers, as can be found by further separation of coordinates as in Section 3.13.

Upon substitution from (3.170) in (3.169), we find:

$$\omega_0^2 T_0''+\left(1+\lambda^2\right)T_0=0 \tag{3.171}$$

If we now choose

$$\omega_0 = \sqrt{1+\lambda^2} \tag{3.172}$$

then the multiplicative factors drop out from (3.171), and its solution is:

$$T_0(\tau) = A_{ij}\cos\tau + B_{ij}\sin\tau \tag{3.173}$$

as usual. Note that we index the coefficients. From (3.162) we have $\tau \approx \omega_0 t$ to the present approximation of the frequency. So, in real time the frequency of oscillation according to (3.172) depends on the eigenvalue λ obtained as a solution to (3.170). The fact that we use double course index ij implies that we are considering a case in two dimensions, more precisely we will consider the square of side π.

Given this, the solution to (3.170) is straightforward:

$$S_0 = \sin(ix_1)\sin(jx_2) \tag{3.174}$$

Before substituting from (3.173)-(3.174) in (3.168) we note that by a suitable choice of the origin for the time scale we can always make the second coefficient of (3.173) zero, so we get:

$$y_0 = A_{ij}\cos\tau\sin(ix_1)\sin(jx_2) \tag{3.175}$$

as a general solution to (3.166).

The remaining coefficient A_{ij} will be determined by the elimination of secular terms in the next differential equation (3.167). Note that (3.175) works for any integers i and j, provided we choose $\omega_0 = \sqrt{1+i^2+j^2}$, which affects the frequency in the solution (3.175) in terms of real time t.

It should be noted that, (3.160) being nonlinear, superpositions of different solutions of the type (3.175) do not work. Through an advanced use of two-timing, which is beyond the scope of this book, it was shown by Keller and Kogelman (at least in one dimension) that asymptotically the solution goes to just one mode, the rest becoming unstable. Which mode it goes to depends on the initial conditions.

Accordingly, for reasons of illustration, suppose that we just deal with the basic mode with $i = j = 1$. Hence we can delete the subscripts on the amplitude in (3.175) and note that we now have $\omega_0 = \sqrt{1+1^2+1^2} = \sqrt{3}$.

Then (3.175) reads:

$$y_0 = A\cos\tau \sin x_1 \sin x_2 \tag{3.176}$$

so we can go ahead calculating its derivatives and substituting in the right hand side of (3.167).

First, we get from (3.176) by differentiation once:

$$\omega_0 y_0' = -\omega_0 A \sin\tau \sin x_1 \sin x_2 \tag{3.177}$$

of which we also need the third power:

$$(\omega_0 y_0')^3 = -\omega_0^3 A^3 \sin^3\tau \sin^3 x_1 \sin^3 x_2 = \tag{3.178}$$

$$-\omega_0^3 A^3 \left(\tfrac{3}{4}\sin\tau - \tfrac{1}{4}\sin 3\tau\right)\left(\tfrac{3}{4}\sin x_1 - \tfrac{1}{4}\sin 3x_1\right)\left(\tfrac{3}{4}\sin x_2 - \tfrac{1}{4}\sin 3x_2\right)$$

Further, differentiating (3.176) twice we have:

$$2\omega_0\omega_1 y_0'' = -2\omega_0\omega_1 A\cos\tau \sin x_1 \sin x_2 \tag{3.179}$$

Substituting from (3.177)-(3.179) into (3.167):

$$\omega_0^2 y_1'' + y_1 - \nabla^2 y_1 = \omega_0 A\left(\tfrac{27}{64}\omega_0^2 A^2 - 1\right)\sin\tau \sin x_1 \sin x_2 \tag{3.180}$$

$$+\tfrac{3}{64}\omega_0^3 A^3 \sin\tau\left(\sin 3x_1 \sin 3x_2 - 3\sin 3x_1 \sin x_2 - 3\sin x_1 \sin 3x_2\right)$$

$$-\tfrac{1}{64}\omega_0^3 A^3 \sin 3\tau\left(\sin 3x_1 \sin 3x_2 - 3\sin 3x_1 \sin x_2 - 3\sin x_1 \sin 3x_2\right)$$

$$-\tfrac{9}{64}\omega_0^3 A^3 \sin 3\tau \sin x_1 \sin x_2 - 2\omega_0\omega_1 A\cos\tau \sin x_1 \sin x_2$$

Fig. 3.8. Two-term perturbation solution for nonlinear wave.

There are several things to note. In the right hand side we have terms which have the basic frequency in space and time along with the next odd harmonic, in all combinations. So, though we discarded the higher modes for the basic solution, we still get the third harmonic (in space and time coordinates), and, had we continued the process of deriving and solving higher order equations in the perturbation series, we would next get the fifth, the seventh, and so forth. The higher order modes are hence present in the nonlinear system, but they arise internally from the structure of the equation, not by superposition and fitting to initial conditions as in the case of linear partial differential equations. In the nonlinear case superposition does not work.

As the first and third modes are present in the right hand side of (3.180), they must also be present in the left hand side. Concentrating on the first (spatial and temporal) mode we note the presence of the first and last terms in equation (3.180). Solving to get a particular solution with these terms present would lead to secular terms where time enters multiplicatively with the trigonometric terms. This violates the assumption of a periodic or even

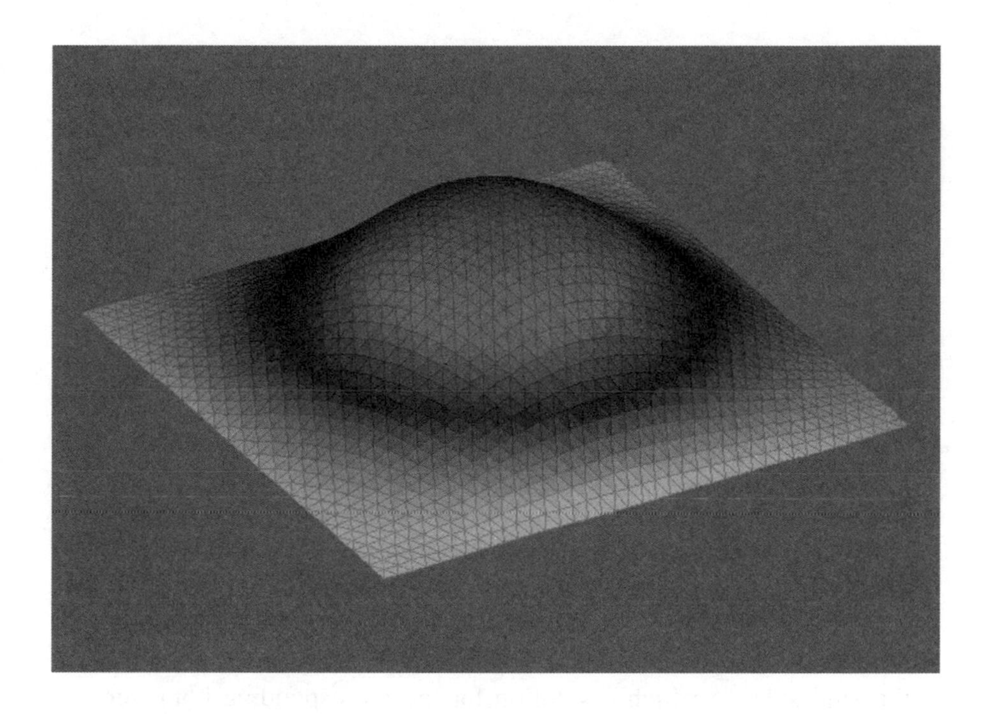

Fig. 3.9. Simulated solution to nonlinear wave.

bounded solution. Again, the only way to avoid this is to make the coefficients of the first and last terms of (3.180) zero.

There is only one reasonable way of achieving this as A should be nonzero, and as we know that $\omega_0 \neq 0$. This way is putting:

$$\omega_1 = 0 \tag{3.181}$$

and

$$A = \frac{8}{9} \tag{3.182}$$

where we take account of $\omega_0 = \sqrt{3}$.

Accordingly, the amplitude is determined by eliminating the secular terms, and so is the next frequency correction, which happens to be zero, so that the first frequency approximation still holds. In the continuation of the process we of course get further nonzero correction terms.

Substituting from (3.181)-(3.182) in (3.180) simplifies the latter a lot, and we obtain:

$$\omega_0^2 y_1'' + y_1 - \nabla^2 y_1 = \tag{3.183}$$

$$= \tfrac{3}{64}\omega_0^3 A^3 \sin\tau(\sin 3x_1 \sin 3x_2 - 3\sin 3x_1 \sin x_2 - 3\sin x_1 \sin 3x_2)$$

$$-\tfrac{1}{64}\omega_0^3 A^3 \sin 3\tau(\sin 3x_1 \sin 3x_2 - 3\sin 3x_1 \sin x_2 - 3\sin x_1 \sin 3x_2)$$

$$-\tfrac{9}{64}\omega_0^3 A^3 \sin 3\tau \sin x_1 \sin x_2$$

for which we can get a particular solution in the frequencies present in the right hand side, to which a solution for the corresponding homogeneous equation (in the basic frequency) has to be added. This complementary function accommodates undetermined coefficients to be fitted to the initial conditions.

Assuming the solution y_1 to be a sum with undetermined coefficients of the three terms on the right in (3.183), taking note that $\omega_0^2 = 3$, substituting, and matching coefficients we find the particular solution:

$$y_1 = \tfrac{\sqrt{3}}{162}\sin\tau(\sin 3x_1 \sin 3x_2 - 6\sin 3x_1 \sin x_2 - 6\sin x_1 \sin 3x_2) \tag{3.184}$$

$$+\tfrac{\sqrt{3}}{486}\sin 3\tau(2\sin 3x_1 \sin 3x_2 - 3\sin 3x_1 \sin x_2 - 3\sin x_1 \sin 3x_2)$$

$$-\tfrac{\sqrt{3}}{81}\sin 3\tau \sin x_1 \sin x_2$$

To this particular solution we add the complementary function. It is the solution to the homogeneous equation, which is identical in form with (3.166), so the solution is of the form (3.176) whose undetermined coefficient can be fitted to suit the initial displacement profile.

Neglecting this complementary function, which can be any sine or cosine of the basic frequency, and putting the perturbation parameter $\varepsilon = 1$, we can just add (3.184) to (3.176), substituting for $A = 8/9$ from (3.182). In this way we get a two term approximation for the nonlinear wave equation (3.160). The result is displayed in Figure 3.8.

For comparison we can use simulation. Macsyma software has a recent PDE solver which is quite efficient for nonlinear problems. A snapshot of an evolving wave is shown in Figure 3.9, where we also see the triangulation used for the numerical solution.

4 Iterated Maps or Difference Equations

4.1 Introduction

In Chapter 2 we saw that the behaviour of differential equations could be studied in terms of first return maps for points on the Poincaré section to itself. In this way the maps are related to differential equation systems.

This, however, is not the only link. At any computer simulation of differential equations the system is discretized: From one point the next is computed as a map, by a set of rules in the computer program, be it through Euler's method or the Runge-Kutta method. Moreover, there is not only a mathematical relation between iterated maps on one hand and dynamical systems described by differential equations on the other. They are also alternatives to be used in the modelling process itself.

In Economics there is a tradition of formulating the dynamical systems as processes in discrete time from the outset, as difference equations instead of as differential equations. The cobweb model, the Cournot duopoly model, and the Samuelson-Hicks business cycle model are all examples of this tradition. Economists used difference equations to such an extent that there were even developed methods for obtaining closed form solutions, which is really instructive as it so clearly shows the parallels and the differences between the two methods.

The order of a difference equation is the largest span of the time lags. If we for instance have:

$$x_t = f\left(x_{t-1}, x_{t-2}\right) \tag{4.1}$$

then the system is of the second order. As in the case of differential equations we can decompose it in two first order equations by defining a new variable $y_t = x_{t-1}$.

In this way (4.1) may be written:

$$x_t = f(x_{t-1}, y_{t-1}) \tag{4.2}$$

$$y_t = x_{t-1} \tag{4.3}$$

For the case of iterated maps we will start with the first order system and end with the second order. Already the link through first return maps on Poincaré sections means reducing the dimension by one. So for instance the map for an autonomous van der Pol oscillator is one dimensional, the map for the forced Duffing equation two dimensional.

Differential equations in one dimension can only have fixed points, attractors and repellors. In two dimensions, the fixed points become more various. There are nodes, stable and unstable, spirals, stable and unstable, and saddles. In addition we have limit cycles, again attracting and repelling, but the Poincaré-Bendixon theorem says that we do not have anything else. So, there is no quasiperiodicity and no chaos in two dimensions. For those phenomena we need three dimensions. The simplest examples of chaotic models, such as the Lorenz system or the forced oscillators, indeed are three dimensional.

The fact that Poincaré sections have one dimension less than the full systems would seem to indicate that chaos occurs for maps in dimension two, but it occurs already in dimension one. So, maps are even more unstable in terms of outcomes than expected compared to differential equations.

The perfect example to start with is the logistic iterated map, which is the most studied system in the chaos literature. Its counterpart in continuous time we already encountered repeatedly. The differential equation could always be solved in closed form, and there was nothing mysterious about it. In discrete time, despite all the literature on the topic, everything is not even known about it yet.

4.2 The Logistic Map

The logistic map was studied by the population ecologist Robert May in 1976:

$$x_{t+1} = \mu(1 - x_t)x_t \tag{4.4}$$

We first note that it obviously has two fixed points defined by:

$$x = \mu(1-x)x \tag{4.5}$$

The solutions are

$$x = 0 \tag{4.6}$$

$$x = \frac{\mu-1}{\mu} \tag{4.7}$$

As for stability we differentiate the right hand side of (4.5). Putting:

$$f(x) = \mu(1-x)x \tag{4.8}$$

we have

$$f'(x) = \mu(1-2x) \tag{4.9}$$

For stability of a map we require:

$$|f'(x)| < 1 \tag{4.10}$$

i.e.

$$|\mu| < 1 \tag{4.11}$$

for the fixed point at origin, and

$$\left|\mu\left(1-2\frac{\mu-1}{\mu}\right)\right| = |2-\mu| < 1 \tag{4.12}$$

for the nonzero fixed point. We see that the fixed point (4.7) is stable if and only if $1 < \mu < 3$.

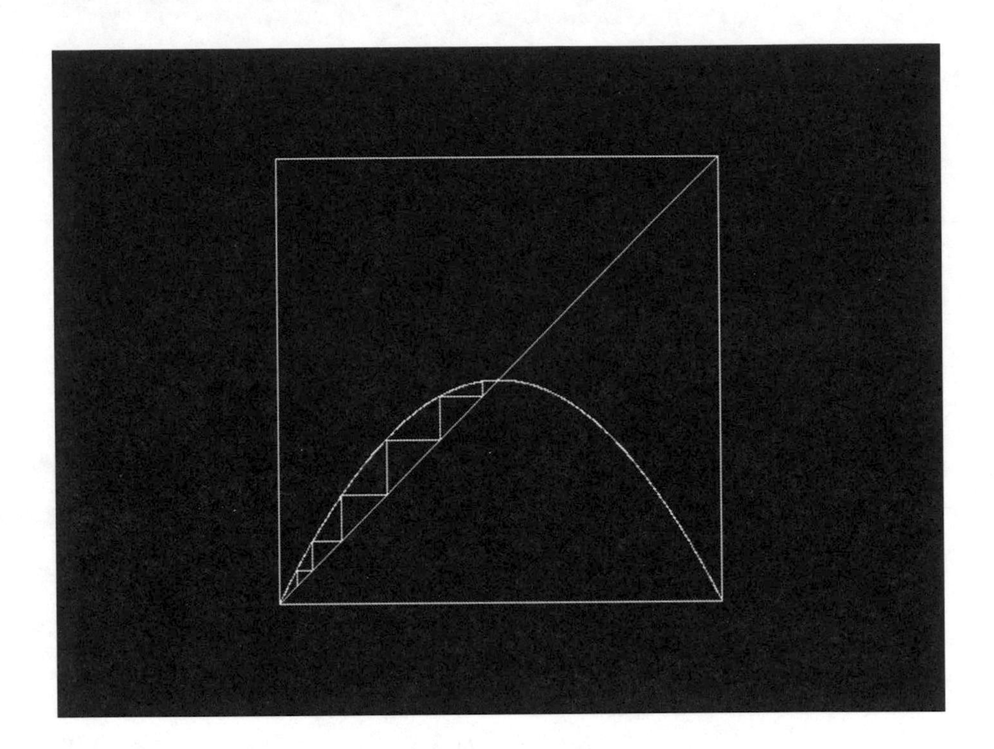

Fig. 4.1. Convergence upon fixed point of the logistic iteration.

For lower parameter values, in the interval zero to unity, we see from (4.11) that the zero fixed point (4.6) is stable instead. So suppose we have $\mu = 1.5$. Then starting at any value $x_0 \in (0,1)$ we get a convergent sequence towards 1/3. Suppose $x_0 = 0.500$. Then $x_1 = 0.375$, $x_2 = 0.351$, $x_3 = 0.342$, and $x_4 = 0.337$, as can be obtained with any pocket calculator. The asymptotic approach to the stable fixed point is unmistakeable.

There is a graphic method of visualizing the process, which is particularly familiar to economists, as it is similar to the diagram method used in the cobweb model, and even more so, to the way convergent multiplier processes were shown with the expenditure function and a 45 degree line for shifting the previous value of the expenditure function as a new argument for it at the next iteration.

This is displayed in Fig. 4.1. The function $f(x) = \mu(1-x)x$ is a parabola turned upside down. For any value of the parameter μ it passes through the origin (0,0) and through the point (1,0). The maximum is always located at

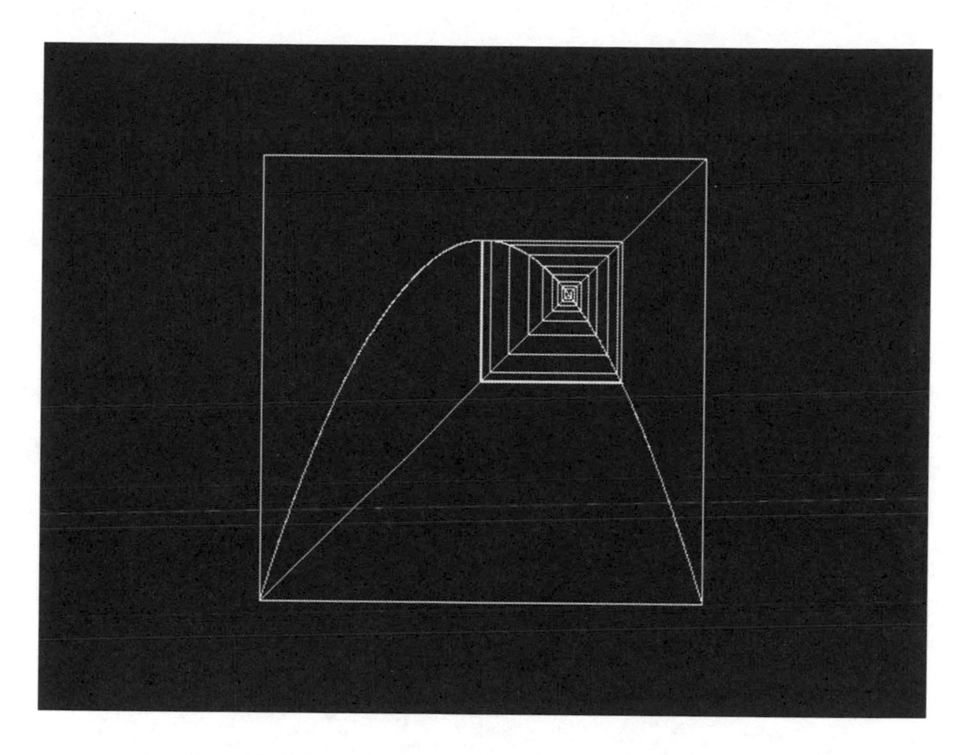

Fig. 4.2. Divergence from fixed point to periodic cycle.

$x = 1/2$, but the maximum value is $\mu / 4$ and hence dependent on the parameter, which actually just scales up the whole function in proportion. Fig. 4.1 also contains the 45 degree line spoken of already. The fixed points are the intersection points of the line with the parabola. We may note that if $\mu = 1$, then the derivative of the parabola at the origin according to (4.9) is unity, and so the parabola and the line are tangential for this parameter value. There are hence not two distinct intersection points, and the same holds true for $\mu < 1$, when there is just one (positive) intersection point.

Once we have $\mu > 1$, there are two distinct fixed points. From (4.12) we see that the positive fixed point turns stable the moment it arises, and from (4.11) that the zero fixed point loses stability at the same moment.

In Fig. 4.1 we also see the convergent process on the stable fixed point. From (4.12) we see that this situation continues for quite a considerable interval. But at $\mu = 3$ the stability condition is no longer fulfilled. The loss of stability according to (4.12) has to do with the slope of the parabola at the intersection point with the 45 degree line. If the slope of the parabola becomes steeper than the unit slope of the straight line, then it is obvious that

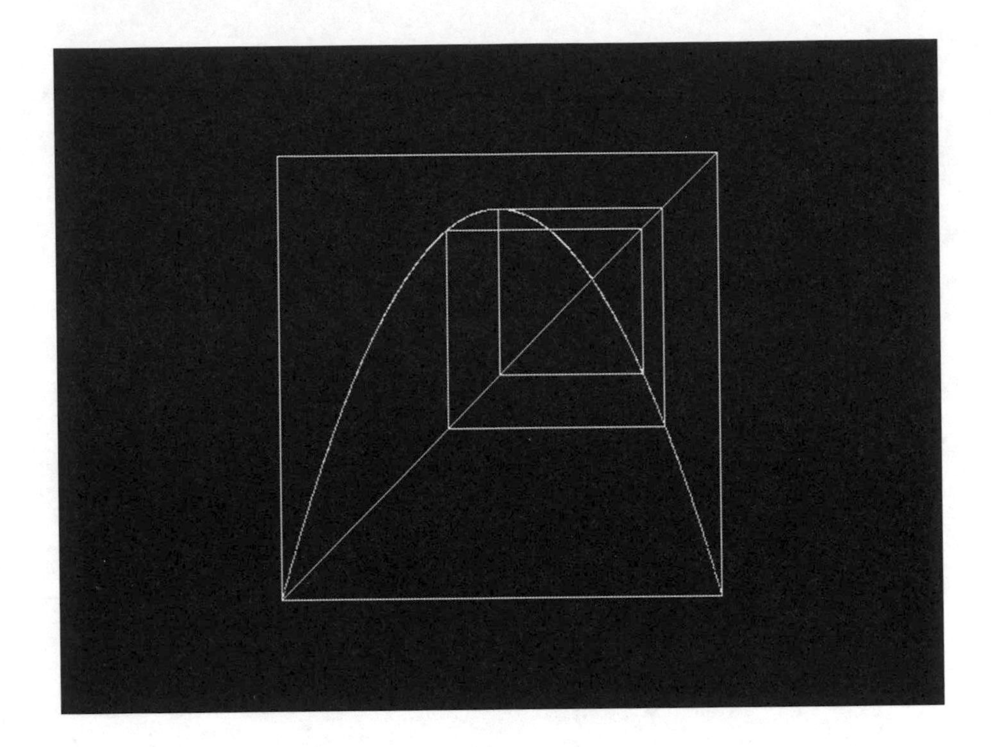

Fig. 4.3. Four period cycle of the logistic map.

the stepwise graphic process in the neighbourhood of the fixed point spirals away from it rather than being attracted by it.

This time it is obvious, however, that there emerge no new fixed points. Once they have become two they remain two. So the process must go to something else than a new fixed point. Fig. 4.2 shows what happens. The process goes to a periodic alternation between two values, i.e. to a two period cycle.

There is now a whole sequence of further period doubling when the parameter increases. Let us check the stability of the two period cycle. Two consecutive iterations of the map (4.8) result in:

$$f^2(x) = f(f(x)) = \mu^2(\mu x^2 - \mu x + 1)(1 - x)x \tag{4.13}$$

The fixed points of the second iterate are obtained from:

$$f^2(x) = x \tag{4.14}$$

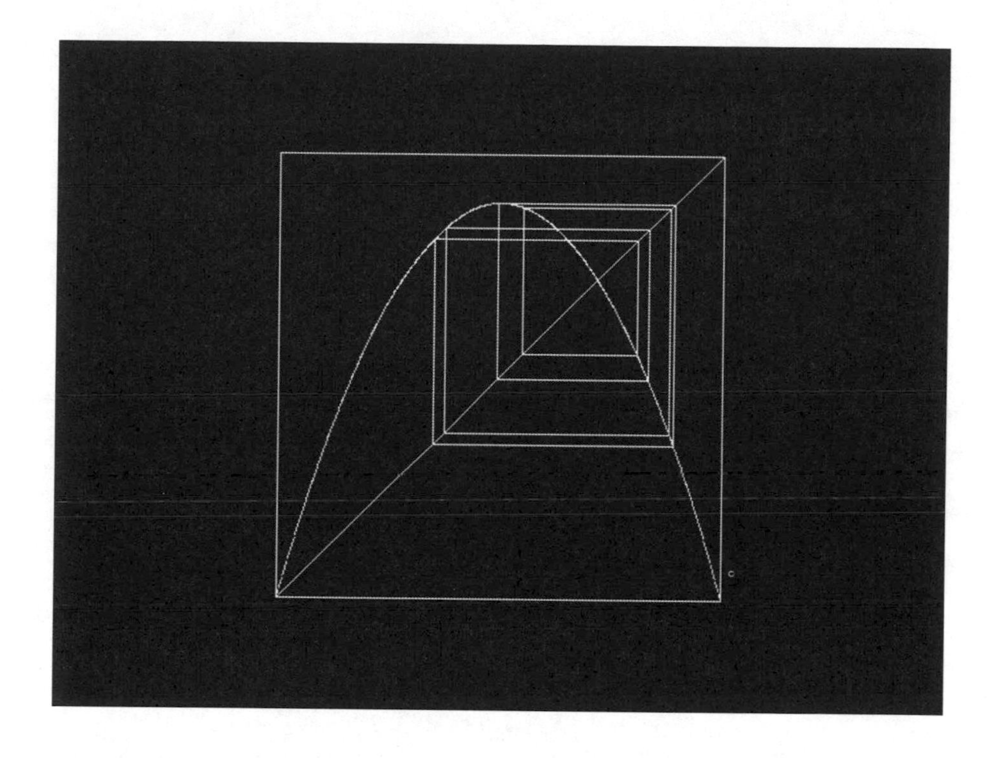

Fig. 4.4. Eight period cycle of the logistic map.

or from (4.13), dividing through by x and a making slight rearrangement:

$$\mu^2(\mu x^2 - \mu x + 1)(1 - x) - 1 = 0 \tag{4.15}$$

This expression easily factorizes into:

$$(\mu + 1 - \mu x)(\mu^2 x^2 - \mu(\mu + 1)x + \mu + 1) = 0 \tag{4.16}$$

The first factor if put equal to zero yields the solution (4.7). This is natural, as the fixed point by definition of course also is a two period cycle, or a cycle of any period. But it is not interesting when it is unstable. So we focus interest on the second factor:

$$\mu^2 x^2 - \mu(\mu + 1) + \mu + 1 = 0 \tag{4.17}$$

which is easily solved and yields the roots:

$$x_1, x_2 = \frac{\mu+1}{2\mu} \pm \frac{\sqrt{\mu+1}\sqrt{\mu-3}}{2\mu} \tag{4.18}$$

We note that they are real when $\mu > 3$, i.e. when the fixed point has lost stability. So, the two period cycle does not even exist before that happens. The two roots (4.18), of course, are the two values between which the process oscillates after the initial transient has faded away.

So, what about stability? To this end take the derivative of (4.13):

$$Df^2(x) = \mu^2(1-2x)(2\mu x^2 - 2\mu x + 1) \tag{4.19}$$

If we substitute any of the roots from (4.18) we get:

$$Df^2(x_i) = 2(\mu+2) - \mu^2 \tag{4.20}$$

For stability of the periodic cycle the absolute value must be less than unity, i.e., we must have:

$$\left|2(\mu+2) - \mu^2\right| < 1 \tag{4.21}$$

For the positive interval of parameter values which is of interest for us the inequality is satisfied for:

$$3 < \mu < 1 + \sqrt{6} \tag{4.22}$$

The lower critical value indeed corresponds to that at which the fixed point loses stability. At the upper, which approximates to 3.449, we may expect that the two period cycle loses stability. Figs. 4.3-4.4 show cycles of period 4 and 8, obtained for parameters equal to 3.5 and 3.54 respectively. We note that, whereas the stability of the fixed point extended over the parameter interval from 1 to 3, the stability range is much smaller for the two period cycle.

So, it continues, and the ratio of two subsequent stability intervals goes to a limiting value, $\delta \approx 4.6692$. This value is called Feigenbaum's Constant,

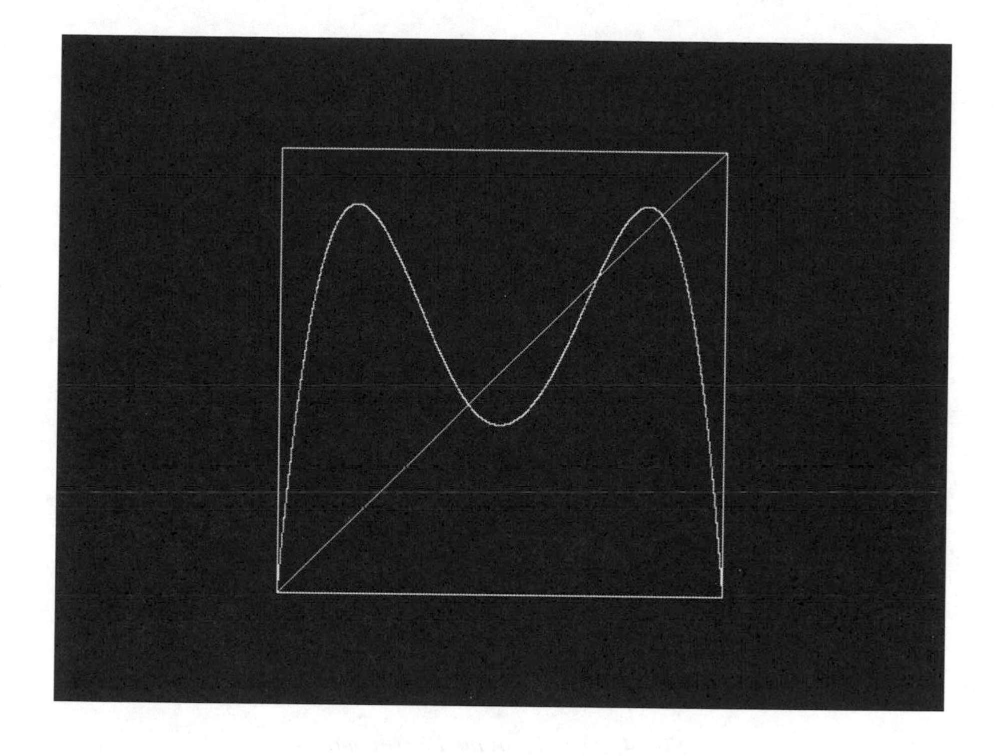

Fig. 4.5. Second iterate and fixed points of the logistic map.

and it is recognised as a new universal constant as it holds not only for the parabola of the logistic iteration, but for all topologically similar one-humped functions.

As the intervals for period doubling are reduced at such a fast pace there is an accumulation point where Achilles in fact overtakes the tortoise, and where we have had an infinite number of period doublings. This point is called the Feigenbaum point. Beyond that something else has to occur. What occurs is chaos.

The Feigenbaum point can in fact be approximated from a given range of stability. We know that the 2-cycle is stable in the interval $\left(3,1+\sqrt{6}\right)$. The convergent series of successively shortened intervals for various higher periodicities is $1+\delta^{-1}+\delta^{-2}+...=\delta/(\delta-1)$, so we would estimate:

$$3+\frac{4.67}{4.67-1}\left(1+\sqrt{6}-3\right)=3.57 \tag{4.23}$$

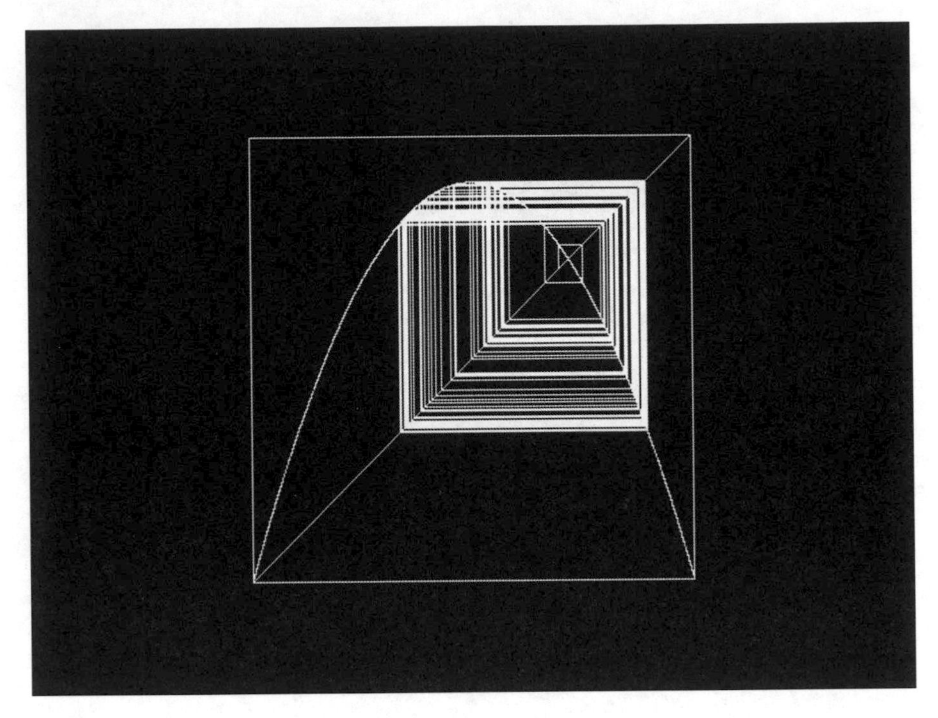

Fig. 4.6. Chaos in the logistic map.

which in fact estimates the true value quite correctly to our limited number of decimals, but had we departed from the stability interval of the 4-cycle, we would have obtained much higher accuracy.

Before looking at the chaotic case, let us present the conclusion for stability of the two period cycle in another form. In Fig. 4.5 we show the twice iterated map (4.13). As we see it is a two-humped function, with three intersections with the straight line. The middle one is the fixed point, the left and right the two roots (4.18). If we could measure the slopes at the intersections accurately, we would find that the absolute value of the slope in the middle intersection exceeds unity, whereas those of the intersections right and left fall short of unity. Had the parameter been slightly lower, then the right hump would not reach up to the line, and we would only have one intersection point. This is the case of the stable fixed point. Would the parameter have been so high that the slopes at all three intersections would exceed unity, then the cycle would not be stable either. In that case we could try a diagram of the fourth iterate, which produces a four-humped curve, and so on.

4.3 The Lyapunov Exponent

In Fig. 4.6 we display a chaotic outcome beyond the Feigenbaum point. The obvious question is: How do we know that the outcome is chaotic? One method is in terms of the Lyapunov exponent which measures the exponential rate of measurement error magnification. Another method is the use of symbolic dynamics. We start with the first method.

The magnification of an infinitesimal difference in initial conditions is measured by the derivative of the mapping (4.9). It is, however not important for the issue of the magnification whether the magnification is in the positive or in the negative direction. Therefore we take the absolute value of the derivative:

$$|f'(x)| = |\mu(1-2x)| \tag{4.24}$$

Moreover, we want to measure the exponential rate of separation of the trajectories. Therefore we measure the logarithm of the preceding expression:

$$\ln|f'(x)| = \ln|\mu(1-2x)| \tag{4.25}$$

Next, it is not necessary that there is exponential separation everywhere in the phase space, so we want to measure the average separation over a long run of iterations, say n:

$$\lambda_n(x_0) = \frac{1}{n}\sum_{t=1}^{t=n} \ln|f'(x_{t-1})| = \frac{1}{n}\sum_{t=1}^{t=n} \ln|\mu(1-2x_{t-1})| \tag{4.26}$$

In practice we always have to end up at a finite, though as large as possible n, but the formal definition of the Lyapunov exponent assumes a limiting value as the number of steps becomes infinite:

$$\lambda(x_0) = \lim_{n\to\infty} \lambda_n(x_0) \tag{4.27}$$

Equations (4.26)-(4.27) provide an operational definition. While we iterate the system (4.4) we can calculate the Lyapunov exponent to a better approximation the larger the number of steps we take. As we measure exponential separation a positive exponent is an indication of sensitive dependence on initial conditions and hence on chaos.

In the lower part of Fig. 4.7 we display the Lyapunov exponent for the logistic map against the value of the parameter. As for the upper part it will be commented soon. Though the formulas (4.26)-(4.27) refer to the initial point x_0, the Lypunov exponent does not depend on this point of departure, once we have left the transient motion and settled on the chaotic attractor, so we can calculate a unique value for each parameter.

As we see there is to the left a considerable section where the Lyapunov exponent is below the zero line, and hence negative. Occasionally it reaches up to the line in isolated humps, which occurs every time a periodic cycle, or the initial fixed point, loses stability. Then, at the Feigenbaum Point, it seems to definitely cross the zero line and remain positive. This signifies the region of chaos, though we see that there are occasional dips to negative values, which means relapse to ordered behaviour. In reality there are infinitely many such dips, though they cannot all, of course, be seen in the diagram.

On top of Fig. 4.7 we display the iterated variable values versus the parameter. We no longer plot all the details of Figs. 4.1-4.4. In the process displayed in Fig. 4.1 there is eventually a fixed point to which the process goes. After the discarded transient process in the beginning there is just one point. Similarly, the process displayed in Fig. 4.2 ultimately oscillates between two values, that in Fig. 4.3 between four values, that in Fig. 4.4 between eight values, and so on. Those values themselves change with the parameter.

So, if we on the vertical axis of the top picture just plot the points which the process ultimately visits for each parameter value, then we see a curve of fixed points, which splits in two when the fixed point loses stability and is replaced by a 2-cycle, further a splitting to a 4-cycle, and an 8-cycle.

We can also infer the location of the Feigenbaum point, and the chaotic regime when an entire interval seems to be visited. Actually we never have intervals, but rather disjoint fractal point sets, it is the finite resolution of the computer screen that deceives us. Fractals are much easier to visualize in two dimensions so we will return to the issue later on.

Note how the splittings occur just above the places where the Lyapunov exponent touches the zero line, indicating loss of stability, and how the chaotic motion sets on when the Lyapunov exponent crosses the line. Finally, note that there are windows in the top picture indicating a relapse to periodicity, just above the dips in the Lyapunov exponent curve. Fig. 4.7 summarizes all the results obtained hitherto in a compact way. There are, of course, infinite numbers of details to find at higher resolution, such as new windows of order, and new dips of the Lyapunov Exponent curve.

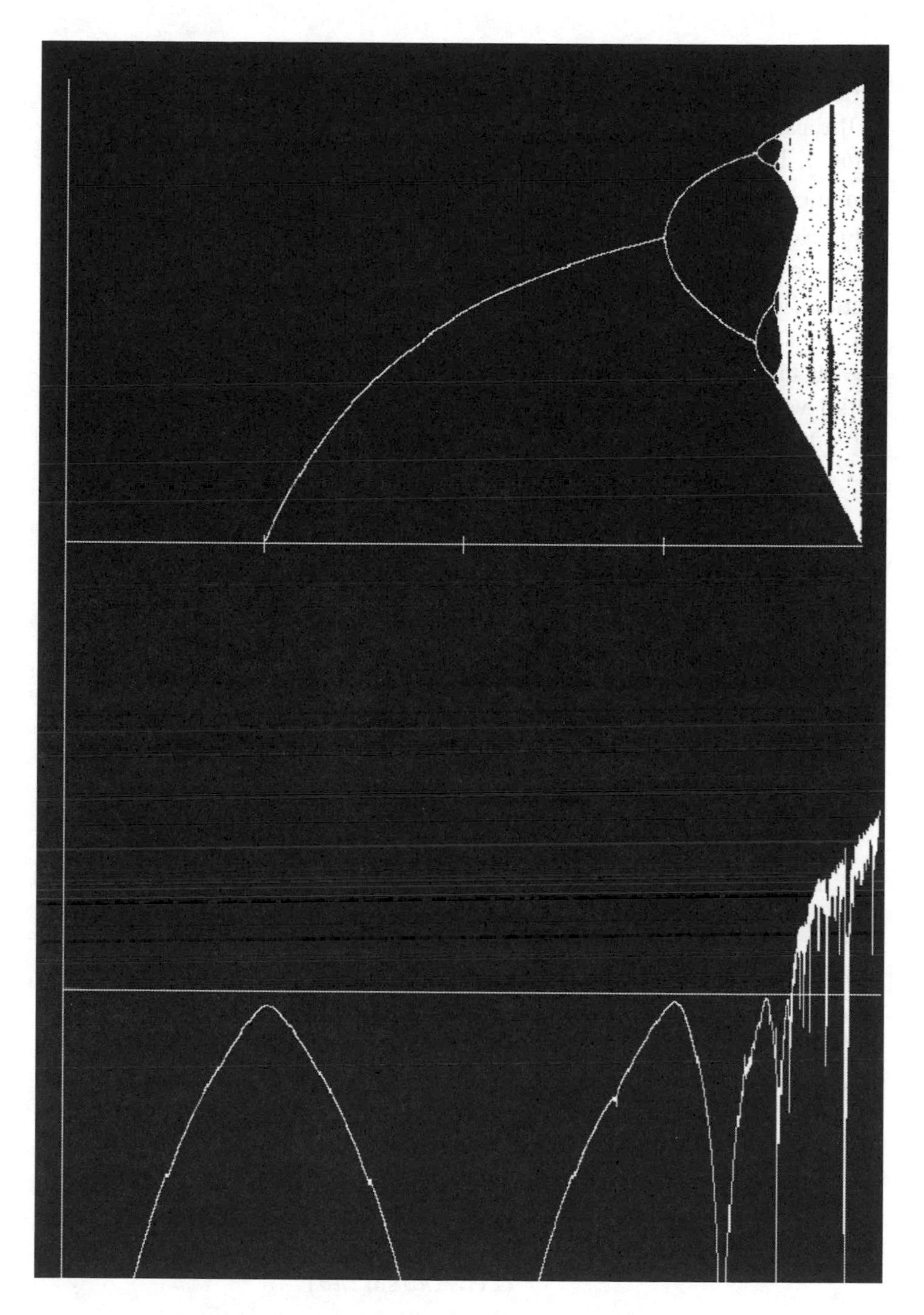

Fig. 4.7. Amplitude and Lyapunov Exponent versus parameter for logistic map.

We also note in the upper part of Fig. 4.7 that the interval on which the attractor is extended when it is chaotic becomes larger the higher the value of the parameter. This situation continues until the parameter reaches the value 4. Then the model just explodes. It is easy to see why. As we saw the maximum value of the parabola (4.8) was $\mu / 4$. If $\mu > 4$ then the maximum value exceeds unity and the parabola protrudes above the unit box that is drawn in Figs. 4.1-4.4. If so, the process would earlier or later end up in that section, with the consequence that the next iterate becomes negative. For iterations in the negative interval the process is always unstable and goes to $-\infty$ at high speed. What has been said also implies that we cannot take initial values for the iteration outside the unit interval (0,1). If we do it explodes.

4.4 Symbolic Dynamics

We indicated that there was another way of formally pinning down the occurrence of chaos, alternative to measuring the Lyapunov exponent. Assume we have $\mu = 4$ in equation (4.8) so that we deal with the boundary case just before the model explodes:

$$f(x) = 4(1 - x)x \tag{4.28}$$

and let us use the coordinate transformation:

$$x = \sin^2 2\pi\theta \tag{4.29}$$

Substituting (4.29) in (4.28) and using $1 - \sin^2 2\pi\theta = \cos^2 2\pi\theta$ we have:

$$\begin{aligned} f(x) &= 4\cos^2 2\pi\theta \sin^2 2\pi\theta \\ &= (2\cos 2\pi\theta \sin 2\pi\theta)^2 \\ &= \sin^2 4\pi\theta \end{aligned} \tag{4.30}$$

We hence see that corresponding to the mapping:

$$x \to 4(1-x)x \tag{4.31}$$

there is an equivalent mapping

$$\theta \to 2\theta \bmod 1 \tag{4.32}$$

The modulus is there because the sine function is periodic. The conclusion is hence that an equivalent mapping to the logistic iteration is the doubling of an angle on the unit circle.

We immediately understand that there is a sensitive dependence on initial conditions such as signifies chaos, because, like the variable itself, differences between two iterated processes are also doubled in each iteration. Thus after n steps the difference $(\theta_0 - \overline{\theta}_0)$ has increased to:

$$(\theta_n - \overline{\theta}_n) = 2^n\theta_0 - 2^n\overline{\theta}_0 = 2^n(\theta_0 - \overline{\theta}_0) \tag{4.33}$$

The Lyapunov exponent of this map is also easily evaluated as the logarithmic separation is constant in each step. The value is just:

$$\ln 2 = 0.69 \tag{4.34}$$

This is indeed the estimate we get for the original logistic iteration, so the shift to new coordinates makes no difference for the Lyapunov exponent.

The doubling of angles on a unit circle has an interesting consequence. If we start from an angle that is a rational number $\theta_0 = m/n$, then, after (n-1) iterations the process would come back to the initial value, and it would have visited m different values on the unit circle. This is so because we have $2^{n-1}(m/n) \bmod 1 = (m/n)$. The result is a periodic trajectory.

As we can take any value on the unit circle as the point of departure we can produce cycles of any periodicity! This is usually taken as an auxiliary condition for chaos, together with sensitive dependence. In the chaotic regime there are periodic cycles of every periodicity, but they are all unstable, so we never find them in simulation. After all the irrational numbers are so many more than the rational numbers. It should be stressed that this is in the chaotic regime. For a different parameter value a cycle may well be an attractor.

As, according to the unit modulus in the map (4.32), all the numbers produced by the iteration are in the unit interval, we can write the numbers as infinite decimal fractions:

$$\theta = 0.d_1 d_2 d_3 \ldots \tag{4.35}$$

where the symbols d_i represent any of the digits 0,1,2,3,4,5,6,7,8,9.

Expressed as fractions, rational and irrational numbers differ by the fact that the former are eventually composed of periods of finite sequences of the same digits in the same order that recur again and again. Everybody has the experience from dividing two integers that the result is such a periodic sequence. For instance: 5/7 = 0 . 714285 714285 ... But the reverse holds too. Take the continued fraction just computed, and multiply it with 1'000'000. Then we obtain the number 714'285 . 714285 714285 ... Note that the fractional part is equal to that of the previous number, due to the fact that the number of digits and sequences is infinite. Thus, subtracting the original number we obtain the difference 714'285. Denoting the original fraction θ, we get: $(1'000'000 - 1)\theta = 714'285$, or $\theta = 714'285 / 999'999 = 5/7$. There hence is a unique correspondence between periodic decimal fractions and rational numbers.

This fact does not depend on what number system we use. We can instead of the decimal system use the binary, writing:

$$\theta = 0.b_1 b_2 b_3 \ldots \tag{4.36}$$

where b_i are now any of the digits 0,1. We now note that the decimal digit 2 is written 10, and that multiplication by that number just shifts all the digits one step to the left. The first digit that is shifted to the left of the point, is then just deleted if it happens to be 1. This is due to the modulus clause in (4.32). So the map (4.32), when working on binary numbers, is equivalent to:

$$0.b_1 b_2 b_3 \ldots \quad \to \quad 0.b_2 b_3 b_4 \ldots \tag{4.37}$$

This is called the shift map, and it can always be defined, though not by such direct means as by the pedagogical trigonometric transformation (4.29) which was possible with $\mu = 4$. The analysis in terms of shift maps is called symbolic dynamics.

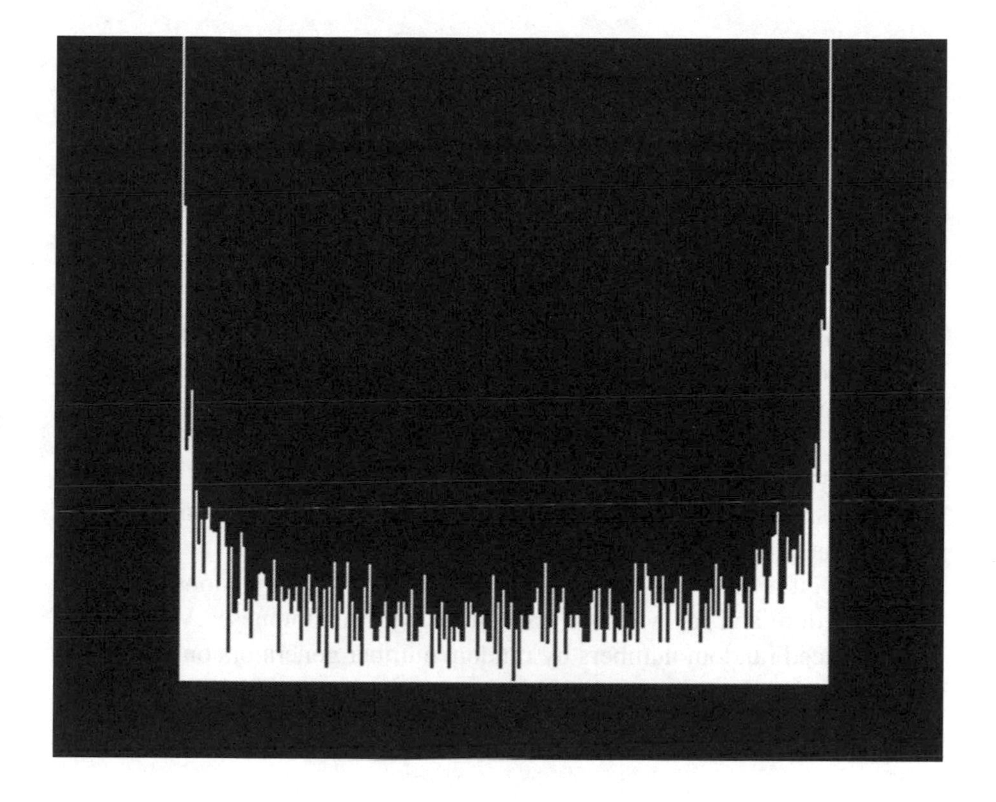

Fig. 4.8. Frequency histogram for the logistic iteration.

The shift map gives some interesting information on sensitive dependence. An initial condition is given by such a binary fraction as on the left of (4.37). In the next run of the iteration all digits become one step more significant. Suppose we have two initial conditions that differ only by digits from the 100th place. After 99 iterations the difference has been shifted to the first place and become the most significant digit, so any closeness of the two originally extremely close trajectories has been lost, and they are as wide apart as any trajectories can be.

The complete future history is thus contained in the infinite sequence of digits in the initial condition. Each of them is shifted so as to become more important, and at a certain stage it becomes the most important digit, after which it is scrapped in the next iteration and so has no more importance at all. As the sequence is infinitely long and has no periodicity it has all the characteristics of a random sequence, and in fact it agrees fully with Sir Karl Popper's definition of randomness.

This also means that the development is completely unpredictable, unless it is periodic, or we know the initial condition with an infinite number of digits. Having one hundred, or one thousand, or one million digits does not help in the long run, because after the corresponding number of iterations we know nothing. As for periodic orbists, we just saw that there exist all imaginable periodic orbits in the chaotic regime, but they are all unstable.

It should, however, be stressed that the randomness involved is very different from that represented by a binomial or a Gaussian distribution. Fig. 4.8 shows a histogram for the logistic iteration, relative frequency plotted against the iterated variable. We see that, far from being bell shaped, the frequency distribution is enclosed on a finite interval, and it has concentrations to the extremely small and extremely large values. Moreover, the whole appearance is anything but smooth with its spiky character. Of course, it is not even continuous, as only variable values net of a Cantor dust are visited, but as always this cannot be seen in finite resolution.

Anyhow, it is clear that chaos has very deep going implications for scientific procedure and for what we understand with randomness. All the actually produced random numbers by random number generators on computers are in fact chaotic series produced by deterministic procedures.

4.5 Sharkovsky's Theorem and the Schwarzian Derivative

Before leaving the topic of the logistic iteration and with that the one dimensional maps we have to shortly state two remarkable results.

The first is Sharkovsky's Theorem. It states that if a map, such as the logistic map (4.8), is just continuous, then there is a certain order of implication for the existence of periodic orbits. By summary:

$$\begin{aligned} 3 &\Rightarrow 5 \Rightarrow 7 \Rightarrow \ldots \\ &\Rightarrow 2^1 \cdot 3 \Rightarrow 2^1 \cdot 5 \Rightarrow 2^1 \cdot 7 \Rightarrow \ldots \\ &\Rightarrow 2^2 \cdot 3 \Rightarrow 2^2 \cdot 5 \Rightarrow 2^2 \cdot 7 \Rightarrow \ldots \\ &\Rightarrow 2^3 \cdot 3 \Rightarrow 2^3 \cdot 5 \Rightarrow 2^3 \cdot 7 \Rightarrow \ldots \\ &\Rightarrow 2^3 \Rightarrow 2^2 \Rightarrow 2 \Rightarrow 1 \end{aligned} \tag{4.38}$$

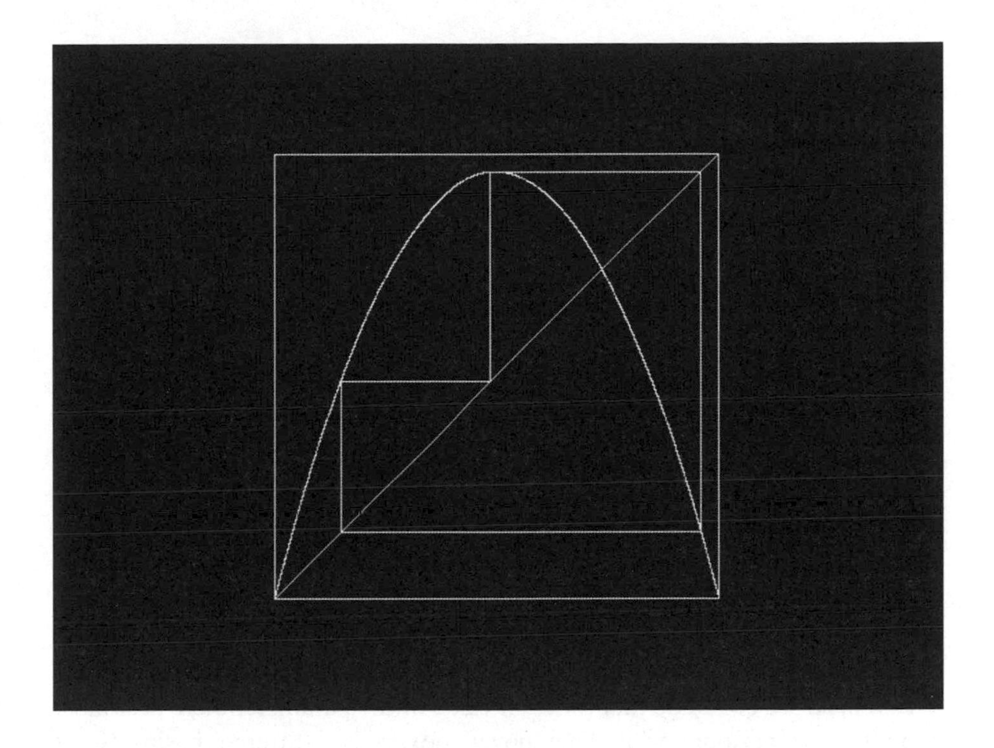

Fig. 4.9. Attracting three-period cycle in the logistic map.

Thus, the existence of a period three cycle implies the existence of cycles of all other periodicities. The logistic mapping indeed is continuous. Continuity, by the way, is a remarkably general assumption for such a strong conclusion. So, if there is a period three cycle there are cycles of all periodicities. This we already know from the discussion in terms of symbolic dynamics. In the chaotic case they were all unstable, but they may turn stable. In Fig. 4.9 we display an attractive cycle of period three, which, by the way can be seen in Fig. 4.7 as the most pronounced window of order and corresponding dip in the Lyapunov Exponent curve.

Similarly there are infinitely many windows of order, where we can locate attractive cycles of the most various periodicities. The period doubling cascade, which, by the way, is one of the most common routes to chaos, is at the lowest level of the hierarchy of periodicities in the Sharkovsky sequence stated in (4.38).

Our second item in this section is the Schwarzian derivative. A negative definite sign for that operator has very far reaching implications some of which will be indicated below.

The formal definition reads:

$$Sf(x) = \frac{f'''(x)}{f'(x)} - \frac{3}{2}\left(\frac{f''(x)}{f'(x)}\right)^2 \tag{4.39}$$

For our mapping $f(x) = \mu(1-x)x$ we have $f'(x) = \mu(1-2x)$, further $f''(0) = -2\mu$, and $f'''(0) = 0$, so:

$$S\mu(1-x)x = -\frac{6}{(1-2x)^2} < 0 \tag{4.40}$$

For instance if the mapping only has a finite number of fixed points and the Schwarzian derivative is negative, then there may be only finitely many different periodic orbits of any given periodicity. Further, there may only exist one attractive periodic orbit. The conditions are fulfilled for the logistic iteration, so we conclude that for each value of the parameter there is at most one attractive periodic orbit. They never coexist with different basins. So, it is generally proved that once the four cycle gains stability, then the two cycle must lose stability, just as we have seen.

For these advanced topics the reader is referred to the authoritative and comprehensive book by Devaney. It is a good reference also for other topics dealt with in this chapter, as is Peitgen, Jürgens and Saupe.

4.6 The Hénon Model

It is now time to turn to two dimensional maps, and an instructive start is the model suggested by the astronomer Michel Hénon 1976. It is a second order process:

$$x_t = 1 - ax_{t-1}^2 + y_{t-1} \tag{4.41}$$

$$y_t = bx_{t-1} \tag{4.42}$$

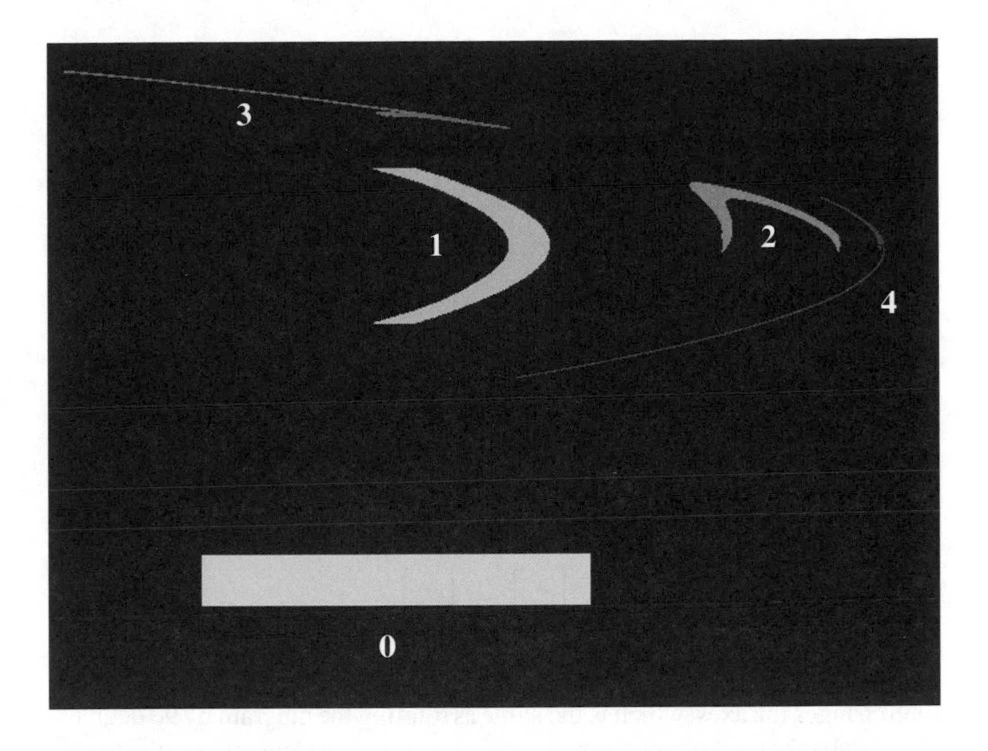

Fig. 4.10. Initial rectangle and four generations of iterates by the Hénon map.

This process could also have been written as one second order difference equation:

$$x_t = 1 - ax_{t-1}^2 + bx_{t-2} \tag{4.43}$$

The system (4.41)-(4.42) takes any points in the phase plane to other points in the phase plane. Its working can be understood by using a strategy that Hénon suggested.

The idea is to study the geometry of this transformation in terms of a composite of three subsequent transformations, operating on a rectangle in the original phase space. First, the map:

$$\begin{bmatrix} x \\ y \end{bmatrix} \rightarrow \begin{bmatrix} x \\ 1 - ax^2 + y \end{bmatrix} \tag{4.44}$$

bends the rectangle over its given base to an upside down parabola shape. Next, the transformation:

$$\begin{bmatrix} x \\ y \end{bmatrix} \rightarrow \begin{bmatrix} bx \\ y \end{bmatrix} \tag{4.45}$$

scales the parabola in the horizontal direction, squeezing it if $b < 1$. Finally, the transformation:

$$\begin{bmatrix} x \\ y \end{bmatrix} \rightarrow \begin{bmatrix} y \\ x \end{bmatrix} \tag{4.46}$$

interchanges the axes, which is the same as rotating the diagram by 90 degrees.

Fig. 4.10 shows the transformation of the original #0 rectangle to the horseshoe #1 by the composition of these three transformations.

Obviously, the new shape is no longer a rectangle, and further transformations, #2 through #4, produce more and more complex shapes. The later generation of iterate is shown in terms of darker shading. We see that further iterates decrease in area, but become more elongated, as they become repeatedly folded, which is not seen in the resolution of the picture.

Ultimately, whatever the shape and size of the point set of initial conditions, provided the system does not explode, it will ultimately settle to the Hénon attractor. This attractor will also be traced by the trajectory starting from any point in phase space, provided again that the system does not explode.

The attractor and its area of convergence are shown in Fig. 4.11. For any initial point in the bright area the process goes to the fractal attractor placed in that attraction basin. For the rest of initial conditions the model explodes.

Within the area of convergence to the chaotic attractor the process is uniformly contractive. We already noted in Fig. 4.10 how the area of a set of initial conditions shrinks. Presently we can even make an immediate numerical estimate of this rate of shrinking.

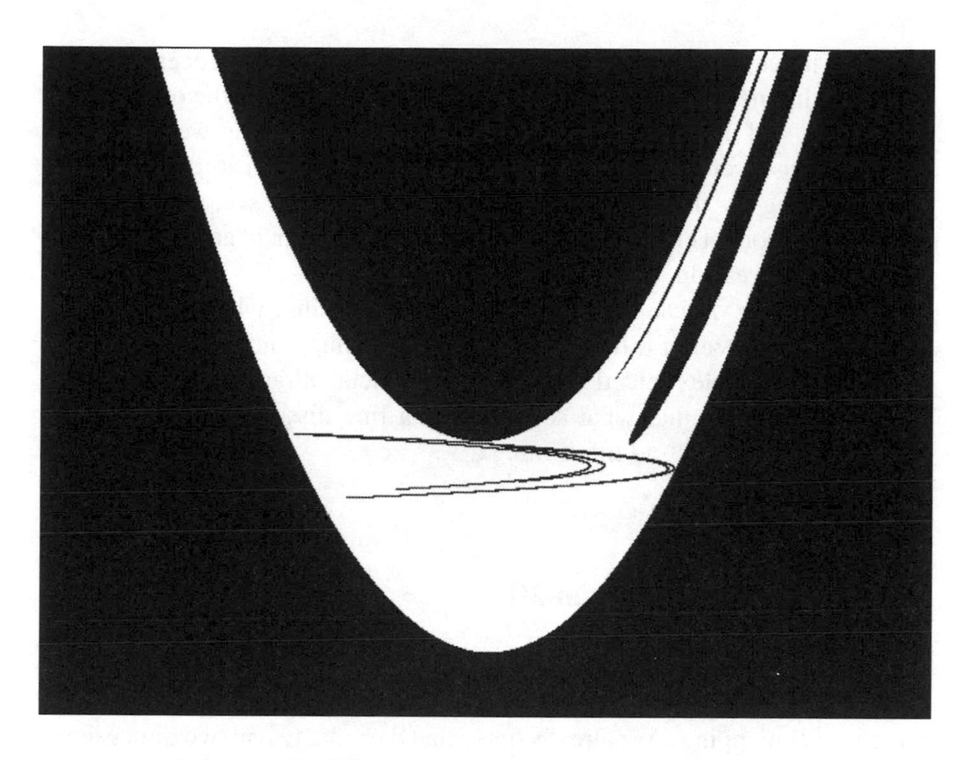

Fig. 4.11. The Hénon attractor and its basin of attraction.

To this end let us rewrite the system (4.41)-(4.42):

$$f(x,y) = 1 - ax^2 + y \tag{4.47}$$

$$g(x,y) = bx \tag{4.48}$$

The change of area by a two dimensional mapping is measured by its Jacobian:

$$\left|\frac{\partial(f,g)}{\partial(x,y)}\right| = \left|\det\begin{bmatrix} -2ax & 1 \\ b & 0 \end{bmatrix}\right| = b \tag{4.49}$$

The absolute value is taken because we are not interested in the reversal of sign due to rotations caused by the mapping. We see that as long as $b < 1$ the system is compressive, as in Fig. 4.11 where $b = 0.3$.

An area, however, has two dimensions, and in a chaotic process there is stretching in one dimension, but a dominant compression in the other so that the process still converges upon the attractor. The stretching we are talking about obviously refers to initial conditions, and in one dimension we measured this stretching by the Lyapunov exponent. As we will see, there are two Lyapunov exponents in two dimensions. One is dominant negative to provide for areal compression, but the other is positive.

The only way we can put an area, though shrinking, which is more and more elongated within a limited space is by folding, and this is what provides for the complex folded structure of the fractal attractor. If we enlarge a piece of it we find that what seems to be a line dissolves in a bundle of thinner lines and folds ad infinitum.

4.7 Lyapunov Exponents in 2D

We now have to deal with the computation of Lyapunov exponents for two dimensional mappings. We already noted that there are two in two dimensions. In general there are as many as the dimension of the space. It is clear that equations (4.26)-(4.27) will not do, because we do not know what the derivative is in two dimensions. The closest we come is the Jacobian matrix of partial derivatives in (4.49), but it has four entries.

Consider again the first iteration in Fig. 4.10. We see that, as mentioned, the initial rectangle is bent, squeezed, and rotated. On purpose, in order to show this, we took a rectangle of considerable base. The deformation, however, would be smaller the smaller we make the initial rectangle. In the limit, with infinitesimal sides, the curved sides become straight and parallel. They, however, are scaled, more precisely squeezed in one direction and elongated in another. Moreover, the pairs of parallel lines delimiting the image of the original rectangle may intersect nonorthogonally. The image of the rectangle is therefore not a rectangle but a rhombic parallelogram.

Equation (4.49) actually measures the ratio of the area of the parallelogram to the area of the original rectangle. We see that our mapping in each step reduces the area in the ratio $b = 0.3$. This may occur in two ways, by scaling down one of the sides, as we know, but also by making the delimiting lines intersect under an acute angle. The area of a parallelogram with finite sides is actually zero if the angle is zero!

What we want to measure by the Lyapunov exponents is the elongation in one direction and the squeezing in the other. We must, however, take in consideration that the orientation of the compressive and expansive directions in space are constantly rotated as we see in Fig. 4.10. Therefore we have to keep memory of the direction in the computations.

Assume a direction θ_t at a certain stage, and define the direction vector:

$$\mathbf{v}_t = \begin{bmatrix} \cos\theta_t \\ \sin\theta_t \end{bmatrix} \tag{4.50}$$

Next take the Jacobian matrix of partial derivatives:

$$\mathbf{J}_t = \begin{bmatrix} -2ax_t & 1 \\ b & 0 \end{bmatrix} \tag{4.51}$$

Then take the matrix product:

$$\mathbf{J}_t \cdot \mathbf{v}_t = \begin{bmatrix} \sin\theta_t - 2ax_t \cos\theta_t \\ b\cos\theta_t \end{bmatrix} \tag{4.52}$$

which is the projection of the partial derivatives on the direction $\mathbf{v}_t$. For our computation of the Lyapunov exponent we now take:

$$\lambda = \lim_{n\to\infty} \frac{1}{n} \sum_{t=1}^{t=n} \ln\left|\mathbf{J}_t \cdot \mathbf{v}_t\right| \tag{4.53}$$

where $\left|\mathbf{J}_t \cdot \mathbf{v}_t\right| = \sqrt{(\sin\theta_t - 2ax_t\cos\theta_t)^2 + (b\cos\theta_t)^2}$ is the Euclidean norm of the vector (4.52). At this stage there is nothing new compared to the reasoning around (4.26)-(4.27). What we still need to do is to state how the direction vectors are updated. The following is the rule:

$$\mathbf{v}_{t+1} = \frac{\mathbf{J}_t \cdot \mathbf{v}_t}{\left|\mathbf{J}_t \cdot \mathbf{v}_t\right|} \tag{4.54}$$

The division by the norm of the vector guarantees that we indeed arrive at a new unit vector, expressible as the cosine and sine of a new angle θ_{t+1}.

We now from (4.51) and (4.53)-(4.54) have all pieces except one: The initial direction vector $\mathbf{v}_0$. The problem is an easy one, we can take any unit vector, it does not matter what the initial direction is. As our process brings elongation in one direction and squeezing in another, the updating rule (4.54) after some time aligns the direction vector along the elongation direction, and keeps it there.

As a consequence of this we understand that the method can only be used to calculate the largest Lyapunov exponent. There are other methods by which we can calculate both at once, but it is the largest exponent that is really interesting as positivity for it indicates chaos. The other then has to be negative, because if both were positive in two dimensions there would be no compression to an attractor, the system would just explode.

In our case we have an indirect method for finding the other Lyapunov exponent. Calculating the largest Lyapunov exponent from the formulas in a simulated process we arrive at the estimate $\lambda = 0.42$. Denote the second Lyapunov exponent by μ. The areal magnification is a result of the combined operation in both dimensions. Therefore with $b = 0.3$ we have:

$$\exp(\lambda + \mu) = 0.3 \tag{4.55}$$

or

$$\mu = \ln 0.3 - \lambda \tag{4.56}$$

Using $\lambda = 0.42$ we get $\mu = -1.62$, negative as conjectured. Using the more advanced methods of determining all Lyapunov exponents at once give the same estimates. These estimates, of course, refer to a specific combination of both parameters: $a = 1.4$, $b = 0.3$, and change as they change.

This is a proper place to stop for a while and consider the operation of the Lyapunov Exponents in a chaotic process. So let us take a three dimensional phase space for illustration though we will not deal with any models of so high dimension. In three dimensions there are three Lyapunov Exponents. Suppose one is positive and the two other negative. Take an initial sphere of initial conditions. In the course of time it is elongated in one dimension but squeezed in two. It hence becomes more and more string like and goes to infinite length and zero crossection. However, compressivity dominates, i.e.

volume constantly shrinks when we do not have an explosive model, and the string is always placed in a finite subspace of the phase space. It hence has to be folded to a complex tangle to fit in that limited space.

As an alternative, suppose we have two positive Lyapunov Exponents and one negative. The latter, of course, again must be dominant in order to produce an overall compression of volume. As there is now one compressive direction and two expansive ones, the initial sphere now becomes a sheet of zero thickness but infinite area. Again it has to be folded and shrinkled in order to fit in the bounded subspace of the phase space. These are the two ways chaos can occur in three dimension, and all that happens is well described by the Lyapunov exponents.

4.8 Fractals and Fractal Dimension

More interest than chaos itself has been attracted by its fingerprints, the fractal attractors. In 1977 Benoit Mandelbrot revolted against the dominance of no lesser person than Euclid. "Mountains are no cones, clouds no spheres, trees no cylinders" Mandelbrot emphatically declared. People had been measuring complicated coast lines, such as the British, and concluded that each better resolution brought in new details and, if the process of resolution enhancement were continued indefinitely, just made the coastline seen as a curve infinitely long.

Mathematicians, such as Giuseppe Peano and Helge Koch in the first decades of the 20th Century actually devised the construction of such curves of infinite length. Mandelbrot suggested a new definition of dimension, fractional dimension. If we halve the linear scale, then a line is divided in $2 = 2^1$ segments of half the original length, a square is divided in $4 = 2^2$ squares with sides half the original length, and a solid cube is divided in $8 = 2^3$ cubes with edges half the original length. The logarithm to the base two of the number of parts equals the dimension of the object, 1 for a curve, 2 for a surface, 3 for a solid.

Maybe a similar scaling law holds true for the British coastline, so that halving the measuring stick results in a number 2^D of parts each half the length of the original measuring unit? The exponent D would be a fraction. If the coastline was of infinite length seen as a normal curve, we would expect a fractional dimension $1 < D < 2$.

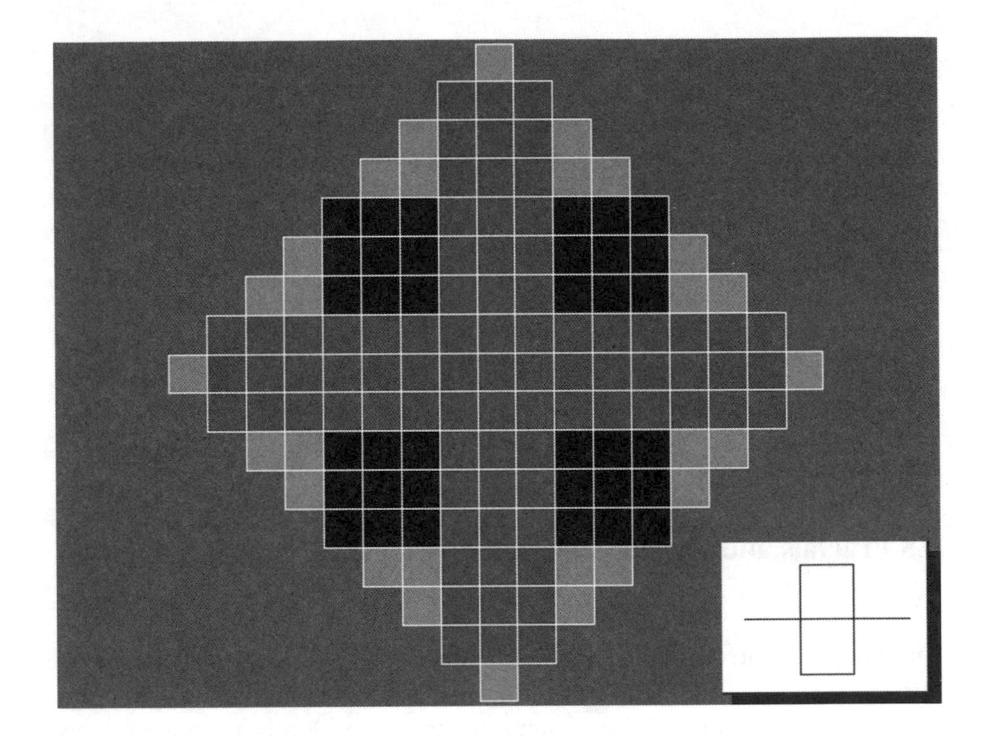

Fig. 4.12. The 2-dimensional Peano curve.

Measurements indeed did show that this was the case, the dimension was estimated to $D = 1.31$. We will return to how the calculation was made. The name fractal was coined for structures with fractional dimension.

Mandelbrot claimed that all natural objects are fractal rather than Euclidean. In the course of time it was detected that deceptive forgeries of objects such as leaves, trees, clouds, and entire landscapes could be created by computer programs producing fractals. Often the algorithms were strangely short and simple in view of the complexity of the objects created. Such fractal programming was used as background in science fiction films, and has today given rise to an entire research field in computer science, called image compression, aiming at the saving of storage space for pictures. In stead of storing a picture pixel by pixel, an algorithm is devised by which the picture is computed and drawn anew each time. The storage space for the programs is by orders of magnitude less than for the pictures. We do not follow this track any longer, we just refer to Barnsley and to Peitgen, Jürgens, and Saupe.

Let us instead look at a few famous fractals. In Fig. 4.12 we show the Peano curve mentioned above. It all starts with the big square indicated by

Fig. 4.13. The Sierpinski gasket. Fractal dimension 1.58.

the dark area. In the next step the four sides of the square are replaced by properly scaled copies of the little picture shown in the inset. From that on we continue the process, replacing each of the segments by the same picture. We note that as each segment in the inset picture is one third of the length of the central line, we scale the picture down by 1/3 in each new step. On the other hand, the inset picture has 9 segments replacing the 3 that are there before on the central line.

If we decrease the measuring unit to 1/3, then a line is divided into 3 pieces, but a square is divided into $3^3 = 9$ pieces. Thus we conclude that the Peano curve does not behave like a curve, rather it behaves like an area, and so we can by theoretical reasoning conclude that its fractal dimension is 2, though it is a curve and has topological dimension 1. Peano in fact derived it as a space filling curve which in the limit comes indefinitely close to any point in the tilted square shown in Fig. 4.12. It is appealing to intuition that space filling curves acquire the fractal dimension of the space in which the curve is placed. Empirical measurements of the fractal dimension of the trachea system in the lungs for instance result in dimensions close to 3.

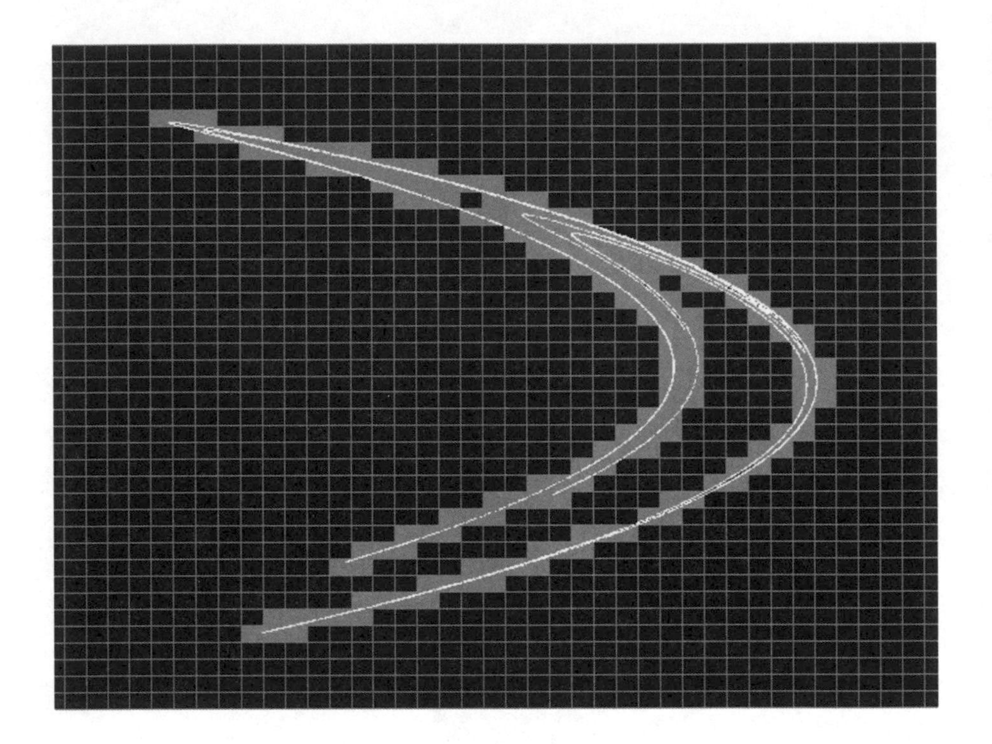

Fig. 4.14. Box counting for the Hénon attractor. 40 by 40 boxes.

Another interesting fractal is the Sierpinski gasket. We take an equilateral triangle, and inscribe a small triangle, which divides the original one in 4, and delete the central triangle. Then we repeat the process over and over, in theory an infinite number of times, in practice a suitable finite number as appropriate for the resolution of the computer screen. At each step the side is scaled to 1/2, and there remain 3 triangles after one is removed. The fractal dimension, reasoning as in the case of the Peano curve is $\ln(3)/\ln(2) = 1.58$.

Both the Peano curve and the Sierpinski gasket are defined in terms of a recursive procedure which, if repeated an infinite number of times, results in exact self similarity. That means, every tiny piece of the structure if scaled up is identical with the whole structure itself. Even if there is not always such exact self similarity in all fractals, it generally holds true that there is an infinite sequence of resolution levels with new details added at each.

We are going to stop this review now. There are lots of interesting fractals to look at in the expanding literature, and some very sophisticated software, such as the Fractint, by which the reader may study innumerable pictures, explore them in tiny detail, and create own variants.

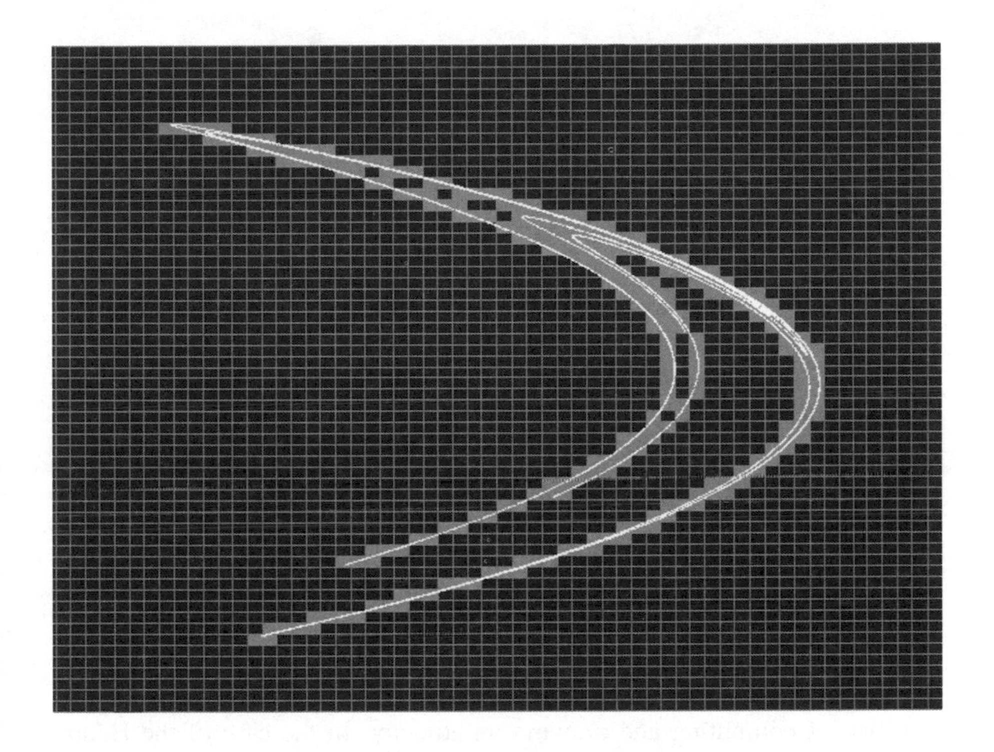

Fig. 4.15. Box counting for the Hénon attractor. 60 by 60 boxes.

Above we estimated the dimension for some very simple fractals created by iterated mathematical processes. It is now time to turn to some more generally applicable dimension definitions. Mandelbrot suggested the so called box counting dimension which would work with any fractal whatever. In Fig. 4.14 we display the Hénon attractor, superposed on a grid of finite mesh. It is proposed to count the number of boxes, coloured gray, which contain any part of the attractor. Suppose the side of the unit cells is denoted s and that the number of boxes is denoted N. Obviously we have $N(s)$, i.e. the number of boxes depends on the side of the cells. The box counting dimension then is defined as.

$$D = \frac{d \ln N(s)}{d \ln\left(\frac{1}{s}\right)} \tag{4.57}$$

In practice we would draw pictures as in Figs. 4.14-4.15 for several mesh sizes, then plot $N(s)$ against $1/s$ on a log-log scale, and estimate the slope of the line. This is quite a complex procedure, but there now exists software which admits to import any picture of a fractal, and, superposing several mesh sizes, makes the count automatically, and then, also automatically, estimates the dimension. The box counting method has for instance been used to make dimension estimates for foaming waves in Hiroshiges's woodcuts. Of course, we always have a limit for the accuracy due to the finite resolution, or if seen numerically, the number of observations.

A slightly different method is the following: Suppose the phase space is three dimensional, and that we centre a sphere of a given radius around any one given point on the attractor. Then, double the radius. Consider the number of points of the attractor contained in the two spheres. If it were a line, then we would expect double the number of points in the larger sphere, if it were a surface, we would expect four times as many, and if it were a solid eight times as many. In general we could measure the numbers of points, compute the logarithm, and divide by the logarithm of 2 if this was the ratio of the radii. Doing this for many points on the attractor and taking an average we get a measure of fractal dimension. The procedure can easily be programmed into that of computing and drawing an attractor. In the case of the Hénon attractor displayed in Figure 4.11 we get the estimate $D = 1.21$.

4.9 The Mandelbrot Set

There is probably no more well known fractal than the Mandelbrot Set. It arises through another two dimensional iterated map. Though the process has no particular application in economics, or any other applied science, the concepts associated with it, the Julia Set and the Mandelbrot Set have a certain general interest. The map, fully written out in Cartesian coordinates, is:

$$x_t = x_{t-1}^2 - y_{t-1}^2 + a \tag{4.58}$$

$$y_t = 2x_{t-1}y_{t-1} + b \tag{4.59}$$

where a and b are two constants. What this mapping does to points in the phase plane is more easily understood if we, as in the Hénon case, decompose it:

$$\begin{bmatrix} x \\ y \end{bmatrix} \to \begin{bmatrix} x^2 - y^2 \\ 2xy \end{bmatrix} \tag{4.60}$$

and

$$\begin{bmatrix} x \\ y \end{bmatrix} \to \begin{bmatrix} x + a \\ y + b \end{bmatrix} \tag{4.61}$$

The transformation (4.61) just translates points by the vector (a, b), but (4.60) is easier to understand if we convert to polar coordinates:

$$\begin{bmatrix} x \\ y \end{bmatrix} = \begin{bmatrix} r\cos\theta \\ r\sin\theta \end{bmatrix} \tag{4.62}$$

Then

$$\begin{bmatrix} x^2 - y^2 \\ 2xy \end{bmatrix} = \begin{bmatrix} r^2\cos^2\theta - r^2\sin^2\theta \\ 2r^2\cos\theta\sin\theta \end{bmatrix} = \begin{bmatrix} r^2\cos 2\theta \\ r^2\sin 2\theta \end{bmatrix} \tag{4.63}$$

Accordingly we have:

$$\begin{bmatrix} r \\ \theta \end{bmatrix} \to \begin{bmatrix} r^2 \\ 2\theta \end{bmatrix} \tag{4.64}$$

as an equivalent to the mapping (4.60). What the mapping does is to double the angular coordinate, and square the radial coordinate. The doubling of angles means a counterclockwise rotation. As for squaring the radius it acts differently on points inside the unit circle and those outside. The first are moved closer to the origin, the latter are mover further out. We note that, in the special case when $r = 1$, only the angle is changed, and then we in fact have exactly an equivalent of the mapping (4.32) above. In summary (4.58)-(4.59) project the points by squaring the radius vector, doubling the angle, and then adding the vector of constants.

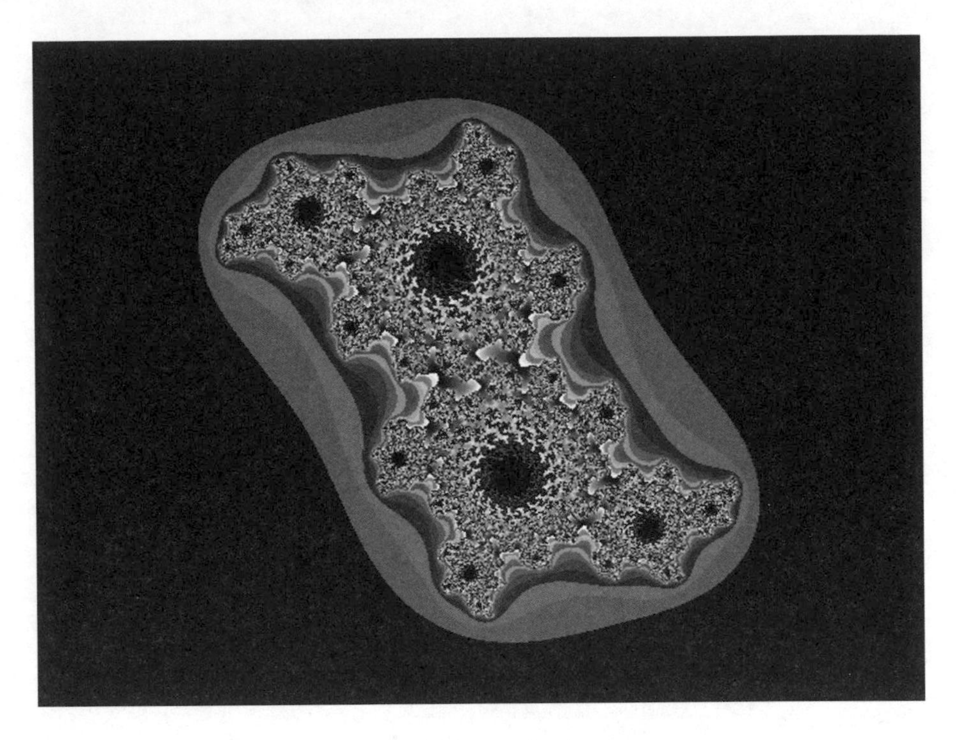

Fig.4.16. A Julia Set.

There is an even more compact way of writing the transformation, by using complex numbers. If we put

$$z = x + iy \tag{4.65}$$

where as usual $i = \sqrt{-1}$, then we have:

$$z^2 = x^2 - y^2 + i2xy \tag{4.66}$$

Further, defining the complex constant:

$$c = a + ib \tag{4.67}$$

we see that (4.58)-(4.59) can be written as the complex analytic function:

$$z_t = z_{t-1}^2 + c \tag{4.68}$$

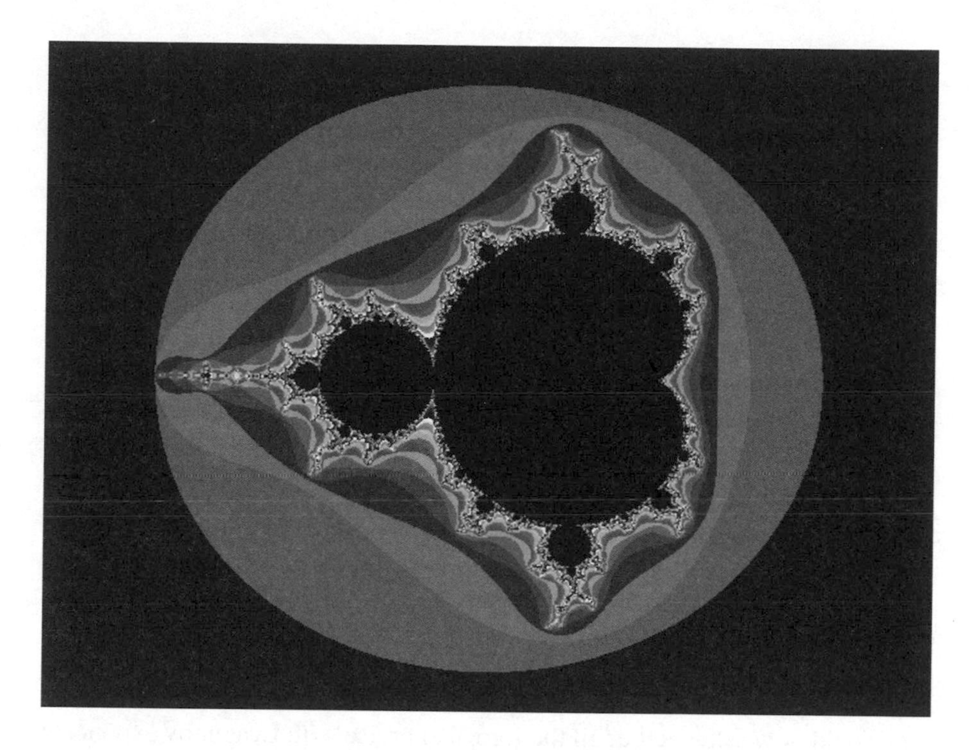

Fig. 4.17. The Mandelbrot Set.

In the 1920es Julia and Fatou studied the behaviour of the iteration (4.68), or written in the more pedestrian style (4.58)-(4.59), especially with respect to the convergence. They realized that the convergence to attractors, which could be several coexistent, or the divergence of the process to infinity from various initial conditions was extremely intricate. If the complex constant is zero, i.e. $c = 0 + i\,0$, then the story is simple. Points inside the unit circle converge in a spiral motion to the origin, which is the unique attractor, points outside it spiral off to infinity, points on the unit circle display a predominantly chaotic movement on the circle, as we know. But with nonzero c the situation becomes infinitely more complex. The deformation of the circle of convergence is usually called the Julia set.

Fig. 4.16 shows the Julia set for an arbitrarily chosen complex constant to be added. It just shows the various attraction basins, and the outside area where the process explodes, but there is a feature added: A shading with respect to how fast the process converges to an attractor or diverges to infinity. The phase space in Fig. 4.16 is, of course, that of initial conditions.

The space shown in Fig. 4.17 is different. It displays the complex constant c or, if we prefer, the coordinates a, b. What is shown is the convergence or divergence of a process starting from the origin in coordinate space. The central body with its satellites and antennas represents convergence. Outside there is divergence to infinity. Here again it is the common usage to show the speed at which the process diverges to infinity in terms of shading.

The quadratic complex iteration is so special that it is not of much use in applications. In particular the complex analytic functions represent a very restrictive group of maps of the plane into itself.

The general ideas of the Julia and Mandelbrot sets are, however, of a much more general interest than these maps. Fractal attraction basins, completed with information about the speed of convergence or divergence, are highly interesting in a more general context. And so is the set in parameter space showing the same properties for a system. There is no need to restrict the ideas to the complex quadratic mapping in the context of which they arouse.

4.10 Can Chaos be Seen?

One could now ask whether all the formal exercise with Lyapunov exponents and symbolic dynamics is really necessary. Could one not, once intuition is opened up for the signs of a phenomenon such as turbulence or chaos, see it with the naked eye? We already touched upon this issue a couple of times.

To find out something more of this consider the following system:

$$x_{t+1} = \left(1 - k\left(x_t^2 + y_t^2\right)\right)\left(x_t^2 - y_t^2\right) \tag{4.69}$$

$$y_{t+1} = \left(1 - k\left(x_t^2 + y_t^2\right)\right)2x_t y_t \tag{4.70}$$

This system is easiest to understand if we introduce polar coordinates: $x_t = r_t \cos\theta_t$ and $y_t = r_t \sin\theta_t$. Substituting these in the right hand sides of (4.69)-(4.70) and computing: $r_{t+1} = \sqrt{x_{t+1}^2 + y_{t+1}^2}$ and $\theta_{t+1} = \operatorname{atan}(y_{t+1} / x_{t+1})$ for the reverse coordinate transformation, we easily obtain:

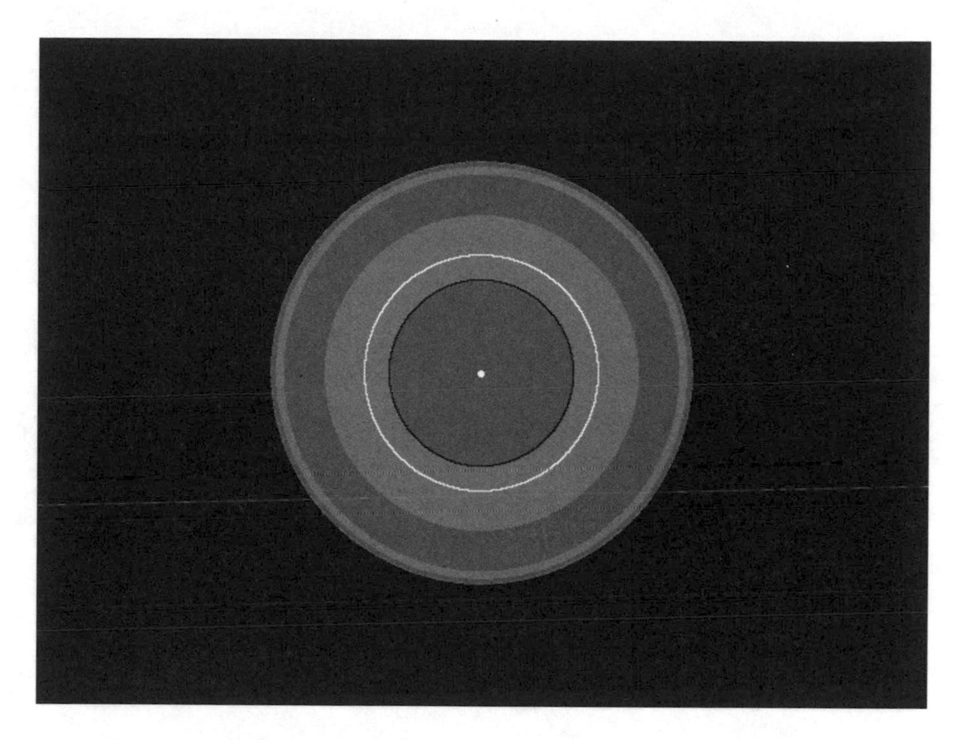

Fig. 4.18. Chaotic limit cycle, fixed point, and three attraction basins.

$$r_{t+1} = r_t^2\left(1 - kr_t^2\right) \tag{4.71}$$

$$\theta_{t+1} = 2\theta_t \tag{4.72}$$

These are two independent iterations in the radial and the angular coordinates respectively. As for the angle it is just doubled in each iteration, according to (4.72). This is already chaotic motion, as we recall from Section 4.4, where we found that the doubling of angles is an image of the most familiar logistic iteration just before it explodes.

As for (4.71), we can, again from the slight similarity to the logistic case, expect the iteration of the radius vector to display a complex dynamics dependent on the parameter k. We get four fixed points for the radius vector, zero and the three roots of the cubic equation: $kr^3 - r + 1 = 0$.

Skipping details, one root is negative. The remaining two roots are either conjugate complex or real and positive. Once they emerge as real numbers, which occurs at a coefficient $k \approx 0.2$, the smaller root creates an unstable

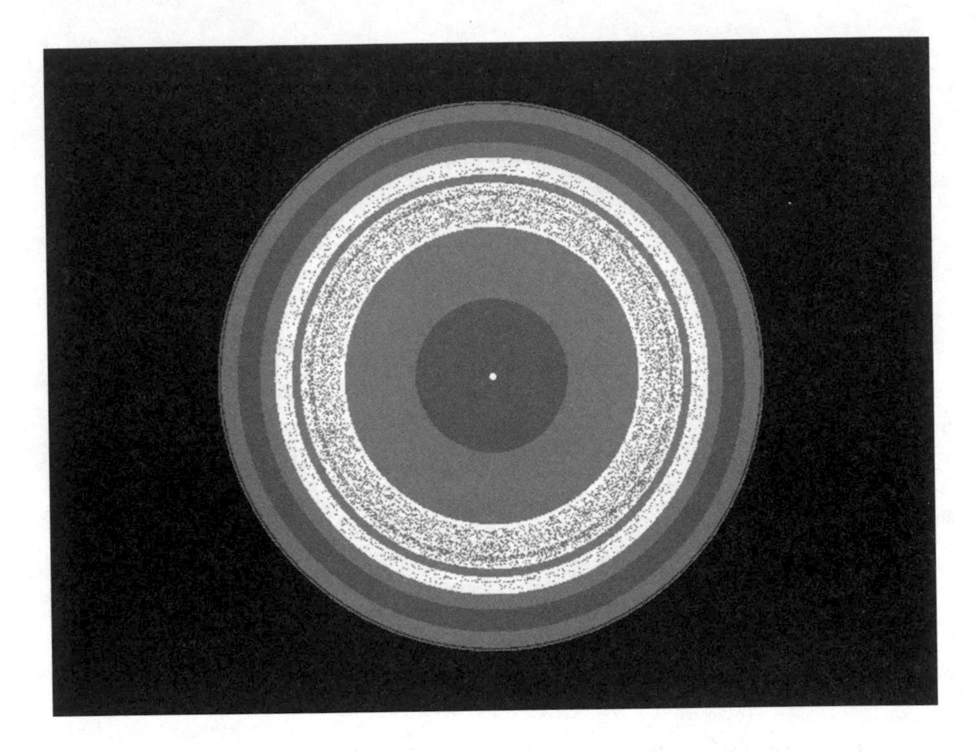

Fig. 4.19. Fractal set of chaotic limit cycles, fixed point and attarction basins.

limit cycle (black) and the larger a stable limit cycle (white). See Figure 4.18. The zero root at origin remains and is always a degenerate but critically stable fixed point. So, the smaller radius separates the attraction basins of the fixed point and the limit cycle.

Consider now the asymptotic motion of (4.71)-(4.72), or, which is the same, the original system (4.69)-(4.70), in the case when there is a stable limit cycle and the iteration starts in the proper attraction basin. It is just an iteration on an invariant circle with a doubling of the angle in each step. In fact a two-dimensional model of the logistic iteration.

What we see at simulation as in Fig. 4.18 just resembles a limit cycle. But the motion is manifestly chaotic. We already know this. If we measure the Lyapunov exponent numerically in the course of simulation, we arrive at the estimated dimension $\ln 2 \approx 0.69$, exactly as we should when the angle is doubled in each iteration.

Moreover, measuring the fractal dimension we arrive at dimension 1, though the "limit cycle" topologically is no curve. There are an infinite number of gaps, corresponding to all the rational angles which result in periodic motion, and hence are never visited in the chaotic motion displayed.

It is interesting to watch the attraction basins. We already noted the existence of the unstable limit cycle that separates the surroundings of the fixed point at origin from the outer ring in which the limit cycle is located. But now we see that the attraction basins are actually nested in several concentric rings, so that initial conditions further out in the phase diagram may again have the origin as attractor. The final attractor, fixed point or cycle, is indicated by the shade of the corresponding ring. There is also an additional feature in Fig. 4.18: The dark area in which the concentric rings are immersed, and which indicates explosion, i.e. convergence to a third attractor at infinity.

Further changes of the parameter, lowering it to about $k \approx 0.1$, cause the "limit cycle" to split in two, and in fact start a period doubling cascade to occur. Finally a chaotic set of limit cycles appears, though still coexistent with the stable fixed point at origin, and with the explosion attractor at infinity. A picture of such a set is shown in Fig. 4.19. We also see that the nesting of the attraction basins remains.

The estimated fractal dimension of the attractor now, of course, exceeds unity. We could also note an interesting fact. The chaotic iteration on the "limit cycle" of Fig. 4.18 was seen to correspond to a positive Lyapunov exponent. Likewise the iteration of (4.67) alone, in a case such as displayed in Fig. 4.19 is, of course, chaotic and hence has a positive Lyapunov exponent. Now, one could expect, as the iteration is a composite of two, each characterized by a positive Lyapunov exponent, that both exponents in the composite map would be positive. This is however not so. The sum of the exponents in the map must be negative, because otherwise the map would not be area compressive, and the process would hence not converge on any finite attractor. The lesson to learn from Fig. 4.18 is that we indeed need the formal methods and definitions for chaos, because there are cases where visual intuition is not of much use.

4.11 The Method of Critical Lines

The method of critical lines for the determination of the so called "absorbing areas" was invented in the 60es already by Christian Mira, though it only recently has gained popularity as a method for the analysis of nonlinear iterations in the plane. An absorbing area, quite like the attraction basin, is an area in phase space delimiting such initial conditions that make the process converge upon the attractor. The two should, however, not be confounded.

Fig. 4.20. Folding of the rectangle by the feedback logistic.

The basin is the set of all those initial conditions for which the system converges to the attractor. As we see in Fig. 4.11 it contains the attractor, but is much larger. An absorbing area, on the contrary is very tightly wrapped around the attractor so as to give an accurate outline of the attractor itself.

The Mira method is most useful for a noninvertible map. To make the distinction clear, consider the Hénon system (4.41)-(4.42). Despite the quadratic term present we can easily solve for the inverse map:

$$x_{t-1} = \frac{1}{b} y_t \tag{4.73}$$

$$y_{t-1} = x_t + \frac{a}{b^2} y_t^2 - 1 \tag{4.74}$$

This means that we can compute unique preimages for the Hénon map and so run the process backwards. This is a rare situation. Normally the maps are not invertible, and so we get multiple preimages. For instance the complex quadratic iteration (4.58)-(4.59) leading to the Julia sets is not invertible.

The reason for the lack of invertibility is the fact that the plane when mapped is normally folded. Consider the iterates of the rectangle by the Hénon map as displayed in Fig. 4.10. They are bent, translated, and rotated, but they are not folded. To be quite exact we did speak of folding in the Hénon case too, so that successively elongated images could be fitted in a limited space, but the images were not folded over themselves so as to produce self-intersection, which is at issue now, because we consider folding the plane over itself.

As a contrast consider the following simple extension of the logistic map (4.4):

$$x_{t+1} = x_t + y_t \tag{4.75}$$

$$y_{t+1} = \mu(1 - y_t)y_t - \sigma x_t \tag{4.76}$$

We just added a new variable which is defined as the cumulative sum of all the outcomes of the logistic iteration, which is itself linearly fed back as an additive term.

Focusing on just one step of the map, we delete the indices and write:

$$u = x + y \tag{4.77}$$

$$v = \mu y - \mu y^2 - \sigma x \tag{4.78}$$

Consider a rectangle in the x, y - plane, drawn in black in the background of Fig. 4.20. Its image in the u, v - plane is shown in gray shade on top of the same picture. We see that, unlike the case of the Hénon map, the image of the rectangle is actually folded. This folding action splits the plane in two parts: Below the fold there are two different preimages for each point, and above there are no preimages at all.

The separating fold itself provides the basic "critical" line in the following construction process. We are going to study its preimage and its various forward images under the iteration (4.77)-(4.78). It is this sequence of critical lines, or rather segments of them that delineate the absorbing area. The fold line is critical in the sense that it is the locus of points with one single preimage (actually two coincident) for the map (4.77)-(4.78), and it provides a boundary between the areas with two different and no preimages at all.

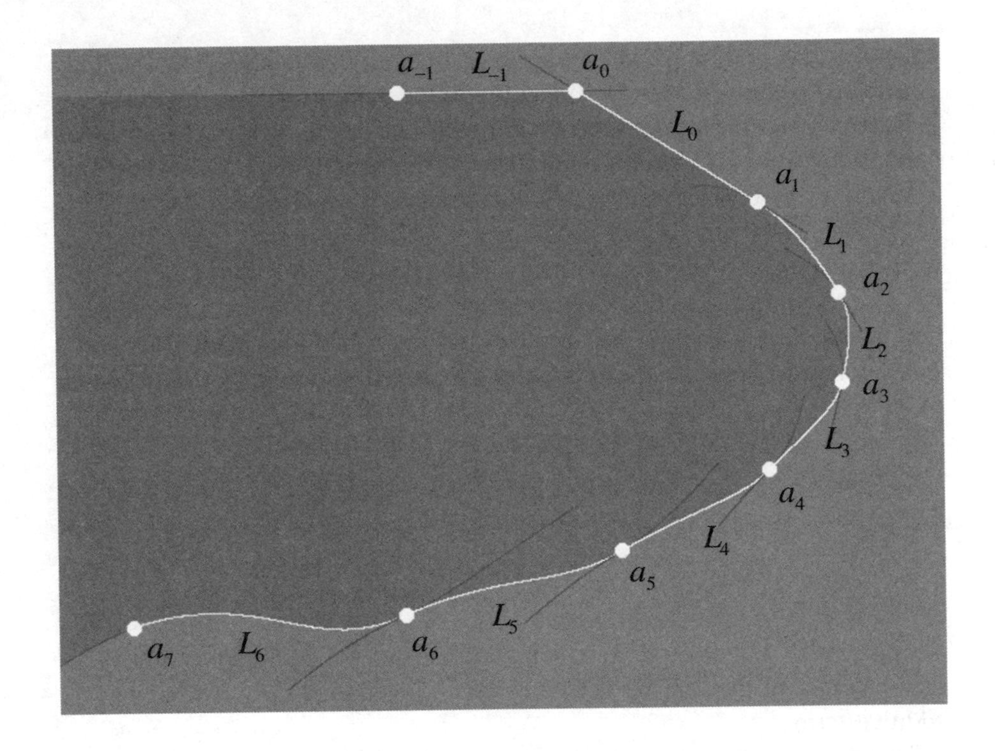

Fig. 4.21.Critical lines and their endpoints.

Note the following interesting fact in Fig. 4.20: The edges of the folded surface, images of the upper and lower horizontal edges of the original rectangle, are parallel with the fold itself. As the edges of the rectangle were taken arbitrary, we conclude that all horizontal lines from the x, y - plane are taken into sloping lines parallel to the fold in the u, v - plane. In particular this holds for the horizontal line which is mapped into the fold itself. Our next task in fact is to find the expression for this particular horizontal line and for its image, the fold. To this end solve (4.77) for x and substitute in (4.78). In this way we obtain:

$$v = (\mu + \sigma)y - \mu y^2 - \sigma u \tag{4.79}$$

Next differentiate (4.79) with respect to y and put the derivative equal to zero:

$$\frac{\partial v}{\partial y} = (\mu + \sigma) - 2\mu y = 0 \tag{4.80}$$

and solve for:

$$y = \frac{\mu + \sigma}{2\mu} \tag{4.81}$$

Finally, substitute y back in (4.79) to obtain:

$$v = \frac{(\mu + \sigma)^2}{4\mu} - \sigma u \tag{4.82}$$

Equation (4.82) in fact defines the fold in the u, v - plane. Likewise, we already got the equation for its preimage under the mapping (4.77)-(4.78). It was given by (4.81). As we already noted we deal with a horizontal line, so there is just a constant vertical coordinate, the horizontal being free. We can check by taking any points located on (4.81), iterating once by (4.77)-(4.78) and checking that their images satisfy (4.82). The facts are illustrated in Fig. 4.21, where we, as in Fig. 4.20, just put the u, v - space on top of the x, y - space. We do not display any rectangles, just the slanting fold and its horizontal preimage. They are denoted L_0 and L_{-1} respectively.

In the sequel we skip the laborious procedure of eliminating, putting the derivative equal to zero, solving, and substituting back as in equations (4.79)-(4.82). It is not even always possible to eliminate and solve. Fortunately, we can just obtain the same result by putting the Jacobian determinant of any system such as (4.77)-(4.78) equal to zero. In this way we obtain the equation for L_{-1} in one single step.

There are various forward images of L_0 in Figure 4.21, the indexing being self-explanatory. We will return to these images, only note that due to the quadratic term in the mapping they become slightly curved, polynomials of ever higher degree in each iteration.

We also display various intersection points between these lines, starting with one particularly interesting point a_0, the point where L_0 and L_{-1} cross. To find its coordinates we identify y according to (4.81) with v according to (4.82):

$$\frac{\mu + \sigma}{2\mu} = \frac{(\mu + \sigma)^2}{4\mu} - \sigma u \tag{4.83}$$

and solve for:

$$u_0 = \frac{(\mu+\sigma)(\mu+\sigma-2)}{4\mu\sigma} \tag{4.84}$$

The vertical coordinate we get from (4.81), identifying v with y:

$$v_0 = \frac{\mu+\sigma}{2\mu} \tag{4.85}$$

Equations (4.84) and (4.85) thus give us the coordinates of the intersection point a_0.

In addition to a_0 we want the coordinates of its preimage a_{-1}. To this end note again that a_0 is located on both L_0 and L_{-1}. Now L_0 is the image of L_{-1}. Hence a_{-1} too must be located on L_{-1}. Its vertical coordinate is hence given by (4.85). To get the horizontal we use (4.77) for one iteration, substituting the right hand side of (4.85) for y, and equate to the right hand side of (4.84):

$$x + \frac{\mu+\sigma}{2\mu} = \frac{(\mu+\sigma)(\mu+\sigma-2)}{4\mu\sigma} \tag{4.86}$$

or, solving for:

$$x_{-1} = \frac{(\mu+\sigma)(\mu-\sigma-2)}{4\mu\sigma} \tag{4.87}$$

As already stated we have:

$$y_{-1} = \frac{\mu+\sigma}{2\mu} \tag{4.88}$$

We are now done with the computations. In Fig. 4.21 we see the critical lines L_0 and L_{-1} and the points a_0 and a_{-1}. In addition we see six forward images of the iterated line L_0. The system maps a_{-1} to a_0. Likewise it maps a_0 to a_1,

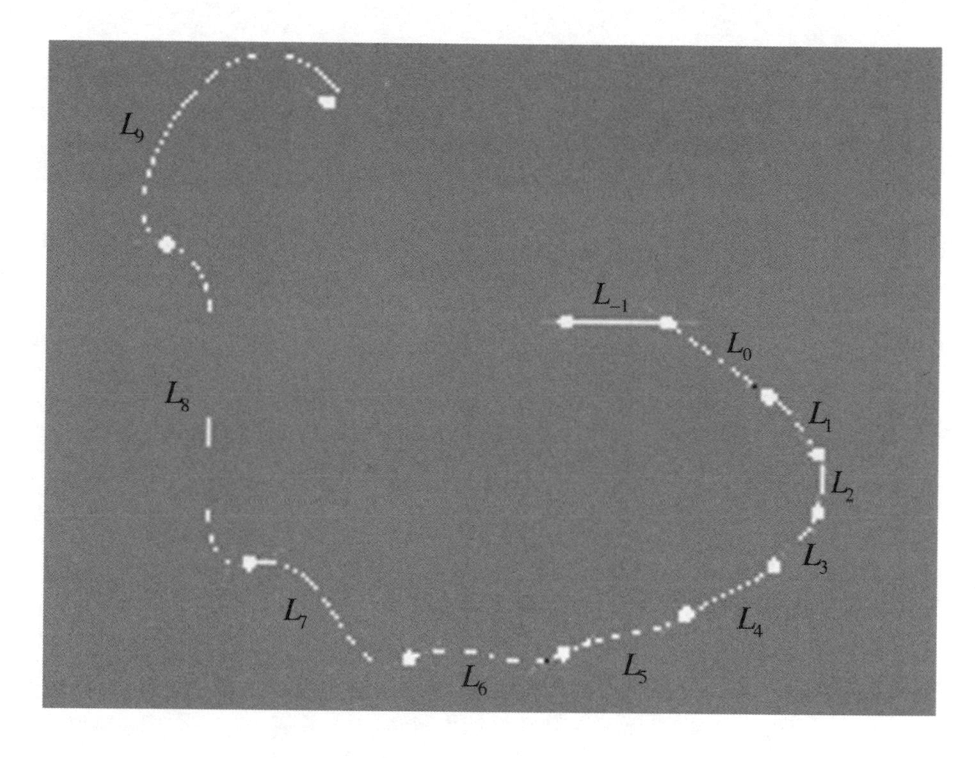

Fig. 4.22. Selection of 10 critical lines.

and so forth. We do not make any further calculations. All the various curve segments are successive forward images of the interval (a_0, a_1) of L_0, just as it itself is an image of the interval (a_{-1}, a_0) of L_{-1}. Note how neatly they fit together in a train of segments.

There is one particularity to the way the segments fit together. Consider the point a_1. It is the image of a_0. As the latter is a point of transverse intersection between L_{-1} and L_0, the point a_1 is a point of tangency between the straight line L_0 and the curve L_1.

The reason for tangency is that the plane is folded in the line L_{-1}, and any curve, such as L_0, which cuts it is mapped to a curve which has a tangency with its image curve, which is L_0 itself. To see this, draw any two intersecting lines on a transparency, then fold it smoothly along one of the lines. The other line will then be seen as a curve, tangent to the fold. These tangencies make the continuation of the lower order critical curves spiral off from an area bounded by the train of critical segments.

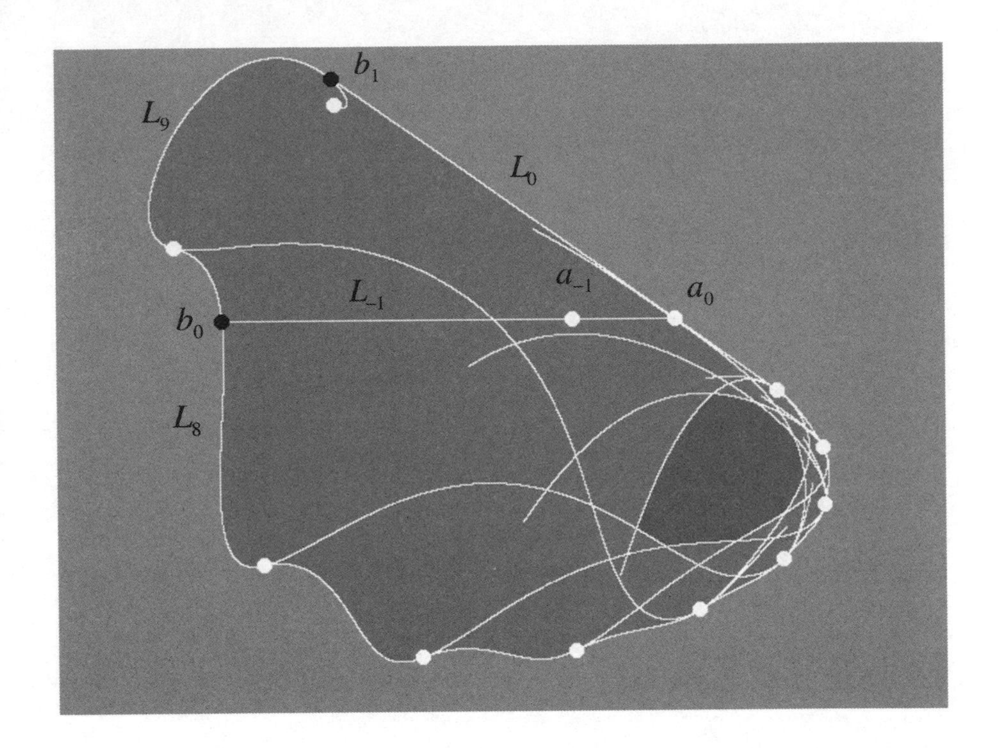

Fig. 4.23. Absorbing area.

This increasing curvature of the boundary also is the reason why the train of segments forms a trapping area for the iterative process. Points inside this area are never moved outside it by the iterated map.

All this is shown on a bigger scale in Fig. 4.22, where a total of 10 curve segments are displayed. This number of iterates of the segment is obviously not sufficient to delimit a closed absorbing area in the phase space. There are two things we can do to close the area: take more iterates, or increase the segment (a_{-1}, a_0) of L_{-1}

Both procedures work, but to capture the absorbing area in as few iterations as possible it is better to enlarge the segment. Figure 4.23 shows a picture where L_{-1} is extended to the left to the point denoted b_0. This point is the intersection of the lines L_8 and L_{-1}. What was said above about the intersection of L_0 and L_{-1} again applies. A transverse intersection with L_{-1} is mapped into a tangency with L_0. Accordingly we see how L_9 meets L_0 at the point of tangency b_1.

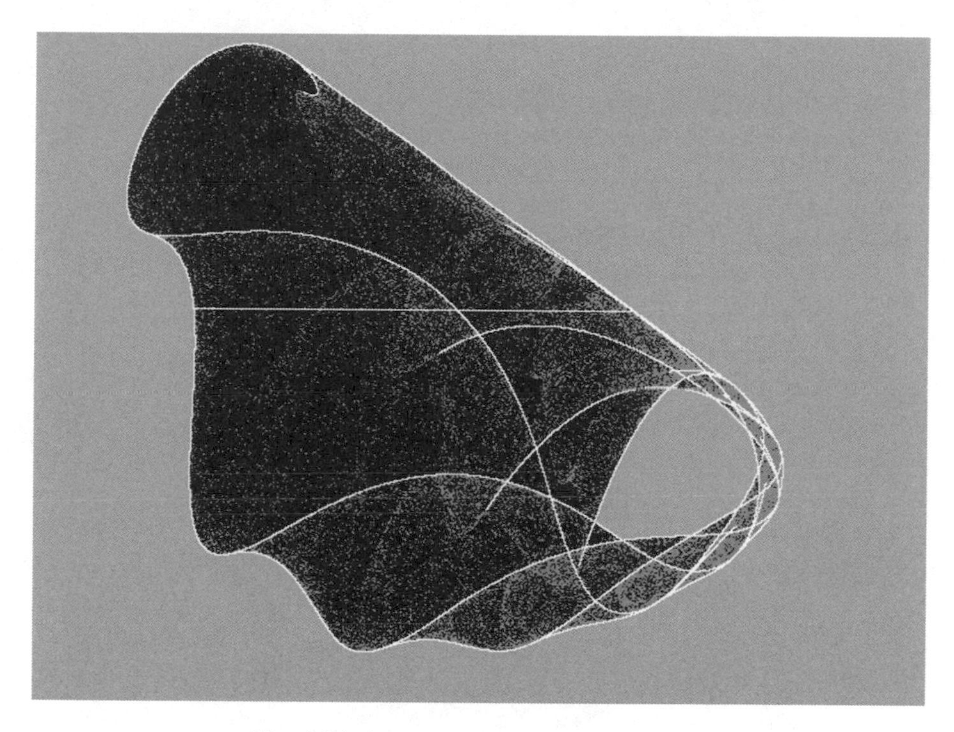

Fig. 4.24. Attractor within absorbing area.

But, just because L_9 meets L_0, the absorbing area becomes closed. Note that as the eighth iterate again meets the line L_{-1}, we need precisely L_0 through L_9 in order to get the complete boundary. To this end we can, however, not work only with the segment (a_{-1}, a_0). We need all of (b_0, a_0). All this becomes clear if we compare Figure 4.23 to Figure 4.22. Note also that the hole in the attractor, indicated by a darker shade becomes outlined in Figure 4.23.

In Figure 4.24 we put the actual attractor inside the absorbing area, and may see how nicely the critical line method works by outlining the exact boundaries of the attractor. Note also the following fact: In most cases the point density varies over a chaotic attractor. It now seems that the points accumulate precisely along the train of critical curve segments. This is no accident. The segments, being various forward images of the fold, which, as we know, separates points with two and those with no preimages, accordingly also separate areas of few and many points in the iteration process.

In Fig. 4.25 we finally see the difference between the absorbing area and the basin of attraction. The latter is shown in black around the absorbing area, and we see that the two actually come quite close to each other along a

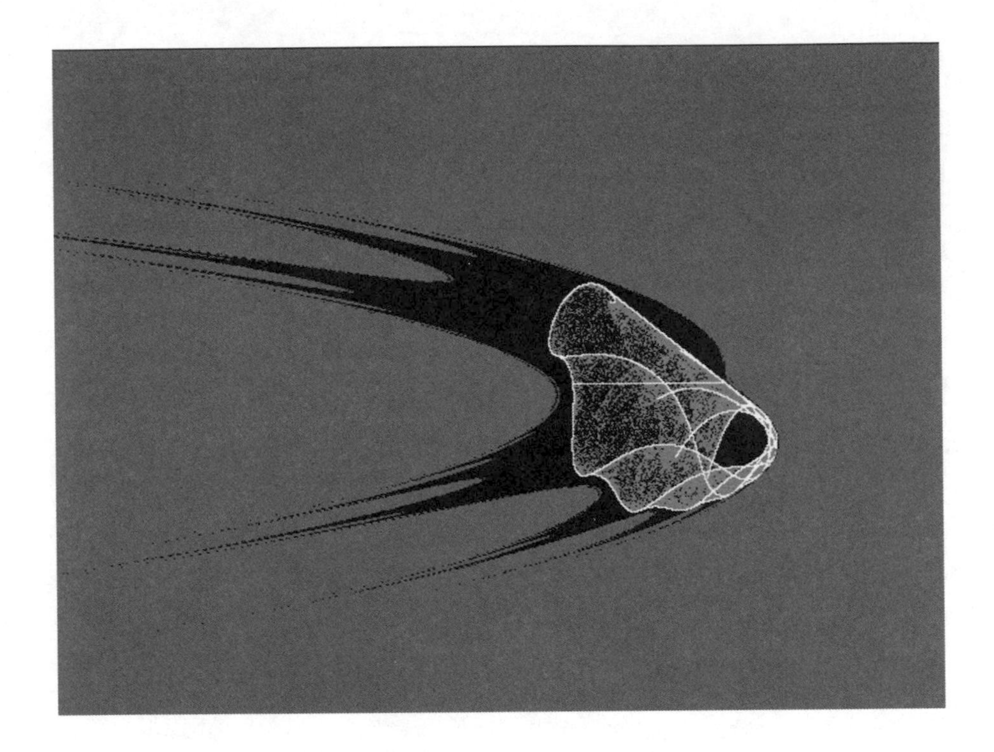

Fig. 4.25. Absorbing area and basin of attraction.

few segments of their boundaries. The black basin is surrounded by a gray area for those initial conditions for which the process explodes, i. e. goes to the attractor at infinity.

The critical curve method can be used to get simple outlines of the attractors, and to understand various bifurcations, in particular the turning of a chaotic attractor into a chaotic repellor. This happens when the absorbing area shoots tongues out into the basin at infinity. Fig. 4.25 is not too far from that point. In many cases, such as a case studied later in Chapter 10, we actually need much fewer curves in order outline the absorbing area and the attractor. For the reader interested in experiment the parameters used for drawing the Figs. 4.20-4.25 were $\lambda = 0.70$ and $\sigma = 0.55$.

We end the discussion of critical curves and absorbing areas here, but the reader can find a lot more of information on the method in the visual and relatively elementary exposition in Abraham, Gardini and Mira (1997), and in the more detailed one by Mira, Gardini, Barugola, and Cathala (1996). The first book also contains a CD-rom with illustrative bifurcation sequences.

5 Bifurcation and Catastrophe

On several occasions we already encountered the concept of bifurcation. In Sections 2.5-2.6 we for instance discussed the Hopf bifurcation from fixed point to limit cycle, and the saddle-node bifurcation, which again broke up the continuous cycle into a discrete set of fixed points.

But, without naming it explicitly, we also touched upon the phenomenon in numerous other places. The discussion in Section 2.3 on structural stability for instance focused phenomena such as the splitting of a "monkey" (six sector) saddle point into two ordinary saddle points, and the further splitting of the "heteroclinic" saddle connection between them. Systems such that small changes to it may lead to bifurcation phenomena were defined to be structurally unstable, and structurally stable systems were defined as such that do not bifurcate. Structural stability, by the way, was touched upon already in our introductory discussion of the frictionless pendulum and of how the phase portrait changed once friction was added.

Finally, the discussion of relaxation cycles in the Rayleigh oscillator in Section 2.9, involved a contrasting process of a slow building up of energy when the system moved along the "characteristic" and fast discharges of energy when the system jumped from one stable branch of the back-bending characteristic to another. With two different time scales involved, we can always take a perspective in which the fast process is regarded as being in constant equilibrium, whereas the slow process is represented by external parametric changes that may cause sudden bifurcations to occur to that equilibrium state.

We will now take a more systematic perspective of all these bifurcation phenomena. Classical dynamical systems theory, such as the Newtonian, the Euler-Lagrange, or the Hamiltonian focused smooth development over time. The basic tool in differential equation theory was the Existence and Uniqueness Theorem for smooth solution trajectories, and there were no general tools to handle sudden jumps, hysteresis (i.e. irreversibility effects), or other bifurcation phenomena. When such phenomena occurred (as in the forced van der Pol or Duffing systems), then the tools for analysis had to be invented ad hoc. Moreover bifurcations, just like the rest of interesting dynamic phenomena, do not occur in linear systems, and so, with a focus on linearity, they became exceptional phenomena which could be dealt with by exceptional methods.

5.1 History of Catastrophe Theory

Around 1970 René Thom felt this situation was deeply unsatisfactory and wrote his seminal work on structural stability and morphogenesis. Thom was particularly interested in modelling the development of shape in living organisms, in particular such phenomena as the sudden change of a bud into a leaf or into a flower. The theory he created was called catastrophe theory, and he aimed at a complete and systematic treatment of bifurcations or catastrophes that could occur in dynamical systems, given the numbers of state variables and exogenous parameters.

The standard forms were worked out as (the simplest possible) polynomials in the state variables with the exogenous parameters as coefficients, and the mathematically deep part of the theory was the proof that any problem with a given number of state variables and parameters could by a smooth coordinate transformation be brought into the corresponding standard form. Unfortunately, the coordinate transformations to be used were extremely hard to find in most cases, which hampered the production of really meaningful applications.

Thom characterized the seven simplest catastrophes involving one to four parameters, four with one state variable (the "fold", the "cusp", the "swallowtail", and the "butterfly"), and three with two state variables (the "elliptic", the "hyperbolic", and the "parabolic" "umblics"). For mathematical detail we refer to Gilmore, Poston and Stewart, or Saunders, which is particularly readable. The decades following Thom's work saw all kinds of applications of catastrophe theory: buckling engineering structures, eating disorders (Anorexia-Bulimia Nervosa), the psychology of attacking dogs, outbursts of riot, and stock market crashes. There were even developed various toys, such as Zeeman's catastrophe machine - a turning wheel with an off centre handle to which two rubber bands are tied, the first with its other end fixed, the second with its other end free to be operated by the player. By moving this other end "in control space" all sorts of unexpected sudden flips of the wheel could then be produced.

Before entering the substance of the topic we should make clear the difference between the concepts of catastrophe and bifurcation. The latter simply is a more general term, referring to any kind of sudden change that can occur in a dynamical system. Catastrophe is limited to so called gradient systems where the fast dynamics acts as if it minimizes some kind of potential. This means that in one dimension there is no difference at all, because we can just integrate the right hand side of a differential equation to obtain the potential (sign reversed). In two dimensions and higher there is a difference. It is far from certain that, for instance, a pair of right hand sides of coupled differential equations can be written as the partial derivatives of one and the same potential function.

Over the history of applied catastrophe/bifurcation theory there was a very pronounced dominance of cases in one dimensional state spaces, so it is not surprising that we find the two concepts used interchangeably. Moreover, whenever we deal with problems based on optimization, as is often the case in economics, the dynamic equations can always be derived from a potential, and so catastrophe theory will do. On the other hand, the periodic motion of an oscillator can, of course, not be derived from a potential, at least not without some mathematical tricks.

The surge of specific catastrophe models has now faded away. This is not so because they are no longer considered as significant, but because catastrophes and bifurcations have been assimilated among all the various phenomena that can occur in nonlinear dynamical systems.

5.2 Morse Functions and Universal Unfoldings in 1 D

Implicit in all catastrophe theory is a fast dynamic process that minimizes some sort of potential such as the process:

$$\dot{x} = -V'(x) \tag{5.1}$$

where $V(x)$ is the potential. As an example take

$$V(x) = \frac{x^2}{2} \tag{5.2}$$

whence we get $\dot{x} = -x$ with the obvious closed form solution $x(t) = x_0 \, e^{-t}$. No matter what the initial value x_0, the process goes asymptotically to zero, which is the unique minimum of the potential.

Now consider various deformations of the potential function and their effect on the dynamical process, more specifically deformations by adding other (integer) powers to the polynomial. The truth is that they make no difference at all. If we add a linear term it just moves the minimum of the potential, left or right from the origin depending on the sign of its coefficient. In a topological sense nothing happens as the unique minimum remains, and as we could even move it back to the origin by a smooth coordinate transformation. The dynamic process remains undisturbed in its approach to this unique minimum.

Now, consider a more remarkable fact: Add a third order term instead. It makes no difference either. Close to the origin the lowest, i.e., in our case the second order term dominates and the character of the minimum, which now even remains where it is, is not altered. The same holds for all higher powers.

Of course, the addition of higher order terms produces results, such as adding new maxima and minima far off from the origin, but this is a global feature, and catastrophe theory is concerned with local phenomena in the neighbourhood of the minimum point.

Given this, the remarkable thing for a potential such as (5.2) is that no addition of terms, whether of orders higher or lower, alters its topological character. It is structurally stable as it stands, or a so called *Morse Function*.

The reader may think that it is fairly restrictive to concentrate on polynomials, but elementary calculus tells us that the very large class of analytical functions, defined as having Taylor Expansions, are expressible in power series. Take for instance the potential

$$V(x) = 1 - \cos(x) = \frac{x^2}{2} - \frac{x^4}{24} + \frac{x^6}{720} - \ldots \tag{5.3}$$

The potential is of the same type as (5.2), and we already know that the higher order powers make no difference, so we can in fact skip all the following terms. Accordingly (5.3) too is Morse.

Catastrophe theory deviates radically from classical calculus. The latter was preoccupied with deriving the remainder term of a truncated series such as (5.3) and showing that it converged to zero in order to prove analyticity. Catastrophe theorists are happy to work with truncated Taylor series, even skipping all higher order terms that are not significant for the local character of the minimum point. Zeeman formulated this shift of focus as not letting "*the Tayl* (the remainder) *wag the dog*" any longer.

Its should now be said that catastrophe theorists are not interested in things as well behaved as Morse Functions, but rather with power functions that lack structural stability on their own and with the ways to stabilize them so that the combined polynomials acquire such stability.

Assume now that we have the potential

$$V(x)=\frac{x^4}{4} \tag{5.4}$$

It looks quite like (5.2), having a single minimum at the origin, but the case is very different from the present point of view. It still holds that there is no difference if we add powers of the fifth order or higher, but lower order powers now make a big difference. As a reminder of the fact that we are not dealing with just polynomials note the equivalent potential

$$V(x)=2(1-\cos(1-\cos x))=\frac{x^4}{4}-\frac{x^6}{24}-\frac{x^8}{480}+\ldots \tag{5.5}$$

where we can again ignore all terms but the first.

Adding lower order terms to (5.4), or (5.5) can now make the minimum split into at most two minima and one maximum, so (5.4) is not one critical point, but rather three coincident critical points. To separate them and produce a structurally stable function we have to add lower order terms. Once this is done the result is called a "universal unfolding". We could, of course, add all lower order terms, linear, quadratic, and cubic, and the result would then be called just "versal" instead of "universal", but one of the lower powers can always be removed by a simple shift of origin. Usually the cubic term is removed and so the universal unfolding reads:

$$V(x)=\frac{x^4}{4}+\alpha\frac{x^2}{2}+\beta x \tag{5.6}$$

For $\alpha=\beta=0$, (5.6) becomes (5.4) with a unique minimum, but in general there are up to three different critical points as solutions to the equation:

$$V'(x) = x^3 + \alpha x + \beta = 0 \tag{5.7}$$

There are two different cases for what (5.6) looks like. Suppose $\alpha > 0$. As both even order terms, the quartic and the quadratic, have a minimum at the origin, they cooperate to produce one single minimum. It is displaced by the linear term, but it remains a unique minimum. The role of the linear term is to break the symmetry that solely even order terms would produce, moving the minimum right or left again depending on the sign of β. In this case (5.7) has one real and two complex conjugate roots. Observe that the single minimum of (5.6) given $\alpha > 0$, unlike the case of (5.4), is structurally stable. No further added term can change the local character of the single minimum.

Suppose we instead have $\alpha < 0$. Then the quadratic term on its own produces a maximum, whereas the quartic still produces a minimum. As the lower order term dominates close to the origin it is the quadratic that makes the combination have a maximum at the origin. However, the quartic dominates further out from the origin, and so produces two minima, one on either side of the maximum.

Note that we are not contradicting the above argument for disregarding added higher order powers. To see this consider $U = x^4/4 + \delta x^2/2$ and $V = x^2/2 + \varepsilon x^4/4$ each with just the same even order terms, though it makes a big difference which term is there first. The extrema for U are located at $x = 0$ and $x = \pm\sqrt{-\delta}$, those for V at $x = 0$ and $x = \pm 1/\sqrt{-\varepsilon}$, so with δ, ε small, the new extrema start out at the origin for U, but at infinity for V. Moreover the extremum for V at the origin is always a minimum whereas for U it flips from minumum to maximum.

Again, with $\alpha < 0$, (5.6) is structurally stable. No further addition of any new powers changes anything in the neighbourhood of any of the critical points.

The dynamics corresponding to (5.1) and (5.7) is

$$\dot{x} = -x^3 - \alpha x - \beta \tag{5.8}$$

Provided $\alpha < 0$, the process now goes to any of the two minima, depending on whether the process starts from one side or the other of the local maximum which acts as a basin watershed for the two attractors. If instead $\alpha > 0$ then the process goes to the single minimum.

As α passes from positive values through zero to negative values there is a bifurcation in (5.8). The single attractor suddenly loses stability and it is replaced by two new attractors one to either side. Any tiny push will put the system in motion from the unstable equilibrium (maximum of the potential), and it can go to either of the new attractors depending on to which side the initial push was.

The bold claim of catastrophe theory now is to have represented all the bifurcation phenomena that can occur with two parameters α, β by the universal unfolding (5.6).

5.3 Morse Functions and Universal Unfoldings in 2 D

We have been occupied with one dimensional state spaces, so what happens if we introduce two state variables x and y? We now have a potential $V(x, y)$ and a pair of differential equations

$$\dot{x} = -\frac{\partial V}{\partial x}, \qquad \dot{y} = -\frac{\partial V}{\partial y} \tag{5.9}$$

As an illustration suppose we have the potential

$$V(x, y) = \frac{x^2 + y^2}{2} \tag{5.10}$$

This is a paraboloid with a unique minimum at the origin, a sort of bowl, and the dynamical system says that a small object put anywhere in the bowl just slides down the walls in the direction of steepest decent. The formal system is simple $\dot{x} = -x,\ \dot{y} = -y$, as are the solutions $x(t) = x_0\, e^{-t},\ y(t) = y_0\, e^{-t}$.

If we now add any powers of the state variables or of products of such powers higher than the second, we again do not alter the potential. This is because sufficiently close to the minimum its character is determined by the lowest order terms which are already there. The only terms of lower order we could add are linear, but again, they only displace the minimum slightly without affecting its character. So, (5.10) is again a Morse function, now in two dimensions.

If we change the potential to $V(x,y) = -(x^2 + y^2)/2$, we still have a Morse Function. The paraboloid is now turned upside down, and there is a unique maximum instead of minimum, but no additional terms can change its local character. The dynamic process $\dot{x} = x,\ \dot{y} = y$ now diverges from the single unstable equilibrium according to the solution $x(t) = x_0\, \mathrm{e}^{t},\ y(t) = y_0\, \mathrm{e}^{t}$. It is noteworthy how closely Morse functions (and hence convexity and uniqueness) are associated with linear dynamic systems.

Suppose next that the quadratic terms have opposite sign, for instance $V(x,y) = (x^2 - y^2)/2$. This is a potential that looks like a saddle, it goes up in the *x*-direction and down in the *y*-direction. The dynamic equations are $\dot{x} = -x,\ \dot{y} = y$ with solutions $x(t) = x_0\, \mathrm{e}^{-t},\ y(t) = y_0\, \mathrm{e}^{t}$. So, the trajectories converge in the *x*-direction but diverge in the *y*-direction. Interchanging the signs in the potential just rotates the saddle by 90 degrees.

It may be a surprise that the potential $V(x,y) = xy$ too is a saddle, only rotated by 45 degrees in comparison to the previous. To see this introduce the coordinates $x = (u+v)/\sqrt{2},\ y = (u-v)/\sqrt{2}$. This coordinate change represents simple rotation, where the square root is necessary to keep the linear dimensions unchanged. Substituting we get $V(u,v) = (u^2 - v^2)/2$.

It so seems that all Morse Functions in 2 D could be written in the form:

$$V(x,y) = \frac{\pm x^2 \pm y^2}{2} \tag{5.11}$$

which in fact is true. The result generalizes to any dimension and is usually called *Morse's Lemma*. The natural question at this stage is: Are all functions of second order of the form $(ax^2 + 2bxy + cy^2)$ Morse? The answer is that this is not true.

Consider the potential

$$V(x,y) = x^2 - 2xy + y^2 = (x-y)^2 \tag{5.12}$$

which is a perfect square. If we introduce the *u*, *v* coordinates, as defined above, we get

$$V(u,v)=\frac{v^2}{2} \tag{5.13}$$

where the u term is not present at all. In one dimension this would have been no problem, but in two dimensions it is. Equation (5.13) says that along the entire v-axis there is an accumulation of local minima, or, in terms of the original potential, along the entire line $x + y = 0$.

This is a structurally unstable phenomenon. Adding just a linear term in the u dimension removes all the minima, adding a quadratic produces a single minimum, or a saddle point.

But there is more to it. Equation (5.13) tells us that there is one dimension, the v dimension which is Morse, so that there is only one dimension, the u dimension, which can be involved in structural instability.

There is a name for this too, the *Splitting Lemma*, which is a complement to Morse's Lemma, stating that in case a function is not Morse it can be split in a Morse part and another involving the dimensions in which instabilities can occur.

As a rule, in any problem with many dimensions of the state space, only a few are involved in structural instability, and so low dimensional state spaces are sufficient to analyse structural change in even high dimensional systems.

This point was emphasized particularly in Haken's "synergetics" school. To see his point let us consider the dynamical system corresponding to (5.13). It is $\dot{u}=0,\ \dot{v}=-v$ and integrates to $u(t)=u_0,\ v(t)=v_0\,\mathrm{e}^{-t}$. We would say that the system in this diagonalized form has eigenvalues 0 and -1. Haken's argument runs as follows: Eigenvalues of any linearised dynamical system in the neighbourhood of a fixed point can be positive, zero, or negative. Positive eigenvalues cannot persist, except for very short periods, because they would blow up the system, so eigenvalues must be on the negative side. Mere probability tells us that only a few eigenvalues among the negative ones are very close to zero. Most are safely clear off the zero value.

Now, all those variables that have clearly negative eigenvalues can be regarded as momentarily damped to their equilibrium values. They can be regarded as "slaved" in the terminology of Haken. The few eigenvalues closest to zero introduce a slow time scale, and they can for a moment pass the zero line and then trigger off bifurcations in the system. But, being few, the bifurcations occur in a low dimensional space, no matter how many the dimensions of the original system. This makes catastrophe theory, though it was developed for systems in few dimensions, retain its value for the analysis of instabilities in high dimensional systems.

There is one warning in place. We should not be so naive as to think that we come across the fast and slow variables directly in whatever model we care to analyse. Neither of the variables x, y in (5.12) can be classified as fast or slow. It is only after defining the u, v variables, which are pure mathematical constructs, that we can identify v as fast, and u as slow (in fact so slow as to be still standing).

Let us now put down the formal condition that was the cause of the structural instability in (5.12):

$$\begin{vmatrix} \dfrac{\partial^2 V}{\partial x^2} & \dfrac{\partial^2 V}{\partial x \partial y} \\ \dfrac{\partial^2 V}{\partial y \partial x} & \dfrac{\partial^2 V}{\partial y^2} \end{vmatrix} = \begin{vmatrix} 2 & -2 \\ -2 & 2 \end{vmatrix} = 0 \tag{5.14}$$

Whenever the determinant in (5.14) is zero we get a perfect square in the second order terms of the Taylor expansion. We can also note that the entries in the determinant of partial derivatives are the coefficients of the linearised right hand sides of the differential equations for the gradient dynamics. As we remember from Chapter 2 the determinant equals the product of the eigenvalues. So one zero eigenvalue is signalled by (5.14) being zero. Next higher order degeneracy occurs if all the entries of (5.14) become zero.

An example of this is given by the potential:

$$V(x, y) = \frac{x^3 - 3xy^2}{3} \tag{5.15}$$

It is easy to see that it has a unique stationary point at the origin, because $\partial V / \partial x = (x^2 - y^2) = 0$ and $\partial V / \partial y = -2xy = 0$ hold only for $x = y = 0$. However, all the second order partial derivatives are zero at the origin, so this third order function is manifestly unstable, and needs several additional terms of lower order for its stabilisation. Geometrically (5.15) defines the picture of the Monkey Saddle which we have encountered several times already as an extreme case of structural instability.

We should now also consider potential functions in 2D not written as polynomials, like we did in the one state variable case, in order to stress the generality of the argument. For instance, we get:

$$V(x,y) = 1 - \cos x \cos y \approx \frac{x^2 + y^2}{2} \tag{5.16}$$

as a function equivalent to the case (5.10), and

$$V(x,y) = 2 - 2\cos(x - y) \approx x^2 - 2xy - y^2 \tag{5.17}$$

as equivalent to the case (5.12). To get (5.15) as a Taylor series we can expand

$$V(x,y) = \frac{1}{6}\sin\left(x + \sqrt{3}y\right)\sin\left(x - \sqrt{3}y\right)\sin(2x) \approx \frac{x^3 - 3xy^2}{3} \tag{5.18}$$

It is easy to see that all the second order partial derivatives to the original trigonometric function (5.18) are zero at the origin, so we can check for degeneracy even before applying the Taylor expansion, as we can in the milder case (5.17) where just the determinant of the second order derivatives vanishes.

Note that (5.15) results in the following pair of differential equations

$$\dot{x} = -x^2 + y^2 \tag{5.19}$$

$$\dot{y} = 2xy \tag{5.20}$$

which we (signs reversed) encountered in Section 2.3. This was the monkey saddle flow, which, as we saw, was split into two disjoint saddles by adding arbitrarily small constants to the right hand sides of (5.19)-(5.20). This would mean adding linear terms to the potential (5.15).

But this is not yet all that can happen to the monkey saddle. Totally it at most hides four singularities, one node and three saddles which remain to be unfolded. They can be produced by adding quadratic terms as well, so that we get the potential

$$V(x,y) = \frac{x^3 - 3xy^2}{3} + \gamma\frac{x^2 + y^2}{2} + \alpha x + \beta y \tag{5.21}$$

It is (5.21) that is the universal unfolding of (5.15). We could add a product term as well, but it makes no difference. The dynamical system then reads:

$$\dot{x} = -x^2 + y^2 - \gamma x - \alpha \tag{5.22}$$

$$\dot{y} = 2xy - \gamma y - \beta \tag{5.23}$$

Putting the right hand sides equal to zero, we can solve for x from the last equation and substitute in the first which then becomes a quartic. The latter as a rule either has two or four real roots, which corresponds to the case of two saddles, or a node and three saddles.

5.4 The Elementary Catastrophes: Fold

We already encountered two of the so called elementary catastrophes in the above discussion, the cusp in equation (5.6) and the elliptic umblic in equation (5.21), but let us now take a more systematic tour around the catastrophes.

The simplest is the *fold*. The potential in its universal unfolding reads

$$V(x) = \frac{x^3}{3} + \alpha x \tag{5.24}$$

We would guess that a quadratic term as well might appear, but again, by choosing a shift of origin, we can easily remove the quadratic, or the linear term, but not both. The normal thing is to skip the highest possible, i.e. presently the quadratic. The fast dynamics associated with the potential (5.24) is given by the differential equation:

$$\dot{x} = -x^2 - \alpha \tag{5.25}$$

If $\alpha > 0$, then (5.25) has no zeros. The process always goes to minus infinity. If $\alpha < 0$, then there are two equilibria for the system, one unstable and one stable, given by the roots $x = \mp\sqrt{-\alpha}$. Depending on whether initially $x_0 < -\sqrt{-\alpha}$ or $x_0 > -\sqrt{-\alpha}$ the system goes to $-\infty$ or to $\sqrt{-\alpha}$.

Accordingly there is a bifurcation when the parameter α passes from negative to positive values. From having had no equilibria at all the system acquires one unstable and one stable equilibrium.

To introduce some formal terms the set of equilibrium points is called the *equilibrium set* and it is obtained by putting (5.25) equal to zero. Due to our gradient dynamics this is the same as:

$$V'(x) = x^2 + \alpha = 0 \tag{5.26}$$

Bifurcation, or catastrophe is now always signalled by the second derivative becoming zero, i.e.

$$V''(x) = 2x = 0 \tag{5.27}$$

This is so because the second order derivative differentiates between minimum, maximum, and inflection point, so, whenever an extremum is about to change character this is indicated by the second derivative becoming zero. The set defined by (5.27) is called the *singularity set.*

For the very simple case introduced we could do without the equilibrium and singularity sets, but this does not hold for the more complicated potentials, so we could as well introduce the terminology now.

The interesting thing is that the two equations (5.26)-(5.27) let us eliminate the state variable x. From (5.27) we gather that whatever the bifurcation, it must occur at $x = 0$. We can then just substitute in (5.26) and obtain:

$$\alpha = 0 \tag{5.28}$$

which is called the *bifurcation set*. It is obtained as the intersection of the equilibrium and singularity sets, and in the present case it is just a point in parameter space.

We already found this out by informal reasoning, so we have only introduced some terminology.

5.5 The Elementary Catastrophes: Cusp

Next on the list is the *cusp* which we already studied, but we state it anew for convenience. The potential was fourth order and it had two control parameters:

$$V(x) = \frac{x^4}{4} + \alpha \frac{x^2}{2} + \beta x \tag{5.29}$$

and its associated dynamics was:

$$\dot{x} = -x^3 - \alpha x - \beta \tag{5.30}$$

We already saw that, depending on whether $\alpha > 0$ or $\alpha < 0$, we may have one stable equilibrium, or two stable equilibria surrounding an unstable one. So there can be a bifurcation when the parameter α takes on negative values.

We should be more precise than this: The equilibrium set is:

$$V'(x) = x^3 + \alpha x + \beta = 0 \tag{5.31}$$

and the singularity set:

$$V''(x) = 3x^2 + \alpha = 0 \tag{5.32}$$

From equations (5.31)-(5.32) we can again eliminate the state variable x, and obtain:

$$4\alpha^3 + 27\beta^2 = 0 \tag{5.33}$$

as the bifurcation set. It is a cuspoid curve, whence the name of the catastrophe. The arithmetical trick to play is to multiply through (5.31) by 3, and (5.32) by x, and subtract. This removes the cubic term so what remains is just a linear equation $2\alpha x + 3\beta = 0$ which we solve for $x = (3/2)(\beta/\alpha)$ and substitute back in (5.32).

We see from (5.33) that a negative α indeed is a necessary, but by no means a sufficient condition for a bifurcation to occur.

With one phase variable and two control parameters we can make a picture in three-space. In Figure 5.1 there is a picture of the cuspoid bifurcation set (5.33) on the bottom sheet, representing the parameter space. Above it is a surface, representing the state variable as a function of the parameters. It is the solution for x from (5.31). As the equation is a cubic it either has just one or three real roots. The geometrical picture of an equation with three roots is a surface which is doubly folded over itself. It is easiest to recognize β for

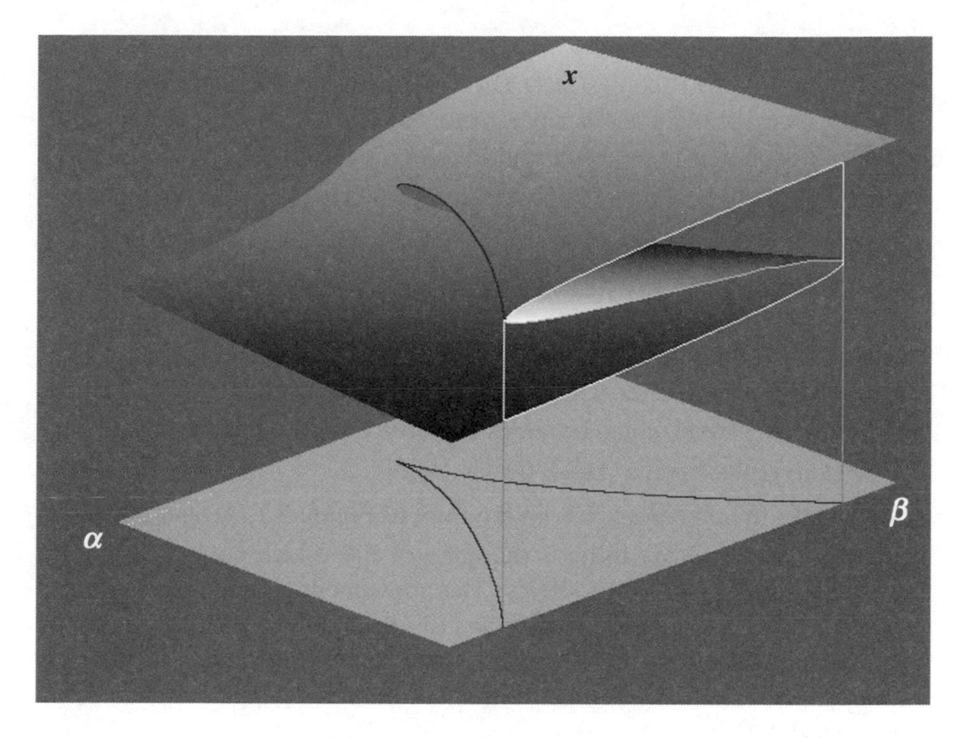

Fig. 5.1. The Cusp Catastrophe.

each given α as a cubic function of x, and the surface in Fig. 5.1 itself as a stack of such curves.

The projection of the fold lines onto the bottom plane in fact provides the cuspoid bifurcation set. Also note that only part of the surface is folded, being in one sheet beyond the sharp cusp point (located at the origin in parameter space.

We can now take an application of the cusp catastrophe to relaxation oscillations in the Rayleigh model of the bowed string represented in equations (2.161)-(2.162) above and shown in Figure 2.19. Remember that the motion was composed by a slow movement along a cubic, and a fast transition between the various branches of the cubic. Let us introduce the new variable $x = \varepsilon y$. Then the system can be restated as:

$$\dot{x} = \varepsilon z \tag{5.34}$$

$$\dot{z} = \frac{1}{\varepsilon}\left(z - \frac{1}{3}z^3 - x\right) \tag{5.35}$$

As in relaxation oscillations ε is very small equation (5.34) becomes slow and equation (5.35) becomes fast. The same holds for the variables: x is so slow that it can be regarded as a parameter, whereas z is so fast that it can be considered as being in constant equilibrium. Accordingly we can consider the parenthesis in (5.35) as being zero

$$z - \frac{1}{3} z^3 - x = 0 \tag{5.36}$$

and x as being a parameter. This is exactly the case stated in (5.31). We have $\alpha = -3$, i.e. negative, which provides for bifurcations. Observe that we have the variable replacements $3x \rightarrow \beta, -x \rightarrow z$.

We display the relaxtion cycle in the front of Figure 5.1. As $\beta = 3x$ moves forth and back, z either follows the edge of the folded surface, or jumps between its upper and lower sheets. This movement also illustrates *hysteresis*, i.e. irreversibility. The jump occurs for different parameter values in its forth and back movement. There are many more illustrations to the cusp catastrophe. Without any competition it was the most popular applied catastrophe model.

5.6 The Elementary Catastrophes: Swallowtail and Butterfly

The reader could now guess that running down the list of new catastrophes the order of the polynomial is increased by one in each step, as is the number of parameters. This is in fact true. Next catastrophe has a fifth order potential and three parameters. It is called *swallowtail*. Its universal unfolding is:

$$V(x) = \frac{x^5}{5} + \alpha \frac{x^3}{3} + \beta \frac{x^2}{2} + \gamma x \tag{5.37}$$

Again we could get rid of one of the powers by using an appropriate linear transformation of the variable, as always choosing the highest (fourth) order. The corresponding dynamics is:

$$\dot{x} = -x^4 - \alpha x^2 - \beta x - \gamma \tag{5.38}$$

Fig. 5.2. The Swallowtail Catastrophe.

which has at most four zeros. There can be two minima and two maxima for (5.37), i.e. two stable and two unstable equilibria for (5.38).

Bifurcations occur when there is loss of stability in any of these equilibria. The equilibrium set is:

$$V'(x) = x^4 + \alpha x^2 + \beta x + \gamma = 0 \tag{5.39}$$

and the singularity set:

$$V''(x) = 4x^3 + 2\alpha x + \beta = 0 \tag{5.40}$$

The system (5.39)-(5.40) furnishes two equations in four variables, one state variable and three parameters. We can hence eliminate the state variable. Doing this we get an implicit function in parameter α, β, γ-space, i.e. a surface. It is a complex self-intersecting sort of surface shown in Figure 5.2 and has a similarity to the tail of a swallow. Hence the poetic name.

Note that this time we cannot produce a picture such as Figure 5.1 displaying the equilibrium state over control space. With one state variable and three control parameters things happen in four-dimensional space. Control space alone is in 3 D, so all we can do is to produce a picture of the bifurcation set in this space. It becomes a surface, not a curve as in the case of the cusp.

Every time the combination of parameters crosses the surface in Figure 5.2 a bifurcation occurs. The formula for the surface, an implicit function in the variables α, β, γ, can again be obtained from eliminating the state variable x from equations (5.39)-(5.40), but it produces an awkward expression, so we do not state it.

Drawing the bifurcation surface in practice, one usually takes the state variable along with one of the controls as parameters. From (5.40) we very easily get $\beta = -4x^3 - 2\alpha x$, and likewise from (5.39) $\gamma = 3x^4 + \alpha x^2$, after substituting for β from the previous expression. In this way the computer can easily generate a picture such as Figure 5.2.

In the present case there is more to it. We have higher order degeneracy, when the third derivative of the potential vanishes along with the two first:

$$V'''(x) = 12x^2 + 2\alpha = 0 \tag{5.41}$$

We can now solve for $\alpha = -6x^2$ from (5.41), then substitute for α in (5.40) and solve for $\beta = 8x^3$, finally substitute α, β in (5.39) and solve for $\gamma = -3x^4$. What we now have are three functions to threespace, parameterized by a single parameter, so this produces no surface but a curve, in fact the cuspoid curve at the tail of the swallow in Figure 5.2. For completeness we should admit that higher degeneracy occurs in the case of the cusp as well, but it just produces a point (the origin in parameter and state space). Similarly there is a point in the present case where even the fourth derivative of the potential vanishes. Differentiating (5.41) one more time we see that this again occurs in the origin of state space, and so from (5.39)-(5.41) we see that this higher degeneracy occurs at the origin of parameter space as well. In Figure 5.2 this point is the cusp point where the tail of the swallow begins.

The most complex of the elementary catastrophes in 1 D is the *butterfly*, with a potential of the sixth order, and four parameters:

$$V(x) = \frac{x^6}{6} + \alpha\frac{x^4}{4} + \beta\frac{x^3}{3} + \gamma\frac{x^2}{2} + \delta x \tag{5.42}$$

The corresponding dynamical system

$$\dot{x} = -x^5 - \alpha x^3 - \beta x^2 - \gamma x - \delta \tag{5.43}$$

produces at most five equilibria, three stable ones, and two unstable ones which function as basin boundaries. The bifurcation set is four-dimensional, so it cannot be portrayed in any simple way even by itself.

5.7 The Elementary Catastrophes: Umblics

This ends the series of one dimensional elementary catastrophes. There are three more in two dimensions, two cases with three and one case with four parameters. It is interesting to note that in 2 D there are no cases with one or two parameters. This is so because, with fewer parameters, the Splitting Lemma lets us separate out one of the state variables as a Morse function, and so what remains of degeneracy is one dimensional.

We already saw this in the case of the potential (5.12) where the determinant of second partial derivatives became zero, so, after a convenient change of coordinates, we obtained one Morse direction and one non-Morse direction. We also saw that the next higher degeneracy, where all second order derivatives vanished, was the monkey saddle which needed three parameters for its universal unfolding.

The monkey saddle in (5.21) in fact represents the *elliptic umblic catastrophe* the universal unfolding of which we restate for convenience:

$$V(x, y) = \frac{x^3 - 3xy^2}{3} + \gamma \frac{x^2 + y^2}{2} + \alpha x + \beta y \tag{5.44}$$

We already stated the dynamical system and noted that there are either one node and three saddles or just two disconnected saddles, so let us skip directly to the equilibrium set, which is defined by zero partial derivatives:

$$V_x(x, y) = x^2 - y^2 + \gamma x + \alpha = 0 \tag{5.45}$$

$$V_y(x, y) = -2xy + \gamma y + \beta = 0 \tag{5.46}$$

Fig. 5.3. The Elliptic Umblic Catastrophe.

where we indicate partial derivatives by the indices. Likewise the singularity set is obtained when the Hessian determinant of partial second derivatives vanishes.

$$\begin{vmatrix} V_{xx} & V_{xy} \\ V_{yx} & V_{yy} \end{vmatrix} = \begin{vmatrix} 2x+\gamma & -2y \\ -2y & -2x+\gamma \end{vmatrix} = \gamma^2 - 4\left(x^2 + y^2\right) = 0 \tag{5.47}$$

In all (5.45)-(5.47) provide us with three equations in two state variables and three control parameters. We can hence again in principle eliminate the state variables and obtain an implicit function in α, β, γ -space, which defines the bifurcation set as a surface, just as in the case of the swallowtail. See Figure 5.3. Whenever the combination of parameters represented as the point α, β, γ crosses the bifurcation set, a bifurcation occurs.

The slow development of parameters can be seen as a curve in α, β, γ -space, and the bifurcations are due to this curve crossing the bifurcation surface. The surface divides the space in three disjoint parts, the two insides

Fig. 5.4. The Hyperbolic Umblic Catastrophe.

to either side of the narrow waist, and the outside. The latter corresponds to cases of two saddles, the two insides to cases of one node and three saddles.

Again, to get at the picture of the bifurcation surface, it is easiest to work with parameterization rather than deriving an implicit function. We can now completely swap the role of variables and parameters, and use the state variables for constructing the bifurcation surface in parameter space. From (5.47) we get $\gamma = \pm 2\sqrt{x^2 + y^2}$, from (5.45) $\alpha = -x^2 + y^2 \mp 2x\sqrt{x^2 + y^2}$ and from (5.46) $\beta = -x^2 + y^2 \mp 2y\sqrt{x^2 + y^2}$.

The *hyperbolic umblic* catastrophe, whose bifurcation set is displayed similarly in Figure 5.4, has the potential:

$$V(x, y) = \frac{x^3 + y^3}{3} + \gamma\, xy + \alpha x + \beta y \tag{5.48}$$

Like its elliptic companion it has either two or four equilibria. The equilibria can be either a maximum, a minimum, and two saddles, or else just one extremum (maximum or minimum) and a saddle point.

The corresponding dynamical system

$$\dot{x} = -x^2 - \gamma y - \alpha \tag{5.49}$$

$$\dot{y} = -y^2 - \gamma x - \beta \tag{5.50}$$

accordingly has either one stable and one unstable node and two saddles, or else just one node (stable or unstable) and a saddle.

Let us conclude with just mentioning the *parabolic umblic* which is fourth degree and has four parameters, like the butterfly, though now in two state variables. Again its bifurcation set cannot be visualized, but it concludes the list of seven elementary catastrophes.

There have been higher order catastrophes analysed in the literature, and we will later use eighth order polynomials, though without invoking any particularities of the deeper parts of catastrophe theory.

6 Monopoly

6.1 Introduction

Traditional microeconomic theory deals with two basic market types: Perfect competition and monopoly. In the case of perfect competition individual firms are assumed to be so small in comparison with the entire market that they cannot noticeably influence market price on their own; they just note the current price and react accordingly with respect to their supply. Only the supply of all the numerous firms together becomes a force on the market strong enough to determine the price in a balance with the demand of all the likewise numerous and small households.

The single monopolist, facing the lot of these households, on the other hand, is assumed to deliberately choose to limit the quantity supplied so as to keep price sufficiently high to yield monopoly profit. If in addition the monopolist is able to efficiently discriminate various submarkets from each other, even different prices may be charged in all those, so as to exploit its monopoly power maximally.

The monopolist must thus know at least the entire demand curve of the market. The information needed is infinitely more complex than that needed by a competitive firm; an entire demand curve instead of just one point on it. To collect this information might be assumed to be costly, and, in view of the variability of the real world, such information collecting would have to be repeated frequently in order to yield a reliable basis for decisions.

It is more likely that a monopolist just knows a few points on the demand function, recently visited in its more or less erratic search of maximum profit. The price policy of any national transportation monopoly could testify to this, though such monopolies could always cloak up any irrational looking behaviour in an alleged pursuit of public benefit.

As to the general form of a demand function, it is assumed to be monotonically downward sloping, demand decreasing as price increases. As

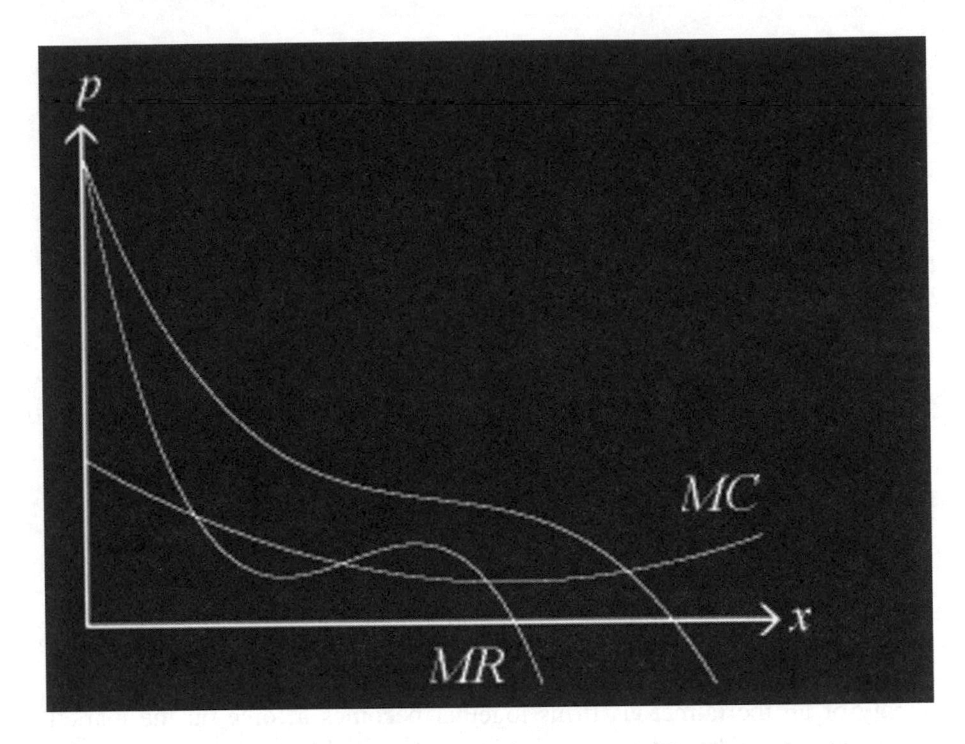

Fig. 6.1. Demand curve, marginal revenue and cost curves.

there are always substitutes for most goods, like driving as an alternative to using the railway, there would be a maximum price above which nobody would demand anything. Likewise, there would be a maximum saturation demand when price goes down to zero. In text book cases the demand curve is just a downward sloping straight line, but this is too simplistic. The elasticity of demand would in general vary over different sections of the demand curve, giving it a convex/concave outline. This is particularly true if the market is a composite of different submarkets with different elasticities, by the way the standard case for the study of price discrimination and dumping.

The full analysis of such cases in terms of graphical reasoning was given in the pioneering work by Joan Robinson in 1933, who recognized the existence of multiple local equilibria.

"*Cases of multiple equilibrium may arise when the demand curve changes its slope, being highly elastic for a stretch, then perhaps becoming relatively inelastic, then elastic again. This may happen, for instance, in a market composed of several subgroups of consumers each with a different level of incomes. There will be several critical points at which a decline in price suddenly brings the commodity within the reach of a whole fresh group of con-*

sumers so that the demand curve becomes rapidly more elastic. The marginal revenue curve corresponding to such a demand curve may fall and rise and fall again, and there will be several points of monopoly equilibrium. The net monopoly revenue at each point would be different, but it is unlikely that any monopolist would have sufficient knowledge of the situation to enable him to choose the greatest one from among them. If the monopolist had reached one equilibrium point there would be no influence luring him towards another at which his gains might be greater."

6.2 The Model

Such a case, illustrated in Fig. 6.1, which could in fact have been cut out from Joan Robinson's book, can be represented by the truncated Taylor series:

$$p = A - Bx + Cx^2 - Dx^3 \tag{6.1}$$

where p denotes commodity price, x denotes the quantity demanded, and A, B, C, D are some positive constants.

In order that the (6.1) function be invertible, i.e. that there be just one quantity demanded corresponding to each price, the curve must be downsloping. The least steep slope is in the point of inflection, defined by:

$$\frac{\mathrm{d}^2 p}{\mathrm{d}x^2} = 2C - 6Dx = 0 \tag{6.2}$$

where the slope is:

$$\frac{\mathrm{d}p}{\mathrm{d}x} = -B + 2Cx + 3Dx^2 \tag{6.3}$$

or, substituting from (6.2) for $x = C/(3D)$,

$$\frac{\mathrm{d}p}{\mathrm{d}x} = -B + \frac{C^2}{3D} < 0 \tag{6.4}$$

The demand curve is down-sloping, as indicated by the imposed sign requirement, provided that the condition:

$$C^2 < 3BD \tag{6.5}$$

holds.

Economists usually analyse the monopolist's behaviour, using the marginal revenue, *MR*, i.e. the derivative of total revenue, $\mathrm{d}/\mathrm{d}x(px) = p + x\mathrm{d}p/\mathrm{d}x$. For the demand curve (6.1) the marginal revenue curve would be:

$$MR = A - 2Bx + 3Cx^2 - 4Dx^3 \tag{6.6}$$

which, in fact, could lack a unique inverse, though the demand curve has one. The marginal revenue curve is displayed along with the demand curve in Fig. 6.1. The lack of an inverse is an interesting case, because it provides an opportunity for multiple equilibria.

Now we want the marginal revenue curve to have a positive derivative at the inflection point. This point is defined by:

$$\frac{\mathrm{d}^2 MR}{\mathrm{d}x^2} = 6C - 24Dx = 0 \tag{6.7}$$

The slope is:

$$\frac{\mathrm{d}MR}{\mathrm{d}x} = -2B + 6Cx - 12Dx^2 \tag{6.8}$$

or, substituting from (6.7) for $x = C/(4D)$,

$$\frac{\mathrm{d}MR}{\mathrm{d}x} = -2B + \frac{3C^2}{4D} > 0 \tag{6.9}$$

The condition for an upward slope at the point of inflection, i.e. for the marginal revenue curve (6.6) not to have a unique inverse is:

$$C^2 > \frac{8}{3} BD \tag{6.10}$$

So, from (6.5) and (6.10) there is in fact a latitude of choice for

$$C^2 \in \left(\tfrac{8}{3} BD, \tfrac{9}{3} BD\right) \tag{6.11}$$

such that the demand curve (6.1) has a unique inverse but the marginal revenue curve (6.6) has not.

Along with the marginal revenue curve, economists use the marginal cost curve, denoted MC. It is the derivative of the total cost as a function of the quantity supplied. Typically the marginal cost curve is assumed to be concave from above, with marginal cost first decreasing with increasing supply and eventually increasing. The truncated Taylor series for the marginal cost would be:

$$MC = E - 2Fx + 3Gx^2 \tag{6.12}$$

where again E, F, and G are positive. The integers are introduced for later convenience at integration.

The standard analysis would involve the condition for maximum profit, $MR = MC$, i.e. from (6.6) and (6.12)

$$(A - E) - 2(B - F)x + 3(C - G)x^2 - 4Dx^3 = 0 \tag{6.13}$$

Whenever the curves look like those in Fig. 6.1 this equation has three distinct real roots. In theory, the monopolist would calculate these by solving the cubic, and then evaluate second order derivatives:

$$-2(B - F) + 6(C - G)x - 12Dx^2 \tag{6.14}$$

to check that the intermediate root is a profit minimum, whereas the higher and lower ones represent genuine maxima. Finally, the monopolist would choose between the two local maxima to identify the global profit maximum.

6.3 Adaptive Search

In reality, however, the monopolist does not know more than a few points on the demand function. The marginal revenue curve, representing the derivative of an incompletely known function, would not be known except in terms of interpolation through recently visited points. The information collected would be of a local character and short lifetime, and the monopolist might not even know that globally there are two distinct profit maxima. As already stressed, the limits on information are due to the difficulty and expense involved in market research, and to the frequency of demand shifts due to changes in the markets of close substitutes.

Given this, the task would be to design a search algorithm for the maximum of the unknown profit function:

$$\Pi(x) = (A-E)x-(B-F)x^2+(C-G)x^3-Dx^4 \tag{6.15}$$

This would be quite easy, provided the monopolist knew that the function was a quartic. The problem, however, is that nothing more except a few points are known.

The simplest algorithm of all, of course, is to estimate the difference of marginal costs and revenues from the two last visited points of the profit function, and, in the vein of Newton, to use a given step length to move in the direction of increasing profits.

Denoting the next last and last visited points by x and y respectively, and the step length by δ, we get the next point as:

$$y+\delta\frac{\Pi(y)-\Pi(x)}{y-x} \tag{6.16}$$

In order to avoid unnecessary numerical instability the denominator in the quotient may be factored out:

$$\begin{aligned} P(x,y) = \frac{\Pi(y)-\Pi(x)}{y-x} &= (A-E)-(B-F)(x+y) \\ &+(C-G)(x^2+2xy+y^2)-D(x^3+x^2y+xy^2+y^3) \end{aligned} \tag{6.17}$$

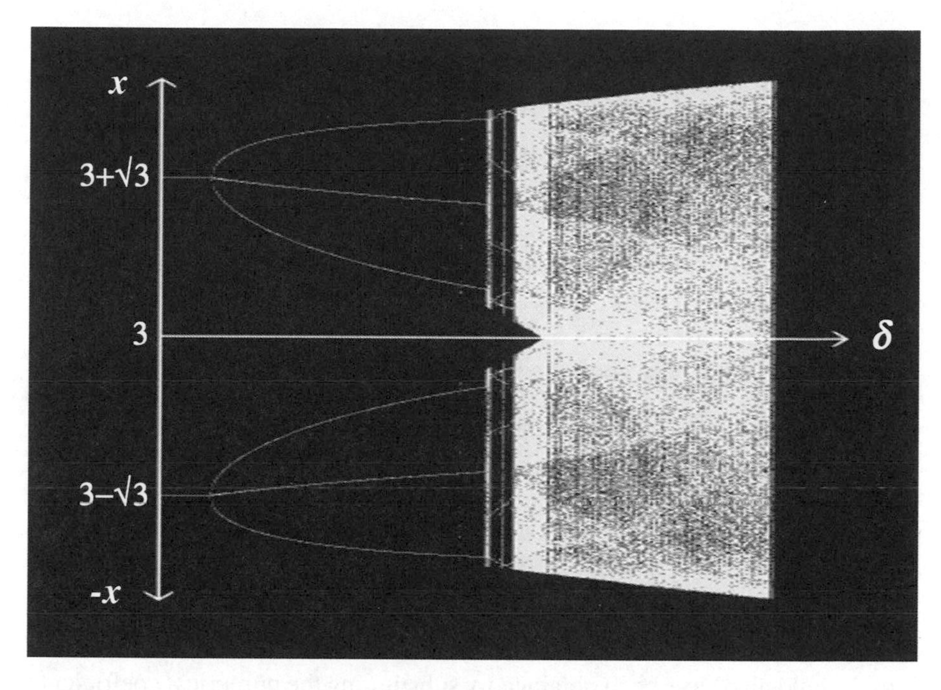

Fig. 6.2. Bifurcation diagram.

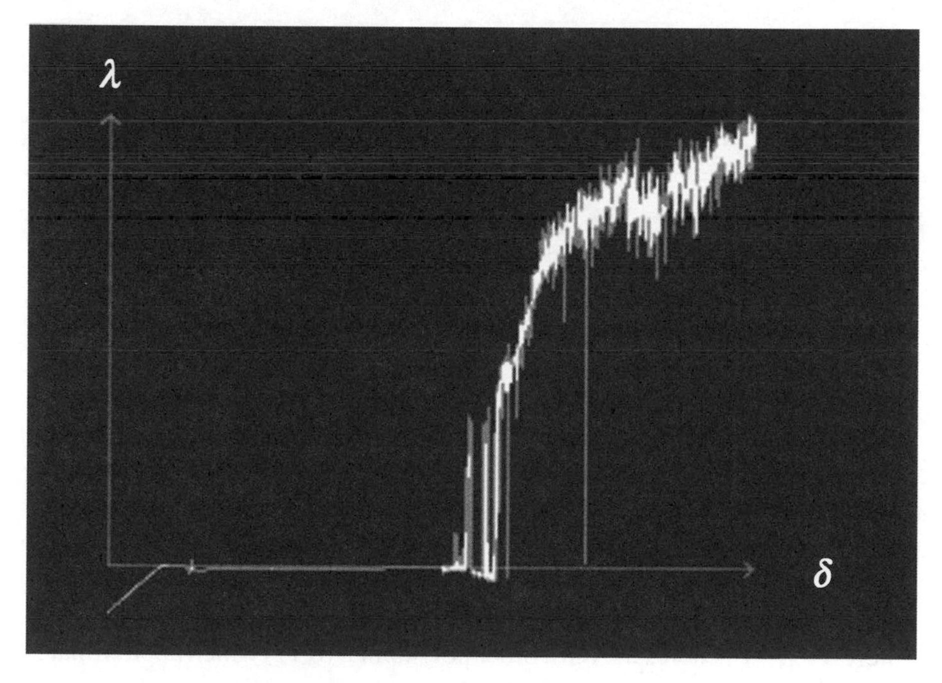

Fig. 6.3. Largest Lyapunov exponent.

6.4 Numerical Results

Iteration of this search process may lead to any of the two local profit maxima, to oscillating processes, or to chaos, depending on the values of the coefficients A through G, and the step size δ. The chaotic attractors may be unique, or else coexist with different attraction basins. In the following computer simulations, as well as in drawing Fig. 6.1, the following values of the coefficients were used: $A = 5.6$, $B = 2.7$, $C = 0.62$, $D = 0.05$, $E = 2$, $F= 0.3$, and $G = 0.02$. With these coefficients the profit function (6.15) would become:

$$\Pi(x) = 3.6x - 2.4x^2 + 0.6x^3 - 0.05x^4 \tag{6.18}$$

Equating its derivative to zero yields the equation:

$$\Pi'(x) = 3.6 - 4.8x + 1.8x^2 - 0.2x^3 = 0 \tag{6.19}$$

which could also have been obtained by substituting the numerical coefficients in condition (6.13) for $MR = MC$. Equation (6.19) has three real roots:

$$x = 3 \text{ and } x = 3 \pm \sqrt{3} \tag{6.20}$$

The search process with the coefficients introduced is defined by the map:

$$x_{t+1} = f(x_t, y_t) \tag{6.21}$$

$$y_{t+1} = g(x_t, y_t) \tag{6.22}$$

where

$$f(x, y) = y \tag{6.23}$$

$$g(x, y) = y + \delta\, P(x, y) \tag{6.24}$$

where $P(x, y)$ denotes the third order polynomial (6.17) which approximates the marginal profit, with the numerical values A through G substituted.

Fig. 6.4. Co-existent four-cycles, and their basins of attraction.

6.5 Fixed Points and Cycles

Before proceeding with the simulations, we note that the fixed points of the iterative mapping (6.21)-(6.22), not unexpectedly, are given by the equations

$$x = y \quad \text{and} \quad 3.6 - 4.8x + 1.8x^2 - 0.2x^3 = 0 \tag{6.25}$$

and that they are the maxima and the minimum of the profit function. To find out the stability of these points we calculate the Jacobian

$$\frac{\partial(f,g)}{\partial(x,y)} = \delta(2.4 + 1.8x - 0.3x^2) \tag{6.26}$$

We have used the information that $x = y$ in the fixed points. Of those we know that the profit minimum $x = 3$ is unstable, but the profit maxima

$$x = 3 \pm \sqrt{3} \tag{6.27}$$

need further discussion. Substituting any of those roots in the Jacobian (6.26), we get:

$$\frac{\partial(f,g)}{\partial(x,y)} = \frac{3}{5}\delta \tag{6.28}$$

Loss of stability occurs for both roots when the Jacobian is unitary, i.e. at the common value:

$$\delta = 5/3 \tag{6.29}$$

This can be seen at the first branching point in the bifurcation diagram in Fig. 6.2 where the phase variable, x (or y) is plotted against the parameter δ.

We see the two alternative fixed points coexist at parameter values no higher than 5/3. After that the fixed points are replaced by cycles, over a quite extensive interval, and chaotic bands, with each its own basin of attraction. After a certain point the attractors merge in a single one. Fig. 6.3 displays the largest Lyapunov exponent, plotted against the step size parameter. The most striking feature is that, once chaos sets on definitely, the Lyapunov exponent remains positive, with no windows of order at all (at least not with the resolution chosen).

The cycles shown in Fig. 6.2 seemingly are of period three, but this is deceptive. For each of the two coexistent cycles three different values are taken on, so let us denote those of one set a, b, and c. They are, however only passed in the sequence a, b, c, b, a, b, c ..., so that one of the values is taken on twice as often as are the other two. Recall that our system is two-dimensional, and that we accordingly deal with a process where the points (a, b), (b, c), (c, b), and (b, a) are visited over and over, so the process is of period four, not three. This is also what we find in the phase diagrams, such as the one displayed in Fig. 6.4.

In Fig. 6.4 we display the approach to each of the coexistent cycles against a background of the attraction basins in different shadings.

To see some details of the periodic solution, let us locate the period four point. After four iterations we are back again, so the following four equations must be fulfilled:

$$z = y + \delta \frac{\Pi(y) - \Pi(x)}{y - x} \tag{6.30}$$

$$w = z + \delta \frac{\Pi(z) - \Pi(y)}{z - y} \tag{6.31}$$

$$x = w + \delta \frac{\Pi(w) - \Pi(z)}{w - z} \tag{6.32}$$

$$y = x + \delta \frac{\Pi(x) - \Pi(w)}{x - w} \tag{6.33}$$

Multiplying through by the denominators, and adding (6.30)-(6.33) pairwise, we arrive at the following two equations:

$$(z - y)(x - y + z - w) = \delta(\Pi(x) - \Pi(z)) \tag{6.34}$$

$$(w - z)(x - y + z - w) = \delta(\Pi(w) - \Pi(y)) \tag{6.35}$$

whereas adding all four of them, (6.30)-(6.33) makes all the profit functions in the right hand sides cancel, and we end up with a perfect square that has to equal zero:

$$(x - y + z - w)^2 = 0 \tag{6.36}$$

Using this last condition in (6.34)-(6.35) we get a pair of useful conditions:

$$\Pi(x) = \Pi(z) \tag{6.37}$$

$$\Pi(y) = \Pi(w) \tag{6.38}$$

So, profits in the sequence x, y, z, w are pairwise equal with a lag of two periods. This may happen in two ways. Either, the points visited are themselves equal, as conjectured for the value visited twice as often as the other two. Or, they may be located sufficiently widely apart as to fall on different rising and falling branches of the quartic profit function.

We now turn to the numerical example to get more substance, and substitute expressions (6.18) for the pairwise equal profits in (6.37)-(6.38). The equations then factor into:

$$(x-z)(x+z-6)\left((x-3)^2+(z-3)^2-6\right)=0 \tag{6.39}$$

$$(y-w)(y+w-6)\left((y-3)^2+(w-3)^2-6\right)=0 \tag{6.40}$$

These give us three possibilities each. Taking x and z for an example, they may be equal, their sum can be equal to 6, or they may be located on a circle centred around the point (3, 3). Among these the middle option is not relevant for a periodic point that moves entirely above or below the line at 3 in Fig. 6.2. Hence any sum of two amplitudes is strictly smaller or larger than 6, depending on which of the alternative cycles we consider. Actually, the middle option refers to the unstable fixed point at 3 of Fig. 6.2.

Obviously, both $x = z$ and $y = w$ cannot hold simultaneously, as we would then no longer deal with a four-period point, but with a two-period point, which, by the way, like the three-period point, can be shown to be impossible. So, what we actually locate are the now likewise unstable fixed points $3 \pm \sqrt{3}$, in addition to the aforementioned one at 3.

In the same way we cannot have both pairs located on the circles $(x-3)^2+(z-3)^2=6$ and $(y-3)^2+(w-3)^2=6$, because, after a somewhat more messy derivation, we end up at the same conclusion, i.e. that we are dealing with one of the unstable fixed points. So, two points in the four-cycle must be equal, say x and z, and two, say y and w, must be on the circle. These considerations, along with (6.36), provide us with three equations:

$$x = z \tag{6.41}$$

$$(y-3)^2+(w-3)^2=6 \tag{6.42}$$

$$x+z=y+w \tag{6.43}$$

In addition we have equation (6.30), which, using the fact that $x = z$ according to (6.41), becomes:

$$x = y + \delta \frac{\Pi(y) - \Pi(x)}{y - x} \tag{6.44}$$

Equations (6.41)-(6.44) are sufficient for calculating the four values of the cycle. For instance with $\delta = 2$, a value for which we according to Fig. 6.2 definitely have a cycle, we get:

$$x = z = 3 \mp \frac{\sqrt{65}}{5} \tag{6.45}$$

$$y = 3 \mp \left(\frac{\sqrt{65}}{5} - \frac{\sqrt{10}}{5} \right) \qquad w = 3 \mp \left(\frac{\sqrt{65}}{5} + \frac{\sqrt{10}}{5} \right) \tag{6.46}$$

The minus sign applies to the lower half of Fig. 6.2, the plus sign to the upper half. These values agree with the results from simulation.

6.6 Chaos

We see from Figs. 6.2 and 6.3 that chaos takes over once the four-period point loses stability. It is therefore of particular interest to locate the parameter value at which this happens. This can be done by analytical means, though the procedure is a little bit messy.

The Jacobian determinant at the periodic point, using the values specified in (6.45)-(6.46), becomes:

$$J = \left(\frac{\delta}{20} \right)^4 \left(\frac{4\sqrt{26}}{5} - 10 \right) \left(\frac{4\sqrt{26}}{5} + 10 \right) \left(\frac{8\sqrt{26}}{25} - \frac{54}{5} \right) \left(\frac{8\sqrt{26}}{25} + \frac{54}{5} \right) \tag{6.47}$$

Fig. 6.5. Co-existent chaotic attractors, and their basins of attraction.

Loss of stability occurs when the Jacobian becomes unitary, i.e. when:

$$J = \frac{1345}{51549}\delta^4 = 1 \tag{6.48}$$

Calculating the critical step length parameter from (6.48) we get:

$$\delta = 2.48813 \tag{6.49}$$

which in fact agrees very well with the results of simulations. Figs. 6.4 and 6.5 are accordingly drawn for parameters on either side of this critical value.

Fig. 6.5 shows two coexistent chaotic attractors against a background of their basins. It can be noted that there is no intricacy in basin boundaries such as in some of the cases of the Newton algorithm studied in the chaos literature, though there are regions in the SW and NE where the "wrong" basins come very close to the final attractors. It is worthwhile to note the symmetry of shapes both of the basins and of the attractors. It is known that for symmetric systems any attractors that are not symmetric in themselves come in pairs which together make up a symmetric picture. This can be seen

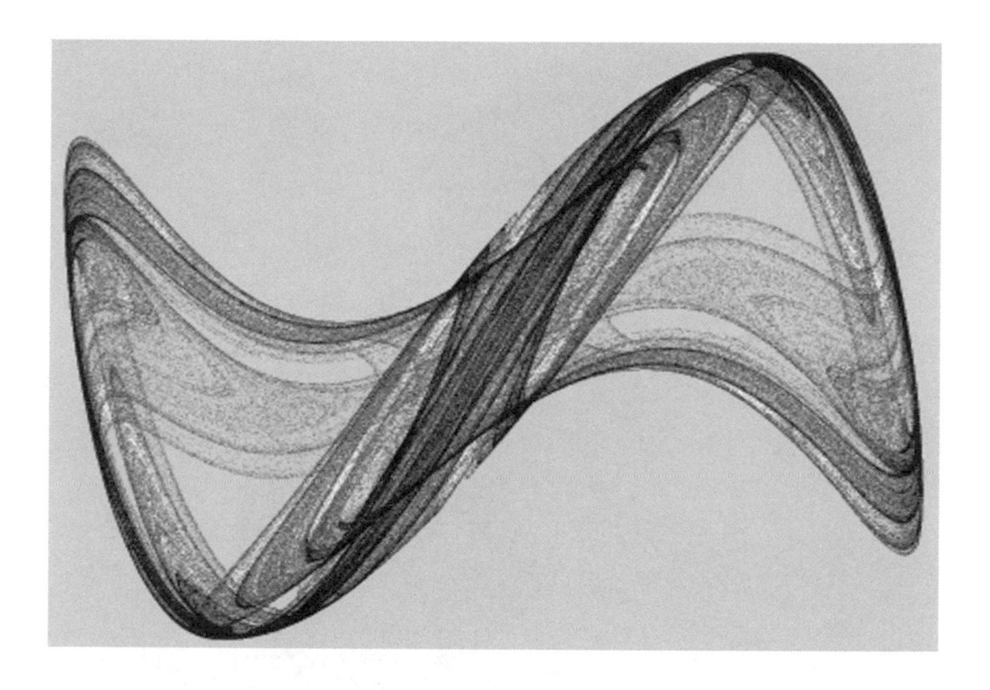

Fig. 6.6. Single chaotic attractor.

as a general consequence of symmetry breaking principles. See Golubitsky for references.

Fig. 6.6, finally, is a portrait of a single chaotic attractor extended over the entire phase diagram, the adjustment step parameter being so large that the process spills over the watershed at the amplitude of 3. As the attractor is there alone it is symmetric as can be expected.

6.7 The Method of Critical Lines

Let us now find out how the attractor in Figure 6.6 could be studied using the concepts of critical lines and absorbing areas. The line along which the mapping becomes folded, L_{-1} in the notation of Section 4.11, is obtained when the determinant of the Jacobian matrix vanishes. In our introduction of the method we used a pedestrian style of variable elimination and differentiation, but the same result can be obtained with just putting the Jacobian determinant equal to zero. Observe that (6.26) and (6.47) as derived

Fig. 6.7. Critical ellipse and two iterates.

above will not do because they were obtained for the fixed point and the four period cycle respectively.

Rather we need the general expression derived from (6.23)-(6.24):

$$\begin{vmatrix} f_x & f_y \\ g_x & g_y \end{vmatrix} = \begin{vmatrix} 0 & 1 \\ \delta P_x & 1+\delta P_y \end{vmatrix} = -\delta P_x = 0 \tag{6.50}$$

Note also that we put the determinant equal to zero, not to unity as before, because we are now interested in the boundary line along which the image is folded, which is an issue different from that of determining the stability of a periodic point. We may also note the interesting fact that the fold (6.50) is independent of the step length parameter δ, as we must have $P_x = 0$.

According to equation (6.17) which defines the polynomial $P(x, y)$, and using the numerical values stated above, we get by simple differentiation and a slight rearrangement:

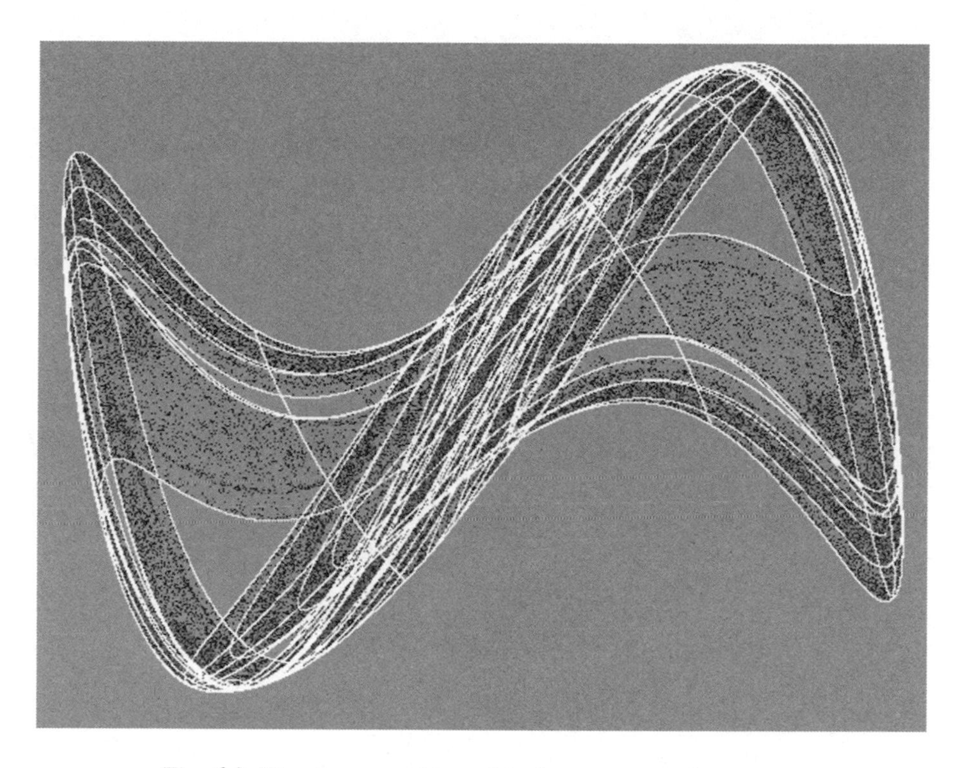

Fig. 6.8. Five iterates of the critical segments and attractor.

$$3(x-3)^2+2(x-3)(y-3)+(y-3)^2=6 \tag{6.51}$$

This is the equation of an ellipse tilted counterclockwise by 45° and centred at the point $x=y=3$. Accordingly, inside the ellipse there are three preimages to the map (6.23)-(6.24), outside there is just one. In Figure 6.7 we display the ellipse and two of its forward iterates. The darker shade is the first iterate, and the brighter the second iterate.

As always we do not use the entire critical curve to delimit the absorbing area, just a segment of it, or rather two segments, approximately a bit more than the white boundary segments of the ellipse of Figure 6.7.

Figure 6.8 displays the result of iterating these segments five times. The attractor from Figure 6.6 is shown in terms of the dark spots, whereas the absorbing area is delineated by the white lines. We see how neatly the attractor fits into these critical lines.

The attractor displayed in Figures 6.6 and 6.8 is quite close to explosion. We find this out if we draw the absorbing area together with the basin of attraction in Figure 6.9. The brighter shade around the absorbing area de-

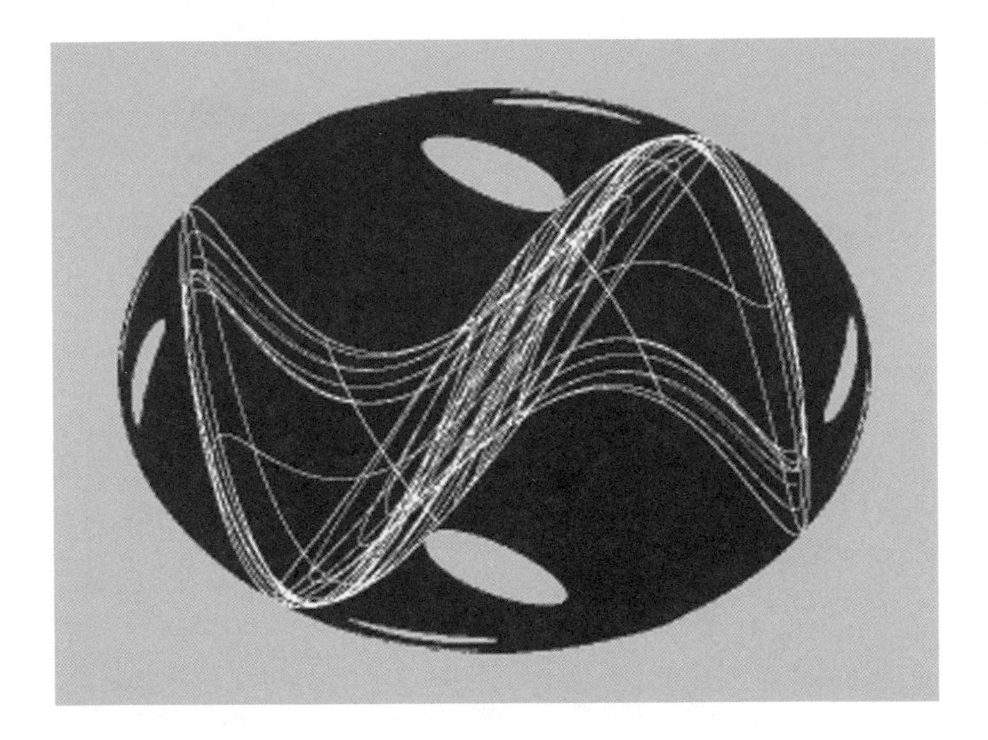

Fig. 6.9. Absorbing area and basins of attraction.

notes the attraction basin of the point at infinity, so we see that it comes pretty close to the absorbing area and hence to the attractor.

Just increase the adjustment step size a little more and the critical curves shoot "tongues" into the basin of infinity as we see in Figure 6.10. We also see that the whole basin and the absorbing area itself become filled with specks which mark initial conditions that make the model explode. The previous chaotic attractor thus loses stability.

We may also discuss the merger of the attractors in Figure 6.5 into one single as displayed in Figure 6.6 by the method of critical lines. First of all recall that the attraction basin boundary in Figure 6.5 shows no intricacy as we already noted. This, however, at least partly, is due to the fact that we do not show the entire attraction regions of the two competing attractors. If we show these basins surrounded by the third basin of the attractor at infinity, which always is there too, then we see that at the boundary to the last attractor there is some intertwining, even if the boundary between the attraction basins of the two finite attractors remains clear-cut.

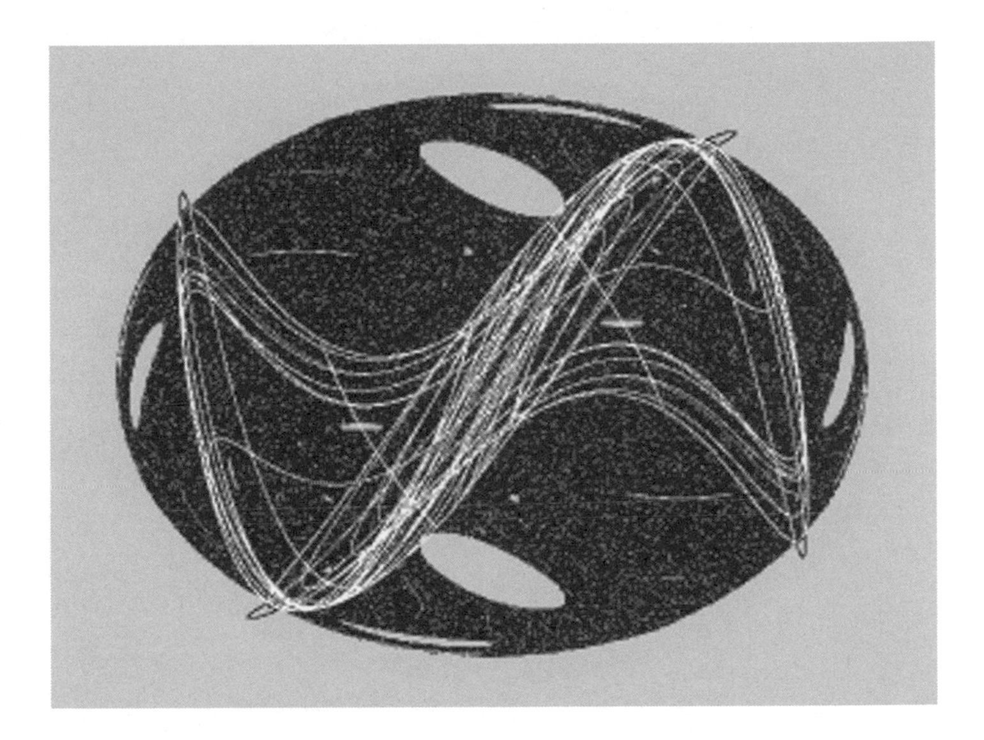

Fig. 6.10. Specky basin and tongues of critical lines.

In Figure 6.11 we show a picture of the absorbing areas to the two finite attractors, their attraction basins, and the basin of the point at infinity. The step length parameter is somewhat larger than in Figure 6.5, so the basins become a bit more twisted, but the two basins are clearly separated and in particular the absorbing areas are neatly set in each its basin of attraction.

Once we get close to the point at which the attractors merge, things, however, become much more intricate. As we see in Figure 6.12, the absorbing area of the lower left attractor shoots tongues into the attraction basin of its twin attractor on the upper right. Accordingly, we see the basins filled of islands, even within the areas delimited by the attractors, belonging to the other attractor.

Hence it is difficult to say from a given initial condition which attractor is the final destiny of the process. Nearness to one attractor now says nothing about this. But yet there exist two different finite attractors.

As the parameter increases even more this may no longer be true. The two basins merge into one as do the attractors, and we have the case displayed in Figures 6.6 and 6.8-6.10

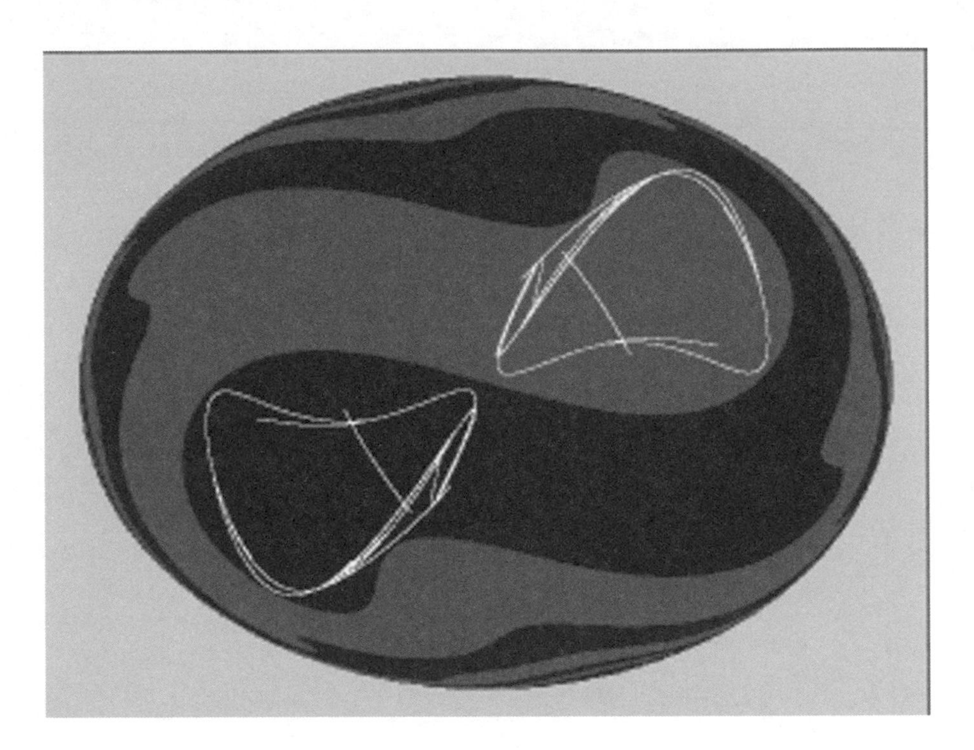

Fig. 6.11. Absorbing areas and basins of two competing attractors.

6.8 Discussion

It may be a bit hard for economists, nourished with textbook monopoly theory, to digest that the monopolist does not know all about the market and may even behave in a way seeming erratic. But we have to admit that it can be prohibitively expensive to get all the information the monopolistic firm needs, and that the knowledge acquired may only be local. The monopolist might have no idea at all of the true *global* outline of the marginal revenue curve, as to how many humps there are and where they are located, as suggested by Joan Robinson.

If the monopolist has not found a local equilibrium, or if he is a little more adventurous than Joan Robinson assumes, then he might devise a search

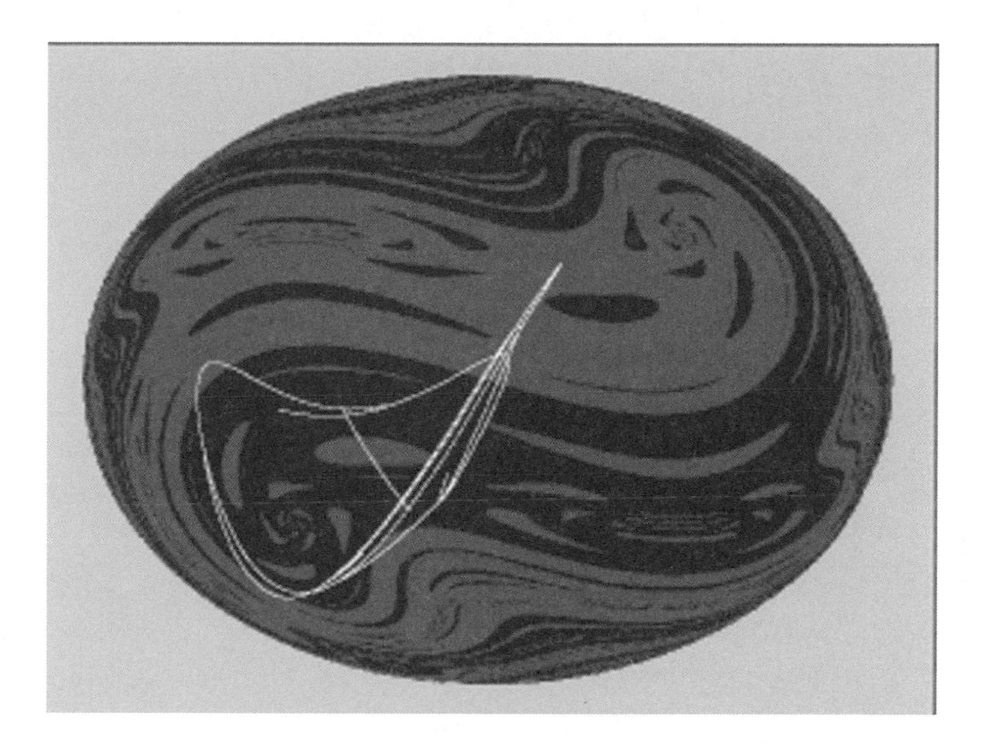

Fig. 6.12. Tongues and basin intricacy just before the merger of attractors.

process like the one described, based on locally estimated marginal revenues. An algorithm like one that was good enough for Sir Isaac Newton can hardly be considered too unsophisticated for a monopolistic transportation utility or the like.

Should the monopolist become suspicious about the fact that he is behaving like a random number generator, the blame could always be on exogenous shifts in the demand function, due to changing prices of close substitutes or to general business cycles. Not even a regularly periodic behaviour need make him suspicious, because the imagery of economic phenomena is crowded with all kinds of regular periodic cycles. Observe also that nothing at all was said about the speed of the process, i.e. the real-time length of the adjustment step. It might accommodate to any degree of conservatism in the monopolist's behaviour we may wish to assume.

Finally, the monopoly profit might on average be quite good, despite this periodic or chaotic behaviour.

7 Duopoly and Oligopoly

7.1 Introduction

Economics recognizes two opposite market forms: competition and monopoly. In the competitive case the firms are very numerous and thus small in relation to the total size of the market. In consequence they consider market price as being approximately given independently of any action they can take on their own with regard to their supply.

In the case of monopoly one single firm dominates the whole market. Its supply influences market price appreciably, and it takes advantage of this to increase its profits by deliberately limiting the supply, thus establishing a monopoly price. It can also apply various discriminatory policies provided it can segregate segments of consumer's demand and charge different prices from different categories of consumers.

Duopoly, though contextually the first step from monopoly towards perfect competition, is analytically not a case of intermediate complexity, but more complicated than any of the extremes. This is so as the duopolists have to take in consideration, not only the behaviour of all the consumers in terms of the entire market demand curve for the commodity. The duopolist must also take account of the behaviour of the competitor, inclusive possible retaliation on his own actions.

The formal theory of duopoly goes as far back as back to 1838, when Augustin Cournot treated the case where there is no retaliation at all, so that in every step each duopolist assumes the latest step taken by the competitor to remain his last. The process was assumed to lead to a steady state, nowadays called Cournot equilibrium, though it is by no means certain that it is stable. As a matter of fact, the following discussion will show how cycles and chaos may arise from very simple Cournot adjustment procedures.

Exactly a Century later, in 1938, Heinrich von Stackelberg made some ingenious extensions of the Cournot model, by assuming that any of the competitors might try to become a "leader", by taking the reactions of the other competitor according to Cournot in explicit consideration when devising his own actions. The tenability of such a situation would, of course, depend on whether the other competitor was content with adhering to his Cournot-like behaviour, i.e., being a "follower".

He might take up the challenge by trying to become a leader himself. The outcome of such warfare would depend on long term production conditions for the competitors, and on their current financial strengths. In the end one of the duopolists might ultimately force the other out of the market, thus becoming a monopolist. They may agree on collusive behaviour, provided law admits this usually forbidden solution. They may also return to the Cournot equilibrium, or to a Stackelberg equilibrium, provided the duopolists tacitly agree to form a leader-follower pair.

At the event of game theory, duopoly was one of the obvious fields of application, and the theory was recast in terms of probabilistic strategies. This, however, falls outside the scope of the present discussion, because the points we want to make do not need anything but the classical deterministic models.

It has been realised that the Cournot model may lead to cyclic behaviour, and David Rand in 1978 conjectured that under suitable conditions the outcome would be chaotic. His purely mathematical treatment does not, however, include any substantial economic assumptions under which this becomes true. In what follows we supply such very simple substantial assumptions.

There exist, of course, other circumstances, such as marginal revenue curves with rising sections and hence several local equilibria, which also fulfil Rand's conjecture, and which have been proposed in the literature. The disadvantage of those cases, however, is that solutions for the "reaction functions" of the duopolists at least are as complex as the roots to cubic equations with a discontinuous choice between branches. The model we suggest is much simpler as it yields explicit solutions for the reaction functions, so that we can easily focus on the dynamics of the process.

7.2 The Cournot Model

Assume an isoelastic demand function, such that price, denoted p, is reciprocal to the total demand. Provided demand equals supply it is made up of the supplies of the two competitors, denoted x and y. Thus:

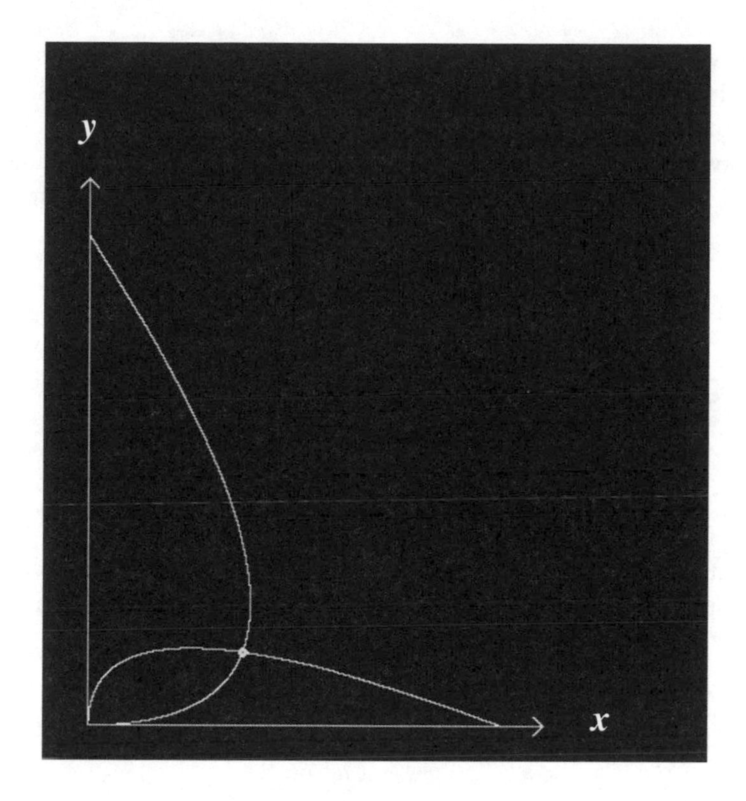

Fig. 7.1. Reaction curves and the Cournot point.

$$p = \frac{1}{x + y} \tag{7.1}$$

This demand function is not unproblematic, because it does not yield a reasonable solution to the collusive case, the reason being that when the possibility of making total supply zero is considered, price can go to infinity and total revenue remain constant. On the other hand, total costs would vanish, and the duopolists could get the entire revenue without incurring any costs. Such a solution is purely formal and does not carry any economic substance. This absurdity does not occur with any of the other solutions, as the presence of a positive supply by the competitor always keeps the price finite. We could also easily remedy this technical problem by adding any positive constant in the denominator of (7.1).

As this would not change any substantial conclusions, but carries the price of making all formulas much more messy, we abstain from this, and just point out this little complication.

Suppose next that the duopolists produce with constant marginal costs, denoted a and b respectively. The profits of the two firms become accordingly:

$$U(x,y) = \frac{x}{x+y} - ax \tag{7.2}$$

$$V(x,y) = \frac{y}{x+y} - by \tag{7.3}$$

The first firm would maximize $U(x,y)$ with respect to x, the second $V(x,y)$ with respect to y. Equating the partial derivatives to zero, we can solve for the reaction functions:

$$x = \sqrt{\frac{y}{a}} - y \tag{7.4}$$

$$y = \sqrt{\frac{x}{b}} - x \tag{7.5}$$

A check of the second order conditions, in fact, testifies that we always deal with local profit maxima, provided quantities are positive as indeed they should be. The reaction functions are displayed in Fig. 7.1. Their general outline is that they start at the origin, have unique maxima, and drop to zero again. The intersection is thus unique.

We can easily solve for the output quantities at the Cournot point, taking (7.4)-(7.5) as a simultaneous system of equations:

$$x = \frac{b}{(a+b)^2} \tag{7.6}$$

$$y = \frac{a}{(a+b)^2} \tag{7.7}$$

This point, of course, is the intersection of the reaction curves as shown in Fig. 7.1. The profits of the duopolists at the Cournot point can be calculated by substituting back from (7.6)-(7.7) in (7.2)-(7.3):

$$U = \frac{b^2}{(a+b)^2} \tag{7.8}$$

$$V = \frac{a^2}{(a+b)^2} \tag{7.9}$$

7.3 Stackelberg Equilibria

We can locate the Stackelberg equilibria as well. Supposing that the first firm is the leader, we can substitute for y from (7.5) into (7.2), which then becomes:

$$U = \sqrt{bx} - ax \tag{7.10}$$

This is a function of x alone, so equating the total derivative of (7.10) to zero we can solve for the optimal output of the leader:

$$x = \frac{b}{4a^2} \tag{7.11}$$

The output of the follower is obtained by substituting from (7.11) in (7.5):

$$y = \frac{2a-b}{4a^2} \tag{7.12}$$

We can finally calculate the profits of the duopolists in the first Stackelberg equilibrium by substituting from (7.11)-(7.12) into (7.2)-(7.3). Thus:

$$U = \frac{b}{4a} \qquad V = \frac{(2a-b)^2}{4a^2} \tag{7.13}$$

As U according to (7.13) can be shown to be larger than according to (7.8), we conclude that it is always preferable for any firm to be a Stackelberg leader rather than to be in a Cournot equilibrium. Both may thus attempt to become leaders, irrespective of the marginal costs, which provide the only difference between the firms in the present model. Marginal costs, however, would play a role for the chance of surviving economic warfare when both persist in their attempts to become leaders.

Let us just write down the equations for the other Stackelberg equilibrium for the sake of completeness. The leadership profit function for the second firm is:

$$V = \sqrt{ay} - by \tag{7.14}$$

and maximizing it yields the optimal solution:

$$y = \frac{a}{4b^2} \tag{7.15}$$

The corresponding passive reaction of the follower then yields:

$$x = \frac{2b - a}{4b^2} \tag{7.16}$$

It remains to calculate the profits in the second Stackelberg equilibrium:

$$U = \frac{(2b - a)^2}{4b^2} \qquad V = \frac{a}{4b} \tag{7.17}$$

Both Stackelberg equilibria are feasible. What is not feasible is if both the duopolists attempt to be leaders simultaneously.

7.4 The Iterative Process

Let us now return to the Cournot case. Once we are interested in the process of adjustment, we have to lag the variables. So, (7.4)-(7.5) must be written:

$$x_{t+1} = \sqrt{\frac{y_t}{a}} - y_t \qquad (7.18)$$

$$y_{t+1} = \sqrt{\frac{x_t}{b}} - x_t \qquad (7.19)$$

This pair of equations is the central piece of the iterative process we are going to study.

7.5 Stability of the Cournot Point

The fixed point of this iteration, of course, is the Cournot equilibrium. To find out something about the stability of it we calculate the derivatives of the functions (7.4)-(7.5):

$$\frac{dx}{dy} = \frac{1}{2}\sqrt{\frac{1}{ay}} - 1 \qquad (7.20)$$

$$\frac{dy}{dx} = \frac{1}{2}\sqrt{\frac{1}{bx}} - 1 \qquad (7.21)$$

Loss of stability occurs when:

$$\left| \frac{dx}{dy} \frac{dy}{dx} \right| = 1 \qquad (7.22)$$

holds, i.e., substituting from (7.6)-(7.7), when:

$$(a-b)^2 = 4ab \qquad (7.23)$$

We observe that this condition can be solved for the ratio:

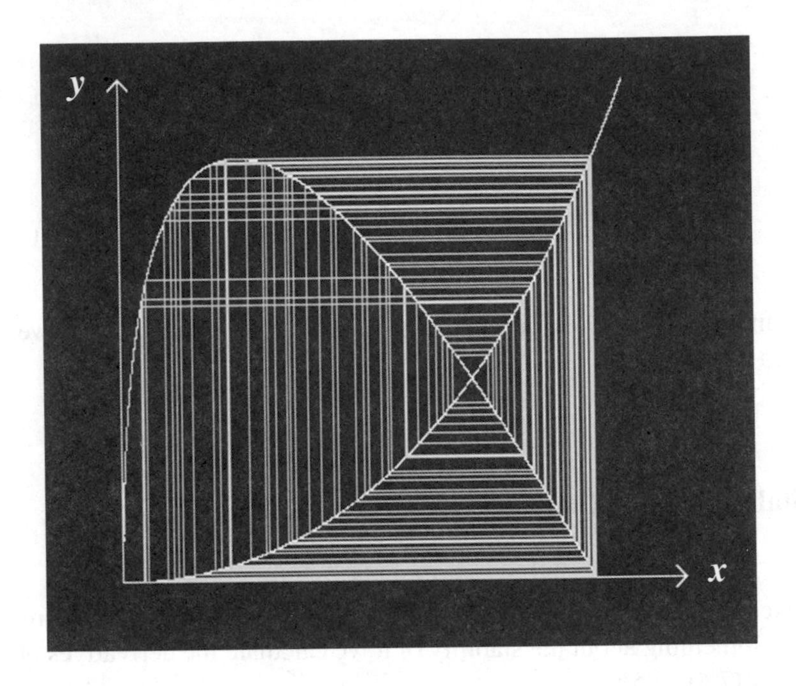

Fig. 7.2. Chaotic Cournot duopoly adjustmens.

$$\frac{a}{b} \quad \text{or} \quad \frac{b}{a} \quad = \quad 3 \pm 2\sqrt{2} \tag{7.24}$$

Thus, whenever one of the ratios of the marginal costs of the duopolists falls outside the interval bounded by the two values specified in (7.24), the Cournot point will not be stable. The two roots happen to be reciprocal, so there is nothing odd in stating this condition for *any* of the ratios. In fact all critical conditions depend on the ratio of marginal costs only, so there is just one single free control parameter.

7.6 Periodic Points and Chaos

As this control parameter passes either of the critical values specified in equation (7.24), the fixed point is replaced by a two period cycle. This should be understood so that x and y each oscillate between two values. The two

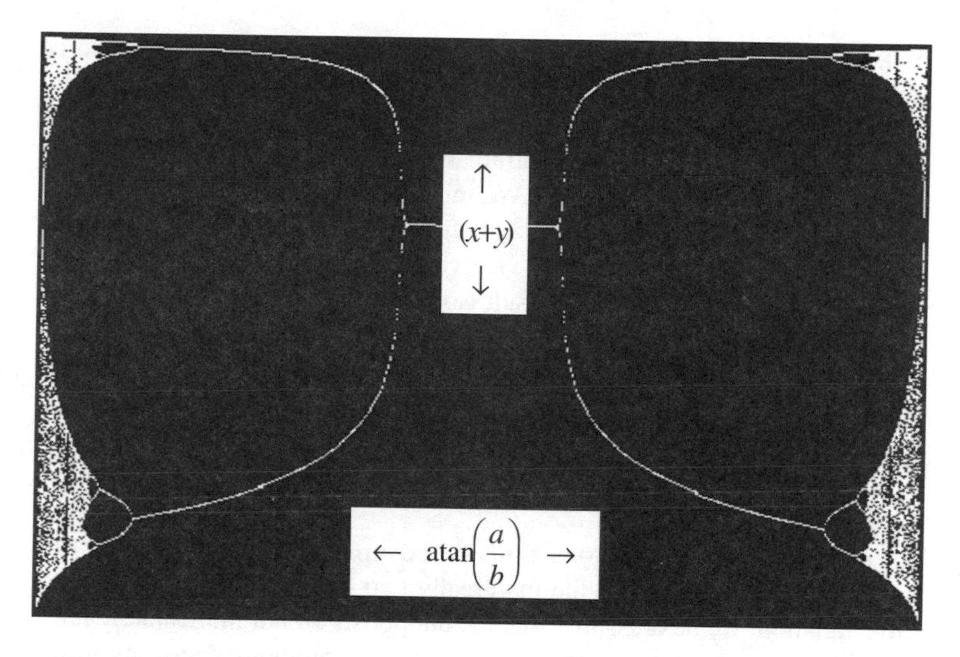

Fig. 7.3. Bifurcation diagram. Sum of amplitudes plotted versus arctangent of marginal cost ratio.

variables themselves, of course, take on different pairs of values in the oscillatory process.

Actually, we are dealing with a pair of independent iterations, which can be seen if (7.19) is substituted in (7.18) or vice versa. We then get iterations of each of the variables alone, without interference of the other one, though the delay is now two periods in terms of the originally introduced lag. For the simplest cycle it then takes four of these periods to return to the initial value.

Implicit in this is also that, contrary to what may be assumed, it has no importance at all whether the duopolists adjust simultaneously or take turns in their adjustments. The only difference is how these two, essentially autonomous, time series are paired together.

After the first cycle appears there is a period doubling cascade to chaos. In Fig. 7.2 we display the cobweb of a chaotic process of adjustments in the same type of diagram as displayed in Fig. 7.1.

In Fig. 7.3 we illustrate the bifurcation diagram. For the sake of symmetry we plot the sum of the phase variables, $x+y$, against the arctangent of the marginal cost ratio a/b. This sum of the phase variables is total supply or, according to (7.1), the reciprocal of commodity price. As mentioned, the arctangent of the cost ratio, rather than the ratio itself is chosen in order to make both ends of the diagram symmetric mirror images. It should also be mentioned, that a substantial section in the middle, corresponding to the stable fixed point, has been removed.

The general appearance of the bifurcation diagram is very like that of the extensively studied logistic map. We might say that this is to be expected from the general look of the reaction functions in Fig. 7.1, but in reality substitution of (7.19) in (7.18) or vice versa results in a two-humped iteration in each variable. Thus, the result is not so obvious.

7.7 Adaptive Expectations

As mentioned, the iterations considered up to now are actually two independent ones. To make the map really two-dimensional, and bring in a little variation, we next assume that the duopolists do not immediately jump to their new optimal positions, but adjust their previous decisions in the direction of the new optimum. Assume:

$$x_{t+1} = x_t + \lambda\left(\sqrt{\frac{y_t}{a}} - y_t - x_t\right) \tag{7.25}$$

$$y_{t+1} = y_t + \mu\left(\sqrt{\frac{x_t}{b}} - x_t - y_t\right) \tag{7.26}$$

With the adjustment speeds λ and μ unitary, we are back at the case already treated. In the contrary case, with those speeds zero, the duopolists will never revise any decision taken. Intermediate values, between zero and unity, bring a host of new possibilities.

The stability of various solutions, such as the Cournot fixed point, and the cycles, now also depend, not on the marginal cost ratio alone, but on the adjustment speeds as well. We can study the stability around any point in phase space by linearising the system (7.25)-(7.26) . The matrix of the linearised system then becomes:

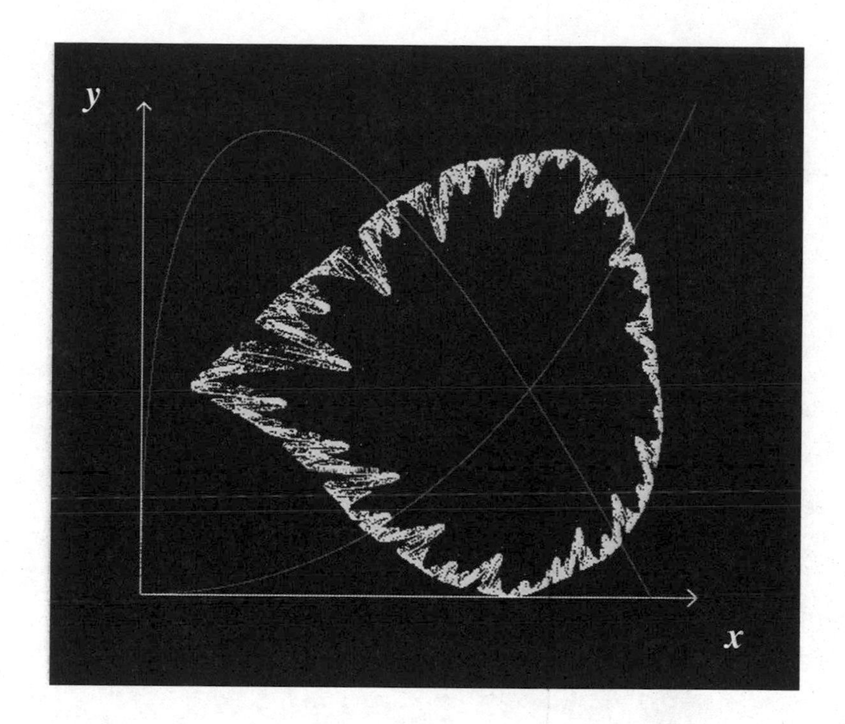

Fig. 7.4. Fractal attractor of the adaptive model

$$\begin{Bmatrix} 1-\lambda & \lambda\left(\dfrac{1}{2\sqrt{ay}}-1\right) \\ \mu\left(\dfrac{1}{2\sqrt{bx}}-1\right) & 1-\mu \end{Bmatrix} \tag{7.27}$$

where we for simplicity deleted the period indices. Considering the Cournot point, we substitute for the phase variables from (7.6)-(7.7). The matrix then becomes:

$$\begin{Bmatrix} 1-\lambda & \dfrac{\lambda(b-a)}{2a} \\ \dfrac{\mu(a-b)}{2b} & 1-\mu \end{Bmatrix} \tag{7.28}$$

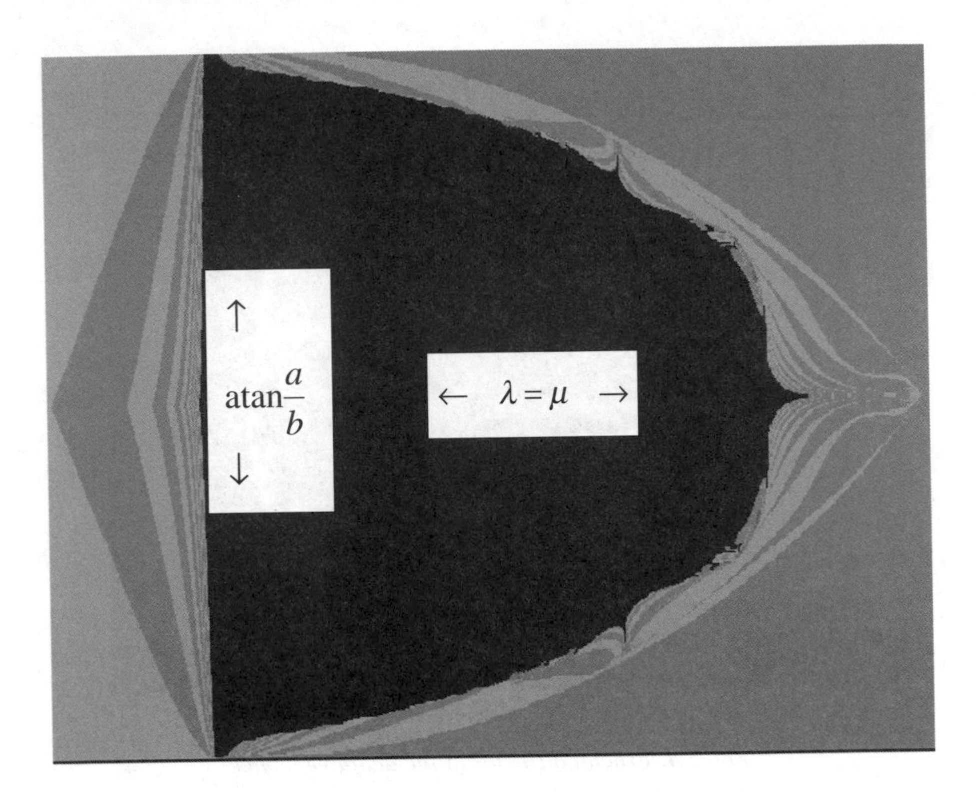

Fig. 7.5. The "Mandelbrot" set for the adaptive model.

At the threshold of loss of stability for the Cournot point the determinant of the matrix (7.28) becomes unitary. Equating to unity and simplifying, we get the condition:

$$(a-b)^2 = 4ab\left(\frac{1}{\lambda}+\frac{1}{\mu}-1\right) \tag{7.29}$$

We note that, with the adjustment speeds unitary, the condition (7.29) becomes identical with (7.23), as it indeed should.

The advantage of the present model, due to its higher dimension, is that we are now able to design fractal attractors in two dimensional embedding space. Such an attractor in the shape of a "leaf" is shown in Fig. 7.4. We can now also display the stability of the model in terms of a "Mandelbrot" set in parameter space. We therefore need two parameters. For the moment we have three, the marginal cost ratio, and the two adjustment speeds. But in the preparation of Fig. 7.5, the adjustment speeds were set equal. The horizontal coordinate is this identical adjustment speed for the expectations, the vertical one is the cost ratio. The dark central body represents stability of the Cournot point, whereas the shaded gray bands indicate the speed at which the model explodes for illicit parameter values. The period doubling cascades and chaos occur in the fractal border fringe of the central body.

The reader should be warned that we preferred to display the entire symmetric "Mandelbrot" set. The feasible region in terms of economic substance is a subset of it, but it includes sufficiently much of the border fringe, including a lot of intricacy at higher resolutions.

We can now investigate the case with the method of critical lines and absorbing areas as indicated in Chapter 4. Consider the fixed points of the iteration (7.25)-(7.26), i.e.

$$x = x + \lambda\left(\sqrt{\frac{y}{a}} - y - x\right) \tag{7.30}$$

$$y = y + \mu\left(\sqrt{\frac{x}{b}} - x - y\right) \tag{7.31}$$

It is obvious that the fixed points are only two, the origin and the Cournot point as given by (7.6)-(7.7) above. The origin is always unstable, whereas we indicated the loss of stability for the Cournot point by equation (7.29).

To outline the shapes of attractors once the Cournot equilibrium loses stability and is replaced by a Hopf cycle and its further bifurcations (of which the chaotic cases are the most interesting in terms of geometry) we need the critical line L_{-1}.

In Section 4.11 we derived it by substituting from one equation in the set corresponding to (7.25)-(7.26), but the same result is obtained by the shortcut of just putting the determinant of the Jacobian matrix (7.27) equal to zero. Observe that we take the general matrix (7.27), not the one at the Cournot point (7.28). Also note that we put the determinant equal to zero not to unity as in deriving (7.29), the reason being that now we just want to locate the line along which the plane is folded.

So, putting the determinant of (7.27) equal to zero we obtain the following expression:

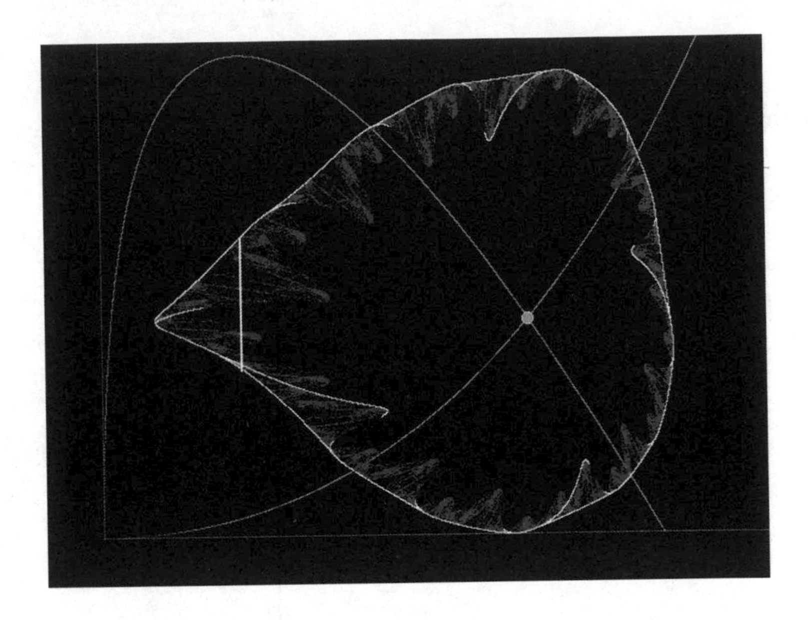

Fig. 7.6. Chaotic attractor and absorbing area with 17 critical lines.

$$\frac{1}{\sqrt{4bx}}+\frac{1}{\sqrt{4ay}}-\frac{1}{\sqrt{4bx}}\frac{1}{\sqrt{4bx}}=\frac{\lambda+\mu-1}{\lambda\mu} \tag{7.32}$$

It is easy to see that this implicit equation results in a pair of hyperbolas in the x, y-plane. Their forward images according to (7.25)-(7.26) are too complex to write down, but they can easily be traced by the computer.

In Fig 7.6 we illustrate a segment of the hyperbola and 17 of its forward images which suffice to outline the outward boundary. The case is that illustrated in Fig. 7.4, and we see the attractor displayed there in gray shade. We also see the reaction functions and the unstable Cournot point at their intersection.

The vertical "line" segment we see on the left actually is part of L_{-1}, i.e. of the hyperbola as defined by (7.32) though it looks like a straight line. We do not see the point a_0 on it, because it is located very high up in the scale we use. This actually illustrates the fact that we get the most accurate pictures of the attractors by just using the parts of the segments that actually intersect the attractor.

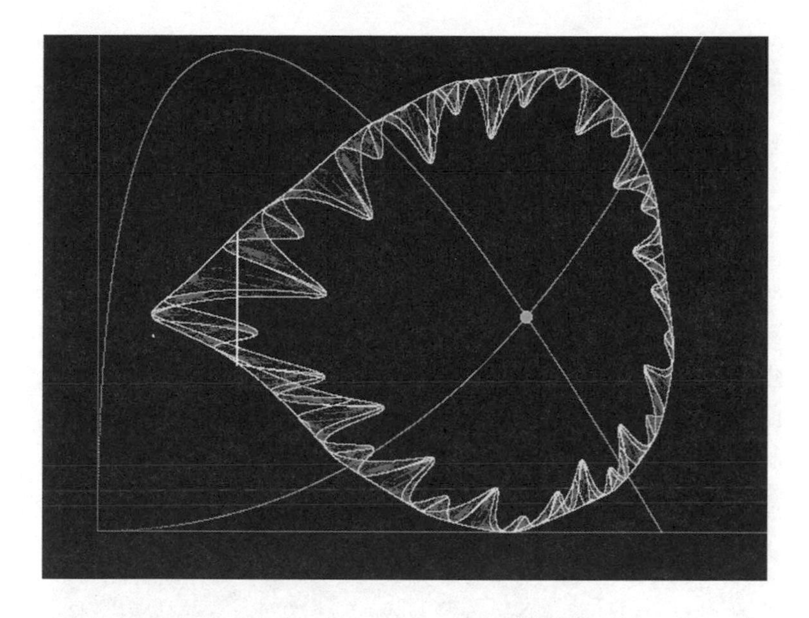

Fig. 7.7. Chaotic attractor and absorbing area with 55 critical lines.

Even more accurate and efficient would have been to exclude the middle part of L_{-1} that, as we see, coincides with the "hole" in the attractor, but we skip this computational complication. Note that we do not need the attractor as advance information to decide which segment or segments of L_{-1} to use.

By using more critical lines as illustrated in Fig. 7.7 we can outline the inner boundary of the attractor as well, and so we can by computer experiment easily find the critical segments to use.

7.8 Adjustments Including Stackelberg Points

Let us now take in consideration that the duopolists might consider trying to become Stackelberg leaders. We noted that Stackelberg leadership, according to (7.11) or (7.15), is always better than the Cournot point, described by (7.6)-(7.7). If thus the process is hunting for a, stable or unstable, Cournot point, according to the reaction functions (7.18)-(7.19), any of the duopolists

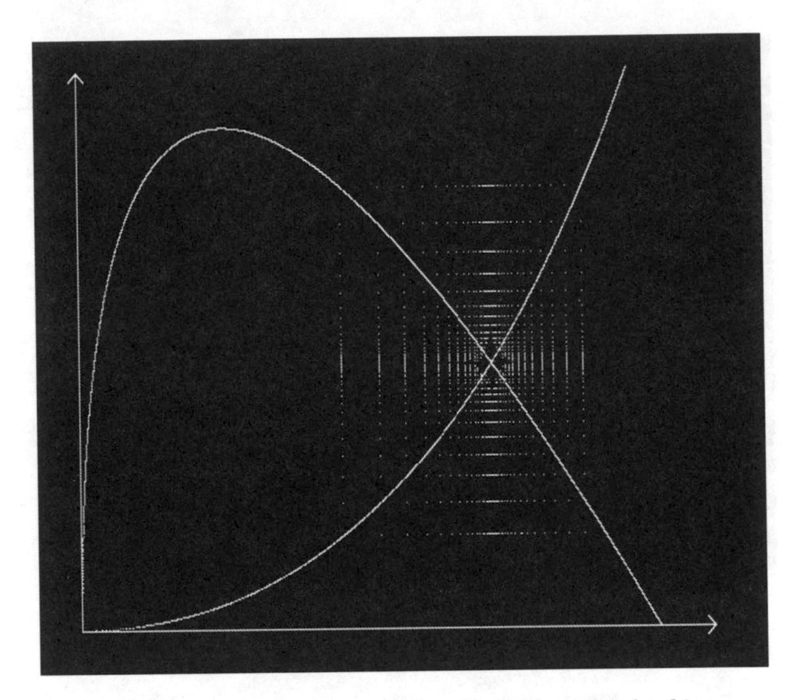

Fig. 7.8. Adjustments including Stackelberg leadership.

might as we saw try Stackelberg leadership at any moment. If one duopolist has chosen to try this strategy, the other might accept this, and this is then the end of the story, but he might as well take up the challenge.

Formalizing, we would then have a generalized dynamical system (a non-linear IFS), where:

$$x_{t+1} = \begin{cases} \sqrt{\dfrac{y_t}{a}} - y_t \\ \dfrac{b}{4a^2} \end{cases} \tag{7.33}$$

$$y_{t+1} = \begin{cases} \sqrt{\dfrac{x_t}{b}} - x_t \\ \dfrac{a}{4b^2} \end{cases} \tag{7.34}$$

and where the choices would be devised on a probabilistic basis. The attractor for such a process does generally not depend on the probabilities, provided they are positive.

We note that the attractor in Fig. 7.8 is just the product set of two independent one-dimensional sets. This is so because we are again back at the situation where the mappings (7.33)-(7.34) are essentially independent, as substitution would show. Such a substitution, of course, results in four alternatives instead of two as in (7.33) and (7.34).

7.9 Oligopoly with Three Firms

The situation becomes much more interesting and complicated with three oligopolists, as it brings in a number of completely new possibilities. Let us keep the notation introduced up to now, but add a third firm, whose profit is denoted W, whose output is denoted z, and whose marginal cost is denoted c. With the unit elasticity demand function introduced, and with the constant marginal costs, the profits of the three oligopolists become:

$$U = \frac{x}{x+y+z} - ax \tag{7.35}$$

$$V = \frac{y}{x+y+z} - by \tag{7.36}$$

$$W = \frac{z}{x+y+z} - cz \tag{7.37}$$

For the Cournot case each oligopolist again equates the partial derivative of its profit, taken with respect to its own control variable, i.e., with respect to its output, to zero. The decision variables of the other firms are, of course, again taken as given. And again we can solve for the control variables and get explicit reaction functions:

$$x = \sqrt{\frac{y+z}{a}} - y - z \tag{7.38}$$

$$y = \sqrt{\frac{z+x}{b}} - z - x \tag{7.39}$$

$$z = \sqrt{\frac{x+y}{c}} - x - y \tag{7.40}$$

As in the duopoly case it is trivial to check that the second order conditions are fulfilled. Note that, due to the simplicity of the demand function, all the formal procedures work with any numbers of oligopolistic competitors.

Next, by taking (7.38)-(7.40) as a system of simultaneous equations, we can easily solve for the quantities produced in the Cournot equilibrium point. The solution becomes:

$$x = \frac{2(b+c-a)}{(a+b+c)^2} \tag{7.41}$$

$$y = \frac{2(c+a-b)}{(a+b+c)^2} \tag{7.42}$$

$$z = \frac{2(a+b-c)}{(a+b+c)^2} \tag{7.43}$$

Next we can substitute back from (7.41)-(7.43) into the equations (7.35)-(7.37) and compute the profits of the three oligopolists in the Cournot equilibrium point:

$$U = \frac{(b+c-a)^2}{(a+b+c)^2} \tag{7.44}$$

$$V = \frac{(c+a-b)^2}{(a+b+c)^2} \tag{7.45}$$

$$W = \frac{(a+b-c)^2}{(a+b+c)^2} \tag{7.46}$$

Not surprisingly, the firm with the lowest marginal cost will make the largest profit. It will, of course make an even larger profit if it becomes a Stackelberg leader, but so will the other firms as well. We will see more about this case in the sequel, but let us for the moment look closer at the dynamic Cournot adjustment process.

First of all we have to find out something about the stability of the Cournot point as defined by (7.41)-(7.43). Unless the Cournot point is unstable, we cannot have chaos or even fluctuations in the long run.

With a third order system we cannot just confine the discussion to checking the value of the determinant of the Jacobian matrix at the Cournot point.

Differentiating the right hand sides of (7.38)-(7.40) we obtain the Jacobian matrix:

$$M = \begin{bmatrix} 0 & \dfrac{1}{2\sqrt{a(y+z)}}-1 & \dfrac{1}{2\sqrt{a(y+z)}}-1 \\ \dfrac{1}{2\sqrt{b(x+z)}}-1 & 0 & \dfrac{1}{2\sqrt{b(x+z)}}-1 \\ \dfrac{1}{2\sqrt{c(x+y)}}-1 & \dfrac{1}{2\sqrt{c(x+y)}}-1 & 0 \end{bmatrix} \tag{7.47}$$

Substituting the coordinates for the Cournot point from (7.41)-(7.43), we get the following simplification:

$$M = \begin{bmatrix} 0 & \dfrac{b+c-3a}{4a} & \dfrac{b+c-3a}{4a} \\ \dfrac{c+a-3b}{4b} & 0 & \dfrac{c+a-3b}{4b} \\ \dfrac{a+b-3c}{4c} & \dfrac{a+b-3c}{4c} & 0 \end{bmatrix} \tag{7.48}$$

It is now that the situation becomes more complicated than in the two-dimensional case.

We have to specify the characteristic equation:

$$\begin{vmatrix} -\lambda & \dfrac{b+c-3a}{4a} & \dfrac{b+c-3a}{4a} \\ \dfrac{c+a-3b}{4b} & -\lambda & \dfrac{c+a-3b}{4b} \\ \dfrac{a+b-3c}{4c} & \dfrac{a+b-3c}{4c} & -\lambda \end{vmatrix} = 0 \tag{7.49}$$

which becomes a cubic in λ. Due to the fact that the eigenvalue appears alone in the main diagonal there is no quadratic term, just the cubic, a linear one, and a constant. The general form is:

$$-\lambda^3 + A\lambda + B = 0 \tag{7.50}$$

Where, due to the composition of (7.49),

$$A = \frac{6\left(a^3+b^3+c^3\right) - 5\left(a^2+b^2+c^2\right)\left(a+b+c\right) + 30abc}{16abc} \tag{7.51}$$

and

$$B = \frac{(a+b-3c)(a+c-3b)(b+c-3a)}{32abc} \tag{7.52}$$

The latter, of course, is the determinant of the Jacobian matrix (7.48). There are two ways the third order system can lose stability. A real eigenvalue can get an absolute value larger than unity, and a pair of conjugate complex eigenvalues can cross the unit circle in the complex plane.

Suppose we put $\lambda = 1$ in (7.50). Then we get:

$$A + B = 1 \tag{7.53}$$

Likewise, suppose we put $\lambda = -1$ in (7.50). Then:

$$A - B = 1 \tag{7.54}$$

We see that the loss of stability due to a real eigenvalue becoming unitary is quite straightforward.

The case of loss of stability due to complex eigenvalues can be settled in the following manner. Equation (7.50) is equivalent to:

$$(\lambda_1 - \lambda)(\lambda_2 - \lambda)(\lambda_3 - \lambda) = 0 \tag{7.55}$$

where $\lambda_1, \lambda_2, \lambda_3$ are the three eigenvalues. If we expand (7.55) and match the coefficients we get:

$$\lambda_1 + \lambda_2 + \lambda_3 = 0 \tag{7.56}$$

$$\lambda_1\lambda_2 + \lambda_1\lambda_3 + \lambda_2\lambda_3 = -A \tag{7.57}$$

$$\lambda_1\lambda_2\lambda_3 = B \tag{7.58}$$

Suppose the first eigenvalue λ_1 is real and the remaining pair complex conjugates, i.e. that $\lambda_2, \lambda_3 = \alpha \pm \mathrm{i}\beta$. Then from (7.56) we have $\lambda_1 = -2\alpha$. Substituting into (7.57) and (7.58) we obtain:

$$A = 3\alpha^2 - \beta^2 \tag{7.59}$$

and

$$B = -2\alpha\left(\alpha^2 + \beta^2\right) \tag{7.60}$$

Now we are ready to introduce the assumption that the complex eigenvalues cross the unit circle, i.e. $\alpha^2 + \beta^2 = 1$. Hence (7.59)-(7.60) become:

$$A = 4\alpha^2 - 1 \tag{7.61}$$

and

$$B = -2\alpha \tag{7.62}$$

Eliminating α between (7.61) and (7.62) we finally obtain:

$$B^2 - A = 1 \tag{7.63}$$

as the proper condition for a pair of complex eigenvalues to cause loss of stability of the Cournot equilibrium for three oligopolists. Equations (7.53)-(7.54) and (7.63) together provide the bifurcation set for all possible losses of stability for the Cournot point. This is displayed in Fig. 7.9 where the bright area, the intersection of the inequalities $A < 1 + B$, $A < 1 - B$, $A > B^2 - 1$, is the region of stability for the Cournot point.

Fig. 7.9 also contains the borderline between real and complex eigenvalues. This is defined by the condition that the discriminant of the cubic (7.50) be zero, i.e., by

$$4A^3 = 27B^2 \tag{7.64}$$

Points in the upper wedge represent cases with three distinct real roots, whereas below there are one real root and a pair of conjugate complex ones.

Of course, we would like to see the region of stability in the space of marginal costs instead. To this end we can substitute from (7.51)-(7.52) in (7.53)-(7.54) and (7.63) to find the boundaries of the region.

We see that the linear coefficient A and the constant B as given by (7.51)-(7.52) only depend on the marginal cost ratios. With three marginal costs there, of course, are two ratios, so choosing a as "numèraire", putting $b = ha$ and $c = ka$, we see that a just drops out from (7.51)-(7.52). We obtain:

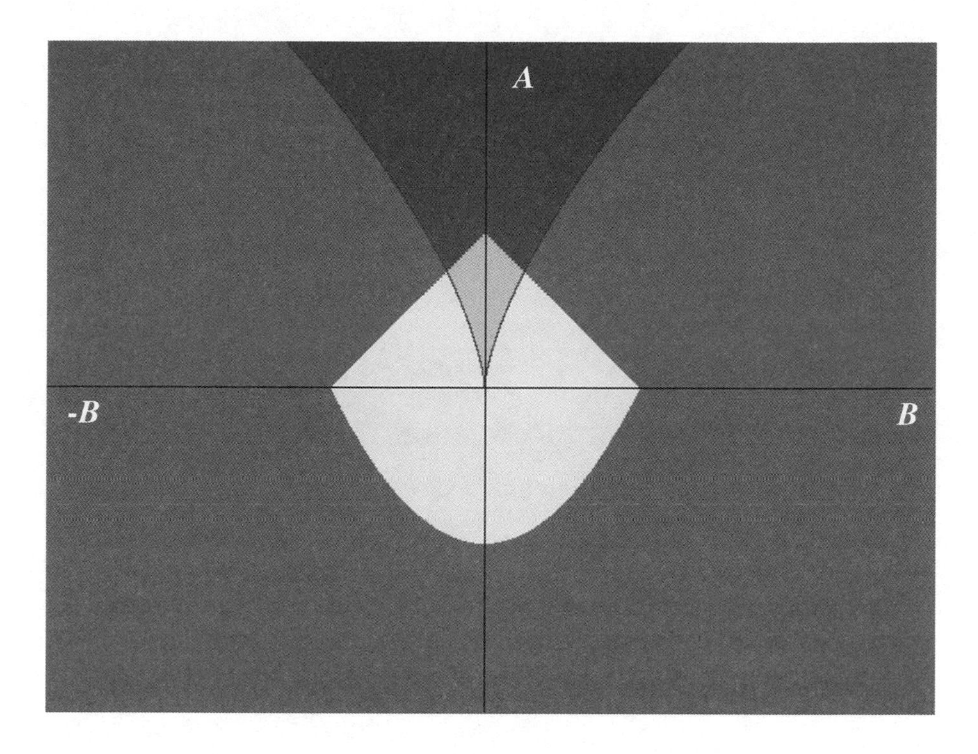

Fig. 7.9. Region of stability for the Cournot point in the A, B-plane.

$$A = \frac{6\left(1+h^3+k^3\right)-5\left(1+h^2+k^2\right)\left(1+h+k\right)+30hk}{16hk} \tag{7.65}$$

and

$$B = \frac{\left(1+h-3k\right)\left(1+k-3h\right)\left(h+k-3\right)}{32hk} \tag{7.66}$$

Accordingly we can display the facts in the two-dimensional h, k - plane, using (7.53)-(7.54), (7.63) and (7.65)-(7.66), though we can expect the computations to become so messy that it is better to let the computer do the job. The result is shown in Fig. 7.10. The bright area represents stability of the Cournot point in h, k-space. The surrounding closed curve is the image of the lower parabola segment of the Figure 7.9. In the positive quadrant of the h, k-space the two other stability constraints are then automatically fulfilled.

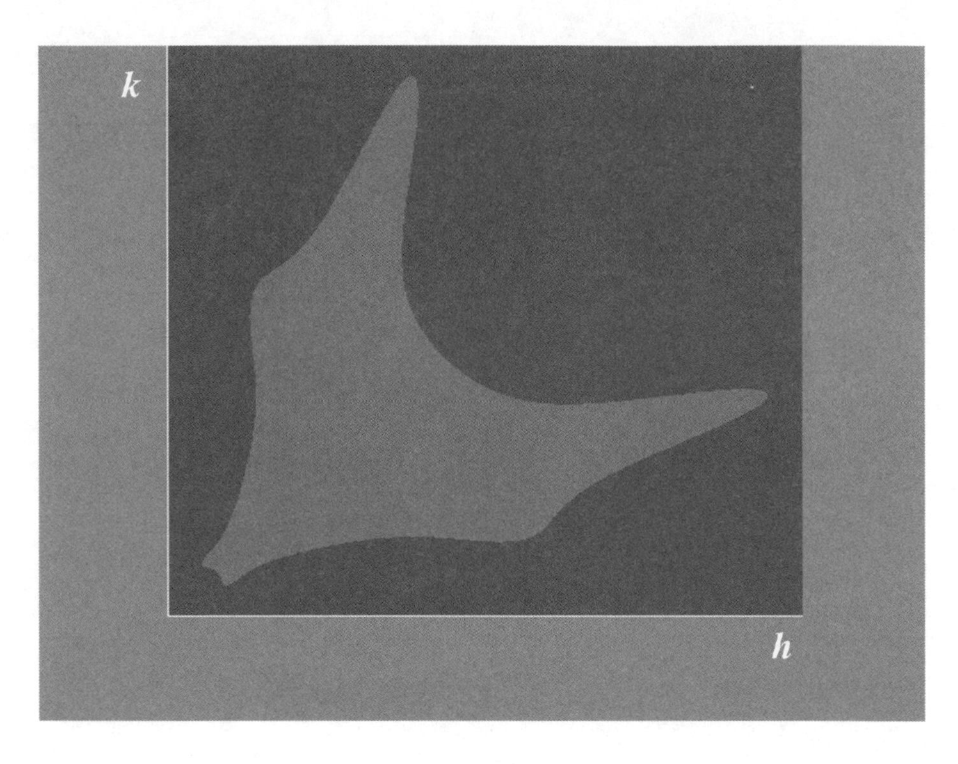

Fig. 7.10. Region of stability for the Cournot point in the h, k-plane.

The bifurcations represented by equations (7.53), (7.54), and (7.63) have their names: When $\lambda_1 = 1$ holds, one usually speaks of a Cusp Bifurcation, whereas the proper name for case of $\lambda_1 = -1$ is Flip Bifurcation. Finally, the case $\lambda_2, \lambda_3 = \alpha \pm i\beta$ with $\alpha^2 + \beta^2 = 1$ is usually called the Chenciner Bifurcation. See Kuznetsov.

Of those the Flip and Chenciner form the borderlines of the stability region. The cusp does not occur. It is the only case we can analyse in closed form. If we put $\lambda = 1$ in (7.50), and substitute from (7.65)-(7.66), then it all boils down to the simple condition:

$$\frac{(h+k+1)^3}{32hk} = 0 \tag{7.67}$$

which obviously cannot be fulfilled for positive *h*, *k*. Negative values would, however, imply that not all marginal costs are positive, which is absurd. We also find that the stability region for complex eigenvalues is completely enclosed in the stability region for the real eigenvalue. Hence it is only the Chencenier bifurcation that is operative in the system.

7.10 Stackelberg Action Reconsidered

Before continuing with the dynamical analysis let us just state the Stackelberg conditions. We should realise that now there are various hierarchies of leadership. It is possible not only to be a Stackelberg leader or else follow the Cournot reaction curve. As each firm now has two competitors, it is also possible to accept the leadership of one competitor, but treat the other competitor as a follower. The mathematical complexity of such cases, with three different levels of behaviour however grows beyond measure. We therefore in the sequel assume only two kinds of behaviour: being a leader, and following the reaction curve.

Supposing the first firm is a leader and the other two followers, we can substitute from the reaction functions (7.39)-(7.40) into the profit function (7.35), which then becomes:

$$U = \tfrac{1}{2}(\sqrt{1+4(b+c)x}-1)-ax \tag{7.68}$$

This leadership profit function depends on the output of the firm itself only, and we can equate its first derivative to zero. The resulting equation in x is readily solved:

$$x = \frac{(b+c)^2 - a^2}{4a^2(b+c)} \tag{7.69}$$

The outputs of the other firms then are obtained from (7.39)-(7.40):

$$y = \frac{(a+b+c)(ac+(a-b)(b+c)}{4a^2(b+c)^2} \tag{7.70}$$

$$z = \frac{(a+b+c)(ab+(a-c)(b+c)}{4a^2(b+c)^2} \tag{7.71}$$

Substituting back from (7.69)-(7.70) into (7.35) we obtain the Stackelberg leadership profit:

$$U = \frac{(b+c-a)^2}{4a(b+c)} \tag{7.72}$$

It is easy to prove that U according to (7.72) is always at least as large as according to (7.44), so any oligopolist may again at any moment try to become a Stackelberg leader. The profits of the two followers in the Cournot point are as easily obtained by substitution, but deriving them is trivial, so we skip it.

7.11 The Iteration with Three Oligopolists

Let us now get back to the Cournot case, and write down the appropriate equations for the iteration by lagging (7.38)-(7.40). We note that we no longer deal with essentially independent mappings. Of course, as always, a system of difference equations can be manipulated to single equations, but they are of higher order, involving several previous values of the single variable, not just a lengthening of the basic period as before. In relation to this it is no longer inessential whether all oligopolists adjust simultaneously, or in a certain order. All different assumptions in this respect now lead to substantially different dynamical systems.

Assuming for the sake of symmetry that adjustment is simultaneous, we have:

$$x_{t+1} = \sqrt{\frac{y_t + z_t}{a}} - y_t - z_t \tag{7.73}$$

$$y_{t+1} = \sqrt{\frac{z_t + x_t}{b}} - z_t - x_t \tag{7.74}$$

$$z_{t+1} = \sqrt{\frac{x_t + y_t}{c}} - x_t - y_t \tag{7.75}$$

7.12 Back to "Duopoly"

Experiments with iterating (7.73)-(7.75), show that, whenever two marginal costs are equal, then the attractor is located in a plane embedded in the three-dimensional phase space of this system. In plain words the outputs of the firms with equal marginal costs tend to become equalized over time. Substituting $b = c$ and $y_t = z_t$ in (7.73)-(7.75) we find $y_{t+1} = z_{t+1}$, so the plane $y = z$ is indeed invariant. As mentioned it even is attracting in simulation, though it is difficult to say whether it is in a strict mathematical sense, but this is not essential for our purpose.

It is therefore possible to study an essentially two-dimensional process, again a kind of "virtual duopoly". To explore this, suppose b=c and assume that the process has already converged, so that $y_t = z_t$ holds true. The system (7.73)-(7.75) then becomes:

$$x_{t+1} = \sqrt{\frac{2y_t}{a}} - 2y_t \tag{7.76}$$

$$y_{t+1} = \sqrt{\frac{x_t + y_t}{b}} - x_t - y_t \tag{7.77}$$

There is a twin equation to (7.77), where z replaces y, but there is no need to reproduce it. The present system is different from the original duopoly case as it cannot be decomposed into independent maps of the first order.

For (7.76)-(7.77) we can compute the Cournot point:

$$y = \frac{2a}{(a+2b)^2} \tag{7.78}$$

$$x = \frac{2(2b-a)}{(a+2b)^2} \tag{7.79}$$

It could also have been found by substitution of b=c into (7.41)-(7.43). To find out about the stability of the Cournot point we can now differentiate the system (7.76)-(7.77) right away, calculate the determinant of its Jacobian matrix and substitute from (7.78)-(7.79). Thus:

$$\text{Det} = \frac{(a-2b)(3a-2b)}{8ab} \tag{7.80}$$

This expression can never equal +1, but it assumes the value -1 for:

$$\frac{b}{a} = 2 \pm \frac{\sqrt{13}}{2} \tag{7.81}$$

Unlike the situation before, the roots are not in reciprocity, so the marginal costs enter asymmetrically into the critical ratio.

Fig. 7.11 displays a sequence of events at a marginal cost ratio near a critical root according to (7.81). On top we see a slow spiralling approach to the stable Cournot equilibrium at a cost ratio close to the critical value. In the middle there is a snapshot just at the moment of bifurcation. On the bottom we see the situation just after the Cournot point has lost stability. What we see is no longer the start of a period doubling cascade, but a Hopf bifurcation from fixed point to cycle.

A cycle in this context of a discrete dynamic process, of course, does not mean that the cycle is traced in a continuous motion. Rather it is like the Poincaré section of a quasiperiodic trajectory wrapped around a torus. The intersection cycle is traced by being hit repeatedly in varying places. There is even a system to how it is hit.

In the spiralling motion we find seven arms, and even the cycle is actually hit in showers of seven shots at a time. We can even see traces of seven corners in the cycle.

These seven corners become interesting for the next bifurcation, which is a global saddle-node bifurcation, involving all the seven corners at once. This is illustrated in Fig. 7.12. On top we see the cycle which has now become more pointed at the corners. There is also a grayscale shading representing the frequency of visiting the points, and we see that the seven corners also are the most frequently visited. On bottom of the picture we see the cycle dissolved in seven disjoint points.

The system has obviously evolved to a regular cycle of period seven. What happened is that at a certain parameter value (marginal cost ratio), there emerged seven coincident pairs of nodes and saddles, which separate upon further parameter change. The saddles become watersheds and the nodes become attractors, so the process converges to this seven period cycle attractor.

Fig. 7.11. Stages of Hopf bifurcation. Top: Critically slow approach to stable Cournot point. Middle: At the bifurcation. Bottom: After the bifurcation to cycle.

Fig. 7.12. Global saddle-node bifurcation.

With further changes of the parameter, the seven points undergo further bifurcations in terms of period doubling cascades which we already know from the case of duopoly. In general, the case of oligopoly with three competitors presents a much richer selection of bifurcations, such as the Hopf and saddle-node, than does pure duopoly.

Of course many chaotic regimes occur, of which we display just one typical attractor in Fig. 7.13.

At this stage we should also add a little technicality in the context of simulations. All the reaction functions, as specified from (7.18)-(7.19) to (7.76)-(7.77), assume negative values for sufficiently large arguments. Such solutions to the optimization problem of the firms are purely mathematical, and, of course, have no factual significance, as it in general is impossible to produce a negative quantity.

In terms of economic substance those cases happen when the competitors supply so much that, in view of its own marginal costs, the firm in question cannot make any profits at all with any positive quantity produced. The consequence then is that the firm produces zero. Thus all the reaction functions should be stated as the maximum of zero and the respective expression stated. For simplicity we did not write this explicitly, but at simulations this fact has to be taken in consideration, otherwise the simulations terminate, as the computer would try to take square roots of negative numbers.

In Figure 7.13 there is one feature added to the picture of the attractor itself: the critical lines.

To find these we again calculate the Jacobian of (7.76)-(7.77) and put the determinant equal to zero. Observe that this does not result in (7.80) which is the Jacobian determinant evaluated at just one point, the Cournot point. The general matrix is:

$$M = \begin{bmatrix} 0 & \dfrac{1}{\sqrt{2ay}} - 2 \\ \dfrac{1}{2\sqrt{b(x+y)}} - 1 & \dfrac{1}{2\sqrt{b(x+y)}} - 1 \end{bmatrix} \tag{7.82}$$

whose determinant becomes zero on either of the conditions:

$$x + y = \frac{1}{4b} \quad \text{or} \quad y = \frac{1}{8a} \tag{7.83}$$

The critical line L_{-1} accordingly consists of two stright lines, one horizontal and one downsloping. This is so because the determinant factorizes. But this is no complication, because we again take the segment of L_{-1} which intersects the attractor as a seed for the further iteration process. In the present case (with $a = 1$, $b = 0.26$) it is a segment of the downsloping curve, visible in the bottom part of the attractor. The horizontal never intersects the attractor at all.

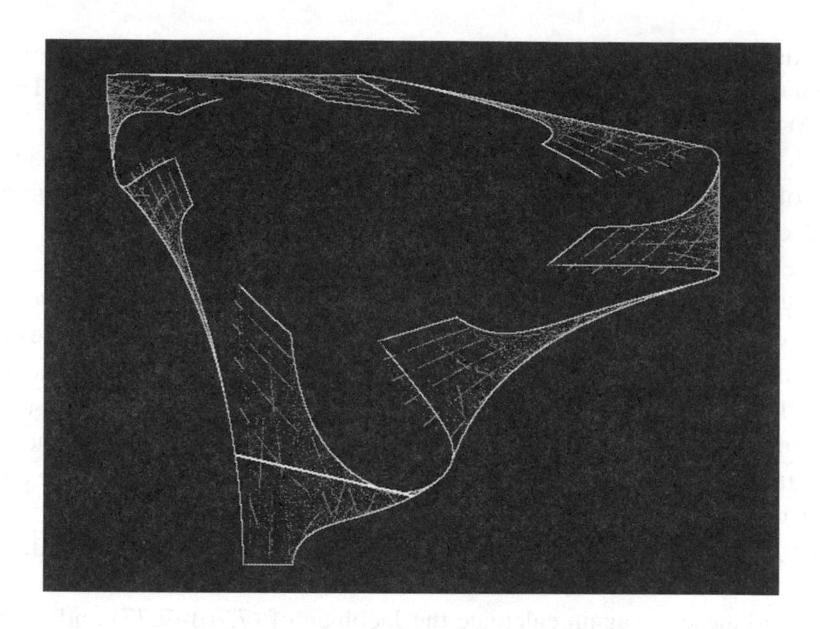

Fig. 7.13. Attractor with part of bondary designed by 14 critical curves.

In Figure 7.13, with 14 iterates of L_{-1} drawn, we see how the outer boundary and part of the inner are clearly outlined. In Figure 7.14, there are 45 iterates drawn, and then we recognise every detail of the actual attractor.

7.13 True Triopoly

After this preliminary revisit to virtual duopoly it is now time to return to the general case where no two marginal costs are equal, and the process hence not confined to take place in any invariant diagonal plane. There is nothing new to simulation techniques, we only have to take some projection in full three-dimensional space to visualize the attractor.

Also, the absorbing area method still works, though it is now a three dimensional solid bounded by critical surfaces, folding space into itself. Again, transverse intersections of any surface with such a critical surface are mapped into tangency contact of their images under the iteration. It is again this repeated tangency which forms the bounded absorbing area.

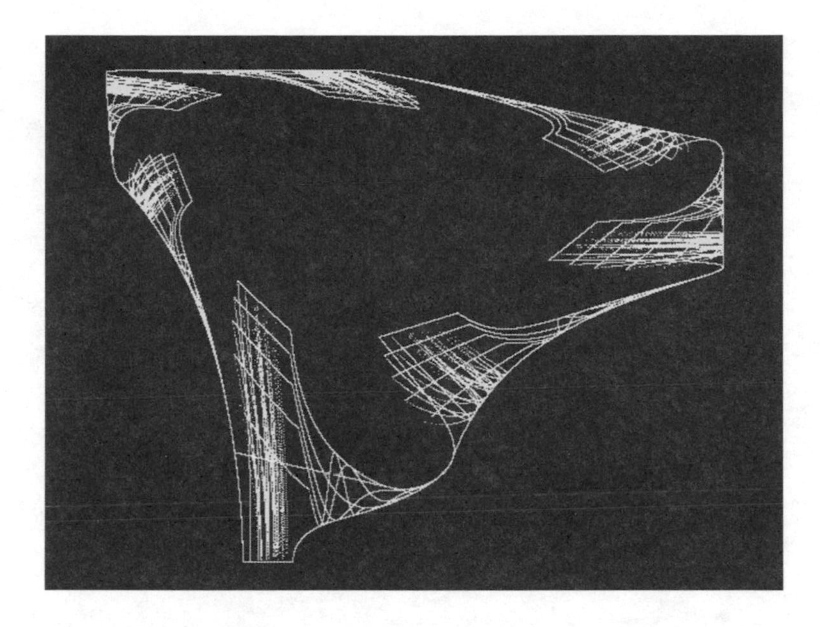

Fig. 7.14. Outline of an attractor with 45 critical curves.

The big difference is that visualisation becomes much more difficult in three dimensions. For this reason, the methods are yet far from as well developed for iterated maps in 3 D as they are in 2 D. Rather, critical surface methods constitute a research area in its first infancy.

To find the critical surfaces for the system (7.73)-(7.75) we take the Jacobian determinant of the matrix (7.47) and put it equal to zero:

$$\begin{vmatrix} 0 & \frac{1}{2\sqrt{a(y+z)}}-1 & \frac{1}{2\sqrt{a(y+z)}}-1 \\ \frac{1}{2\sqrt{b(x+z)}}-1 & 0 & \frac{1}{2\sqrt{b(x+z)}}-1 \\ \frac{1}{2\sqrt{c(x+y)}}-1 & \frac{1}{2\sqrt{c(x+y)}}-1 & 0 \end{vmatrix} = 0 \quad (7.84)$$

Again, note that we are not dealing with the Jacobian just at the Cournot point, and that we put it equal to zero, not to unity.

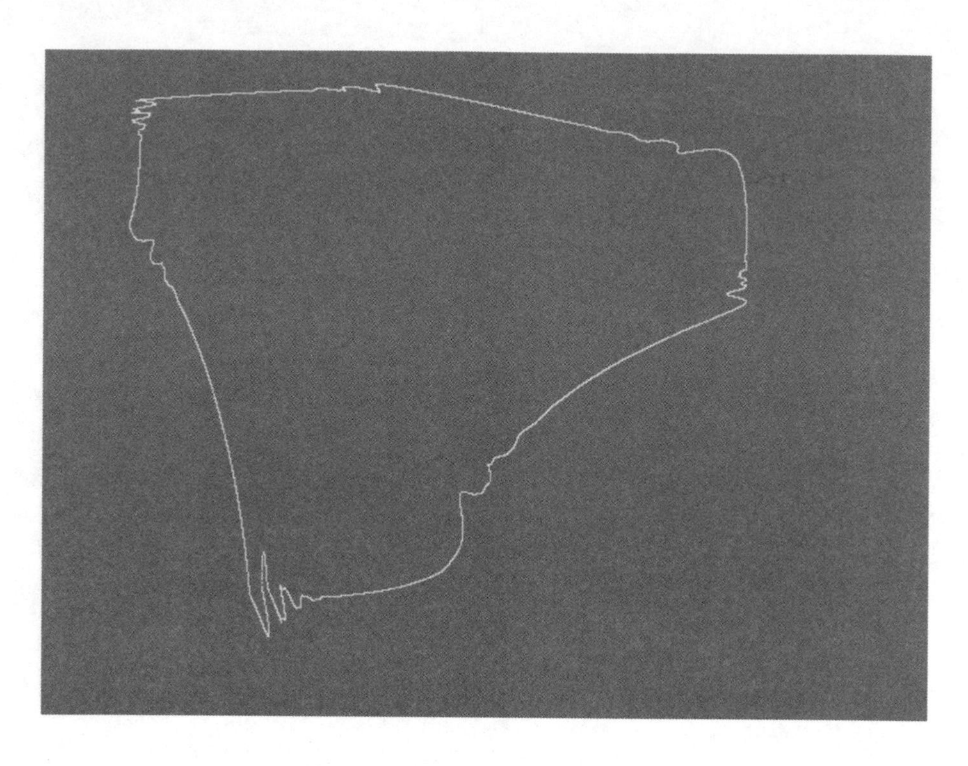

Fig. 7.15. Attractor for a case of Cournot triopoly.

Note also the equal nonzero entries in the first, second and third rows. They can hence be factored out from (7.84), and the remaining determinant becomes equal to the constant 2. Accordingly, the Jacobian determinant becomes zero under the three different conditions that one of the three entries be equal to zero. Simplified these read:

$$y+z=\frac{1}{4a} \tag{7.85}$$

$$x+z=\frac{1}{4b} \tag{7.86}$$

$$x+y=\frac{1}{4c} \tag{7.87}$$

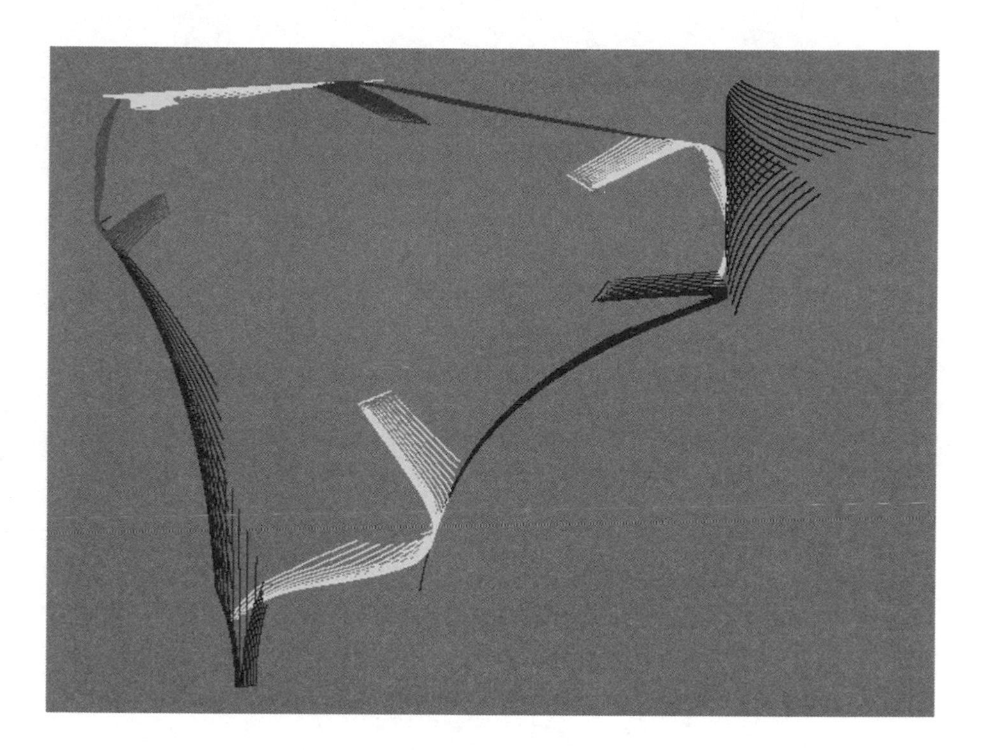

Fig. 7.16. Eight critical surfaces for the attractor in the previous picture.

Obviously they represent three different critical (plane) surfaces in 3D, which together make up the critical surface of order -1. As in previous cases where the critical manifolds were composites of different pieces, only one is operative.

So, let us take the first (8.85), and iterate it once under (7.73)-(7.75). From the first of these equations we find that the critical plane surface of order -1 $y_t + z_t = 1/(4a)$ maps to another plane surface $x_{t+1} = 1/(4a)$, which is part of the critical surface of order 0. Following iterates no longer are plane.

In Figure 7.15 we show a general shape of an attractor. It has passed trough a Hopf bifurcation from the Cournot equilibrium to a cycle with some irregularities due to further bifurcations.

In the companion Figure 7.16 we display a sequence of eight critical surfaces. We do not show the plane of order -1 as it would obscure the picture too much, but the plane of order 0 is shown to the right. It does not seem plane, but it is. The reason for the distorted impression is that the coordinate lines, straight in the plane of order -1, are mapped onto parabolas, which even intersect, in the next plane. The following seven surfaces are, however,

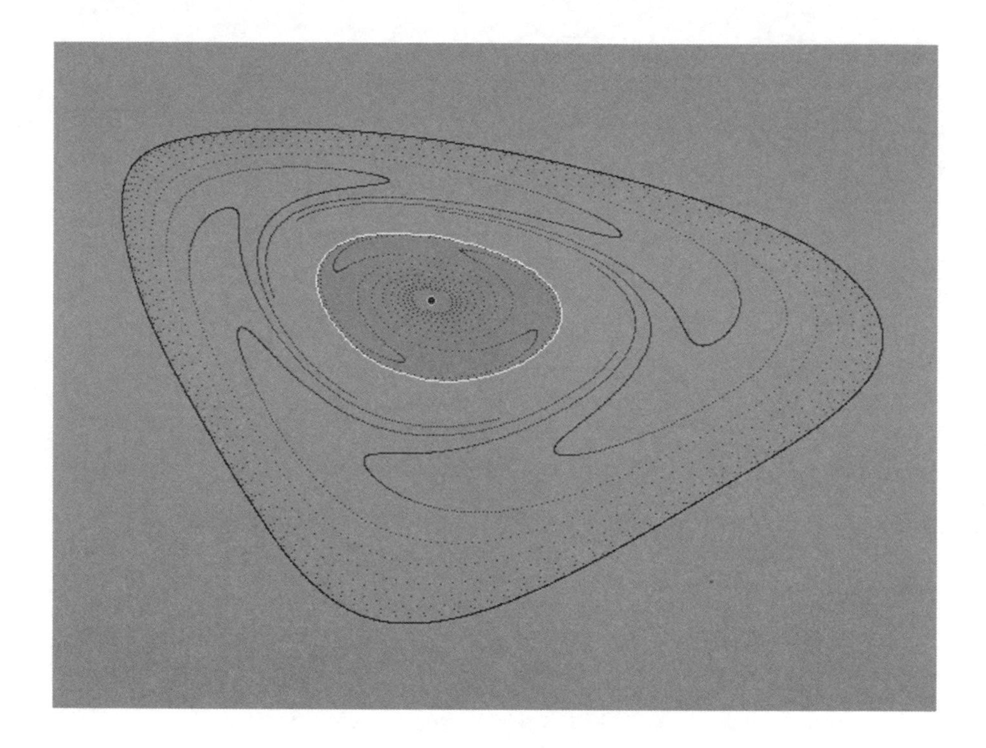

Fig. 7.17. Coexistence of fixed point and cycle.

no longer plane, but strongly folded so as to squeeze parts of the locally almost flat attractor in its folds. We also see the tangencies in the train of critical surfaces.

The reader can just get a slight flavour of how the critical surface method works in 3 D, and also of the visualisation difficulties. Basins, in the case of coexistent attractors, not even attempted to be drawn in the picture, are difficult to show, because holes, belonging to some attractor, may be completely hidden in the interior of that of another one, like the peels of an onion.

Agliari discovered that even the simplified system (7.76)-(7.77) has coexistent attractors. See Agliari, Gardini and Puu. The Cournot equilibrium may thus still be stable for some parameter interval after the cycle has been created, and lose stability first as the attraction basin shrinks upon the Cournot point.

Such a case is shown in Figure 7.17. We find the black Cournot point in the centre along with the black bifurcated triangular cycle. The attraction basins are drawn in different gray shades, and they are separated by the white repelling ellipse shape invariant curve. In each basin there is also shown a spiralling trajectory approaching its proper attractor. Upon further change of the

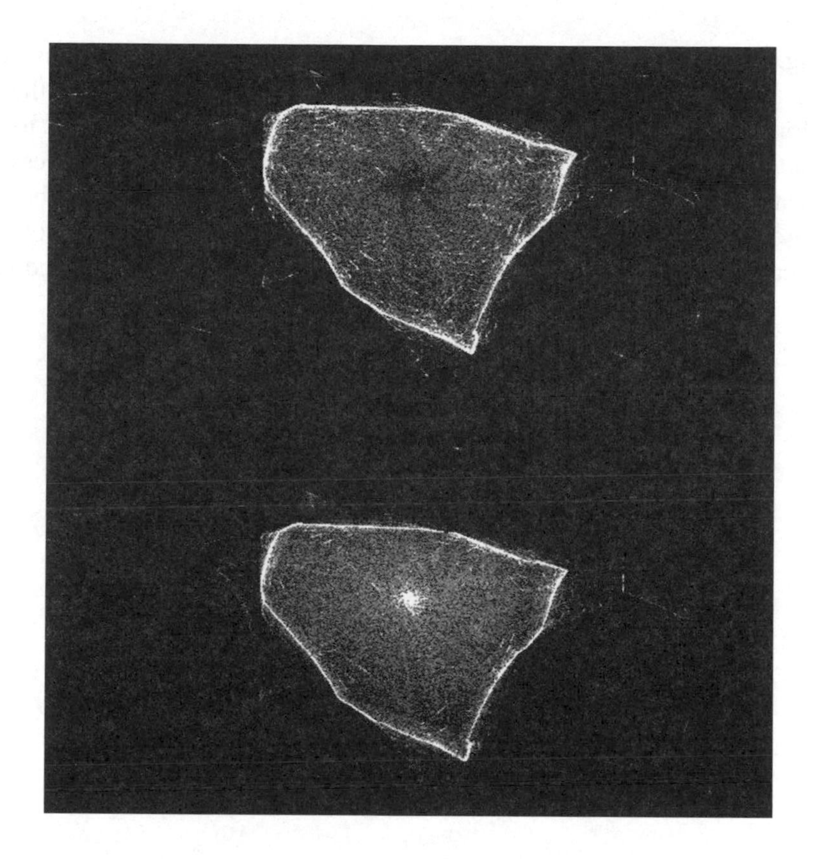

Fig. 7.18. Attractor with probabilistic order of adjustment.

parameters the inner basin may implode upon the Cournot point and hence annihilate its stability. The present digression warns us that the comments around Figure 7.11 were a bit oversimplified as two bifurcations were taken as one.

7.14 Changing the Order of Adjustment

We mentioned that the order of adjustment matters in the case of three competitors. To try this out we can again do it by introducing a certain probability that each competitor abstains from adjusting in any run of the process, just sticking to the old decision.

In Fig. 7.18 we display the effect of the random perturbation of the adjustment order. It is interesting to compare the cases on the top and bottom. On the top randomness is very small. In both cases a new dimension is added to the attractor, which becomes filled and not just curve like. Nevertheless, as indicated by shading, the Hopf cycle is visited most frequently. On the bottom the probabilistic disturbance is somewhat higher with the result that both the bifurcated cycle and the neighbourhood of the Cournot point are visited very often.

8 Business Cycles: Continuous Time

8.1 The Multiplier-Accelerator Model

The invention by Paul Samuelson in 1939 of the business cycle machine, combining the multiplier and the accelerator, certainly was a major event. That two such simple forces as consumers spending a given fraction of their incomes on consumption and producers keeping a fixed ratio of capital stock to output (=real income) combined to produce cyclical change was simple, surprising and convincing at the same time. This model if any qualifies for the attribute of scientific elegance. In passing it should be stressed that the Keynesian macroeconomic outlook was an essential background.

The original model was cast in terms of a process in discrete time, i.e., a difference equation, and was further elaborated with great ingenuity by Sir John Hicks in 1950. Roy Harrod in 1948 was inspired to formulate the process in continuous time, as a differential equation. Instead of arriving at a system generating cycles he arrived at one generating growth, but he clearly realized that in the growth process there was a knife-edge balance immersed in surrounding instability.

This set the tradition for some decades. Business cycles were formulated by difference equations, growth by differential equations. Today we would say that Samuelson and Hicks happened to choose a second order process, whereas Harrod happened to choose a first order one. We would also recognize that this has nothing to do with the choice among discrete and continuous modelling. Dynamical processes of any order can be cast either as difference equations or differential equations.

The choice of model type (discrete versus continuous) can be regarded as a matter of pure convenience. For analytical purposes, when we want to apply theorems from the rich body of literature on differential equations, the continuous outlook is to be preferred. Whenever we want to apply the model to an empirical time series, which necessarily is discrete, we have to discretise.

The same is true when the tools of analysis fail and we are forced to proceed by simulation. Algorithms for differential systems are by necessity discretizations.

Today we would perhaps prefer the continuous outlook a little once the chaotic regime has been discovered. A simple logistic growth process that is well behaved in continuous time has become the most studied prototype of chaotic processes in discrete time. Instabilities that thus occur in discrete first order processes do not arise in continuous processes until they reach third order. To avoid too drastic instabilities merely due to fixed delays and immediate action that are not a true image of reality we might slightly prefer continuous modelling, but when this reservation has been made, the choice still is a matter of convenience.

The choice of the order of the process is the outcome of a different decision. In the golden age of economic dynamics Phillips in 1954 clearly realized that a second order process along the Samuelson-Hicks lines could be formulated as a continuous process, where the time derivatives could be interpreted in terms of exponentially distributed lags. As we want to study a second order process, but want to have access to the analytical tools of differential equation theory, the Phillips model is a convenient point of departure. We will use the very clear outline of it given by Allen in 1956.

8.2 The Original Model

Let us denote income by Y. Savings, S, are a given proportion s of income. Capital stock K is retained in a fixed proportion to income, and the ratio is denoted v. Investments, denoted I, are by definition the rate of change of capital stock. Thus: $I = v\dot{Y}$ and $S = sY$. In equilibrium we would have $I = S$ and this was the point of departure for Harrod. Phillips assumed that there was an adaptive process so that income increased in proportion to the difference of investments and savings, i.e., $\dot{Y} \propto (I - S)$. A similar delay was assumed to occur in the adjustment of investments, so that $\dot{I} \propto (v\dot{Y} - I)$.

In the adjustment equations adjustment speeds would be involved, and for generality Phillips assumed different speeds for the two adaptive processes. The precedence of Samuelson and Hicks who assumed identical unit lags for all kinds of adjustments makes it a licit simplification in modelling to assume the two lags to be equal. Once this is done we only need to assume a suitable measurement unit of time to make the adjustment speeds unitary so that we can dispense with the symbols altogether. We thus have:

$$\dot{Y} = I - sY \tag{8.1}$$

$$\dot{I} = v\dot{Y} - I \tag{8.2}$$

We differentiate the first equation once more, and use the original equations to eliminate investment and its time derivative, thus obtaining the basic equation:

$$\ddot{Y} - (v - 1 - s)\dot{Y} + sY = 0 \tag{8.3}$$

which is capable of producing damped or explosive oscillations, depending on the sign of (v-1-s). We should mention that we have omitted the so called autonomous expenditures (such as government expenditures, exports, or investments not induced by the accelerator). The reason is that they only contribute by an inhomogeneity that can be dealt with by deriving a particular solution to which the solution of the homogeneous equation above can always be added. This superposition principle holds for the original model as well as for all models we are going to present, so we disregard the autonomous term completely. What we have to keep in mind is that a limited negative "income" then is nothing absurd per se as it is measured as the deviation from a (positive) stationary equilibrium.

8.3 Nonlinear Investment Functions and Limit Cycles

This original multiplier-accelerator model clearly displays the limitations of linear analysis. The system, like the linear growth model due to Harrod, is capable of either producing exponentially damped or exponentially explosive change, but nothing else. There is the boundary case where $v = (1 + s)$ producing a standing cycle, but the probability for it is vanishingly small, and it represents a case that in terms of modern systems theory is called structurally unstable. An additional problem is that in the border case the amplitude of change is arbitrary and depends on initial conditions, but not on the system.

As we saw, Arnol'd suggested in his discussion of the mathematical pendulum that friction should be included in the model, no matter what we knew about it from empirical evidence, just because the pendulum without friction would be an unstable boundary case between damped and explosive motion. In modelling reality we should avoid assumptions that are critical in the sense that the behaviour would turn out qualitatively different if those assumptions were not fulfilled. This is something we must consider as we always are committing mistakes in the abstraction process of scientific modelling.

So, disregarding the boundary case, two possibilities remain, either damped or explosive movement. In the first case the model does not generate movement itself, it only maintains it if change is introduced exogenously, but it always tends to damp out any movement and make the system, if left to itself, return to eternal equilibrium. The question is whether such a model is at all dynamic.

The case of explosive motion, in addition to the factual absurdity of producing unlimited change, presents a problem of scientific procedure. In all physical systems, the source of inspiration for mathematical modelling in economics, linearisation is a first approximation, and it is always clear that the approximation holds when the variables remain within reasonable bounds. Linearised relations were never used as components in global models that generated unbounded change, because this would violate the conditions for linearisation, no matter how generously the bounds were set. It seems absurd that economic growth theory was based on such global linearisations that would never have been accepted in physics. Explosive cycles are equally absurd, but those never became tradition in economics as pure growth did, maybe because the former also involve negative infinite values.

The ideal would be a system that neither explodes nor erases movement that happens to be present, but that produces a persistent cyclic movement of finite amplitude, defined by the system itself. Such a cyclic orbit should be an attractor, such that the system from any initial condition within a basin of attraction tends to settle on a movement along it, the system losing memory of the initial conditions.

Hicks realized how this could be arranged within the multiplier-accelerator model. From a factual point of view he argues that the accelerator as a linear relation would be effective only at small rates of change. If income decreased so much that in order to adjust the capital stock, more capital could be dispensed with than disappeared by the process of natural depreciation, capital would not be actively destroyed. This would put a lower limit, the "floor", to disinvestments. Likewise, if income increased very fast, then a moment would come when resources such as labour force or raw materials available would limit production and prevent the firms from increasing capital stock any further. So there would also be a "ceiling" to investments. The constraints were introduced as linear inequalities, but there is no doubt that they were an integral part of the Hicksian investment theory.

The investment function thus became nonlinear. Hicks also observed, by careful calculations with his difference equation model, that the process of change would be limited by the floor and the ceiling, creeping along them for some time, but would turn around and produce a truly cyclic movement within finite bounds defined by the model.

Somewhat later Richard Goodwin in 1951 introduced a smooth nonlinearity in the investment function. The floor and ceiling were reached asymptotically by a an investment function that turned away smoothly from the linear relation between investments and income growth around the origin. By invoking the Poincaré-Bendixon theorem Goodwin made it likely that the model with a nonlinear accelerator possessed a limit cycle.

These represent two possibilities of introducing the nonlinearity in the investment function. As a matter of fact the set of possibilities is unlimited. A linear function is fixed by one numerical value (if it passes through the origin as the investment function does). A nonlinear function has innumerable degrees of freedom, and any choice must be more or less arbitrary.

In the following we will use a power series representation of the lowest degree that represents all the features we want to be represented. Such a power series is the truncated Taylor expansion of any function that interests us. Modern mathematics developed with catastrophe theory encourages us to work with truncated Taylor series. As long as they are determinate, we are instructed to dispense with analyticity and remainders and no longer "let the Tayl wag the dog". This seems particularly appropriate for applications in economics where mathematics was always used to obtain qualitative results.

The reader who wants examples from physics can think of the so called "hard" and "soft" springs which represent the step from the linear Hooke's Law into the world of nonlinearity opened up by Duffing. The same modification of the linear approximation by a cubic term was used by van der Pol in the context of electrical circuits. We choose a nonlinear investment function:

$$v\left(\dot{Y}-\frac{1}{3}\dot{Y}^3\right) \tag{8.4}$$

to replace the linear accelerator $v\dot{Y}$ everywhere. This function possesses the following properties: It passes through the origin having the slope v there, and it then tapers off at large positive and negative rates of change.

We have given mathematical reasons for using the linear-cubic Taylor approximation of the odd nonlinear investment function. In Chapter 10 on discrete time processes we also supply factual reasons for the two turning points of the complete investment function implicit in the cubic by considering public investments distributed contracyclically over time. We do not duplicate these reasons here.

Anyhow, the resulting model is:

$$\ddot{Y} + sY = (v - 1 - s)\dot{Y} - \frac{v}{3}\dot{Y}^3 \tag{8.5}$$

The left hand side is that of a simple harmonic oscillator, which the model becomes if the right hand side is put equal to zero. The right hand side

$$(v - 1 - s)\dot{Y} - \frac{v}{3}\dot{Y}^3 \tag{8.6}$$

is the cubic characteristic which will play an important role in the following.

8.4 Limit Cycles: Existence

The existence of a limit cycle to this model, though obvious by simulation experiments for the case when $(v - 1 - s) > 0$, is not quite trivial to prove. This is so because the obvious tool to use, the Poincaré-Bendixon theorem is not directly applicable. We will present a heuristic outline of a proof that also shows why the theorem referred to does not apply immediately.

Let us first multiply the differential equation (8.5) through by $\dot{Y}$ and identify the products $\dot{Y}\ddot{Y}$ and $Y\dot{Y}$ as half the time derivatives of $\dot{Y}^2$ and Y^2 respectively. Thus:

$$\frac{1}{2}\frac{d}{dt}\left(\dot{Y}^2 + sY^2\right) = (v - 1 - s)\dot{Y}^2 - \frac{v}{3}\dot{Y}^4 \tag{8.7}$$

Provided that

$$(v-1-s)>0 \tag{8.8}$$

is fulfilled the quadratic and quartic terms in the right hand side have opposite signs. Accordingly the quadratic term dominates and the right hand side is positive when $|\dot{Y}|$ is small, whereas the quartic term dominates and the right hand side is negative when $|\dot{Y}|$ is large.

Thus the expression

$$E=\dot{Y}^2+sY^2 \tag{8.9}$$

where the symbol E alludes to "energy" (kinetic and potential), increases for small rates of change of income and decreases for large rates of change. For constant values of E the equation represents concentric ellipses around the origin in $Y, \dot{Y}$ phase space.

We can also say that within the strip

$$|\dot{Y}|<\sqrt{\frac{3(v-1-s)}{v}} \tag{8.10}$$

of the phase plane the vector field defined by our differential equation is directed outward on any such ellipse, whereas outside this strip it is directed inward.

The Poincaré-Bendixon theorem now requires that an annular region of phase space be defined, such that no trajectory entering it ever escapes from it. It would be natural to try the ellipses of constant E as boundaries for an elliptic annulus. This works well for the inner boundary. We just have to choose an ellipse small enough to be contained in the strip defined above. But for the outer boundary there is a problem. Only the part of it lying outside the strip referred to is such that the ellipse captures any trajectory.

There remains a possibility that the trajectory escapes through the openings in the ellipse left by the strip. By expanding the outer ellipse these openings can be made vanishingly small in relation to the whole ellipse, but the holes still are there, and the system might escape any elliptic annulus.

Therefore let us consider how the system would escape such an annulus with a very large outer boundary. In such an escape process it is inevitable that $|Y|$ goes to infinity, whereas $|\dot{Y}|$ remains bounded. We easily calculate

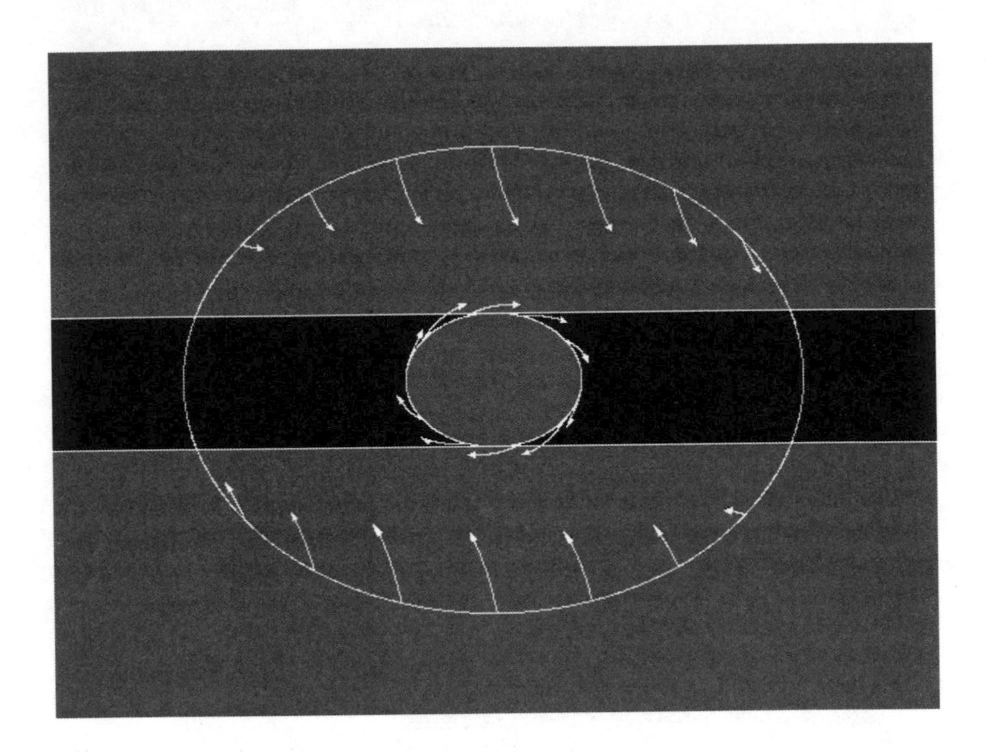

Fig. 8.1. The punctuated elliptic annulus, and the strip of energy gain.

the maximum absolute value of the terms containing the first derivative of income as:

$$K = \frac{2}{3} \frac{(v-1-s)^{3/2}}{v^{1/2}} \tag{8.11}$$

so that the original differential equation implies:

$$\left| \ddot{Y} + sY \right| < K \tag{8.12}$$

If thus the absolute value of Y becomes very large, we conclude that so does its second derivative, which takes the opposite sign. With both the variable and its second derivative very large we can regard the constant K as vanishing, and so the differential equation $\ddot{Y} + sY = 0$ would integrate to $\dot{Y}^2 + sY^2 = E$.

For very large elliptic boundaries the trajectory actually follows the boundary. Accordingly, it becomes more and more unlikely that a trajectory can escape the larger the outer ellipse is. The geometry of the present discussion is illustrated in Figure 8.1.

On the other hand the system only has one equilibrium point at the origin of phase space, excluded by the inner boundary of the elliptic annulus, and so, provided we settle the outer boundary problem properly, the conditions for a periodic solution required by the Poincaré-Bendixon theorem would be satisfied. To settle that problem mathematically properly brings us too far in mathematical detail and fails to be instructive any longer.

There is also another possibility for an existence proof. By differentiating our original equation (8.5) once more and setting $X = \dot{Y}$ we can write it:

$$\ddot{X} + sX - \left(\left(v - 1 - s\right) - vX^2\right)\dot{X} = 0 \tag{8.13}$$

which is the thoroughly investigated van der Pol equation. It is of the Liénard type and Jordan and Smith in 1977 produced a proof for the existence of a unique limit cycle for it. There even exists a neat graphical method of constructing the solution. See Hayashi. What remains, knowing that the derivatives of income change periodically, is to show that income itself does that too, but this is fairly trivial.

8.5 Limit Cycles: Asymptotic Approximation

For the approximation of a limit cycle to the model (8.5) the perturbation methods are most suitable. See Stoker, Hayashi, and Jordan and Smith.

It is unavoidable that the formulas become somewhat complex. As our purpose is to illustrate the use of the method it seems reasonable to simplify the basic equation we deal with by choosing specific parameter values to this end. Thus we first change the time scale by a factor $\sqrt{s}$ to remove the constant in the left hand side. This in substance amounts to choosing a unit of time that makes the period of the cycle produced if the right hand side were zero equal to 2π. If we next choose $s = 1$ and $v = 3$, we get the special case:

$$\ddot{Y} + Y = \dot{Y} - \dot{Y}^3 \tag{8.14}$$

The rate of saving may seem a little exaggerated, but we recall that the investment function was chosen to illustrate qualitative properties only, and we could easily imagine different coefficients of its linear and cubic terms that would require more reasonable rates of saving and still yield the simplified expression we desire. Anyhow, this is no major point.

The basic method still is one suggested by Poincaré and perfected by Lindstedt. In order to apply the method we pretend that the right hand side, containing the nonlinearity, is small. To this end we multiply it by a small parameter ε, obtaining:

$$\ddot{Y} + Y = \varepsilon\left(\dot{Y} - \dot{Y}^3\right) \tag{8.15}$$

We try to find a solution in the form of a power series in ε:

$$Y(\tau) = Y_0(\tau) + \varepsilon Y_1(\tau) + \varepsilon^2 Y_2(\tau) + \ldots \tag{8.16}$$

where

$$\tau = \left(1 + \varepsilon\omega_1 + \varepsilon^2\omega_2 + \ldots\right)t \tag{8.17}$$

is a new time scale. The attempted solution is now substituted into the differential equation, which itself becomes a power series in ε. According to the suggestion by Poincaré we require that the differential equation holds for each power separately, i.e., we equate each of the coefficients to zero. This results in a series of differential equations of increasing complexity, the first ones of which are:

$$\ddot{Y}_0 + Y_0 = 0 \tag{8.18}$$

$$\ddot{Y}_1 + Y_1 = -2\omega_1\ddot{Y}_0 - \dot{Y}_0^3 \tag{8.19}$$

and

$$\ddot{Y}_2 + Y_2 = -2\omega_1\ddot{Y}_1 - \left(\omega_1^2 + 2\omega_2\right)\ddot{Y}_0 + \dot{Y}_1 + \omega_1\dot{Y}_0 - 3\dot{Y}_0^2\dot{Y}_1 - 3\omega_1\dot{Y}_0^3 \tag{8.20}$$

We note that these equations can be solved in sequence. In each equation the right hand side only contains solutions to equations higher up in the list. We expect to get a sequence of solutions that approximate the real solution to a better degree the more terms we retain. Formally, the series solution only works when $\varepsilon \ll 1$, but even when the smallness parameter is as large as unitary, as it is in our case, the series may still converge quite rapidly. In fact it is seen to do.

We have to settle the issue of initial conditions. As the system is autonomous the choice of time origin is arbitrary and we can assume:

$$Y(0) = A_0 + \varepsilon A_1 + \varepsilon^2 A_2 + \dots \tag{8.21}$$

and

$$\dot{Y}(0) = 0 \tag{8.22}$$

When the solution is periodic these initial conditions imply that:

$$Y_i(0) = A_i \tag{8.23}$$

and

$$\dot{Y}_i(0) = 0 \tag{8.24}$$

We can now proceed to solving the equations. Starting with equation (8.18) at the top of the list we immediately see that Y_0 has a solution in terms of pure sine and cosine terms. The initial conditions imply that only the latter should be retained. Thus:

$$Y_0 = A_0 \cos \tau \tag{8.25}$$

where the amplitude is undetermined for the moment. We next note that, given (8.25), we obtain the expressions $\dot{Y}_0 = -A_0 \sin \tau$, $\ddot{Y}_0 = -A_0 \cos \tau$, and $\dot{Y}_0^3 = \frac{3}{4} \sin \tau - \frac{1}{4} \sin 3\tau$.

These expressions are inserted in the right hand side of equation (8.19), next on the list, which thus becomes:

$$\ddot{Y}_1 + Y_1 = 2\omega_1 A_0 \cos\tau - A_0\left(1 - \frac{3}{4}A_0^2\right)\sin\tau - \frac{1}{4}A_0^3 \sin 3\tau \tag{8.26}$$

This equation is linear and easy to solve by several standard methods. We note that terms of the form $\cos\tau$ and $\sin\tau$ in the right hand side of (8.26) would give rise to solution terms of the type $\tau\sin\tau$ and $\tau\cos\tau$. These terms violate the assumption that the solution is periodic as they grow beyond any limit with increasing time, and were called "secular" by Poincaré. Lindstedt's solution was to put the coefficients of any terms that would lead to secular solution terms equal to zero. So we have to require.

$$2\omega_1 A_0 = 0 \tag{8.27}$$

and

$$A_0\left(1 - \frac{3}{4}A_0^2\right) = 0 \tag{8.28}$$

There are two possibilities. One is that $A_0 = 0$, which is not interesting as it means that the amplitude of the periodic solution is zero. The other is that $\omega_1 = 0$ and $A_0 = 2/\sqrt{3}$.

Thus, the amplitude of our first approximation is fixed by the elimination of the secular terms. In the same way we learn that the period is not affected to the next approximation. The differential equation (8.26) that we are presently dealing with is thus simplified to:

$$\ddot{Y}_1 + Y_1 = -\frac{2}{3\sqrt{3}}\sin 3\tau \tag{8.29}$$

and it is easily solved. The general solution reads:

$$Y_1 = A_1\cos\tau + B_1\sin\tau + \frac{1}{12\sqrt{3}}\sin 3\tau \tag{8.30}$$

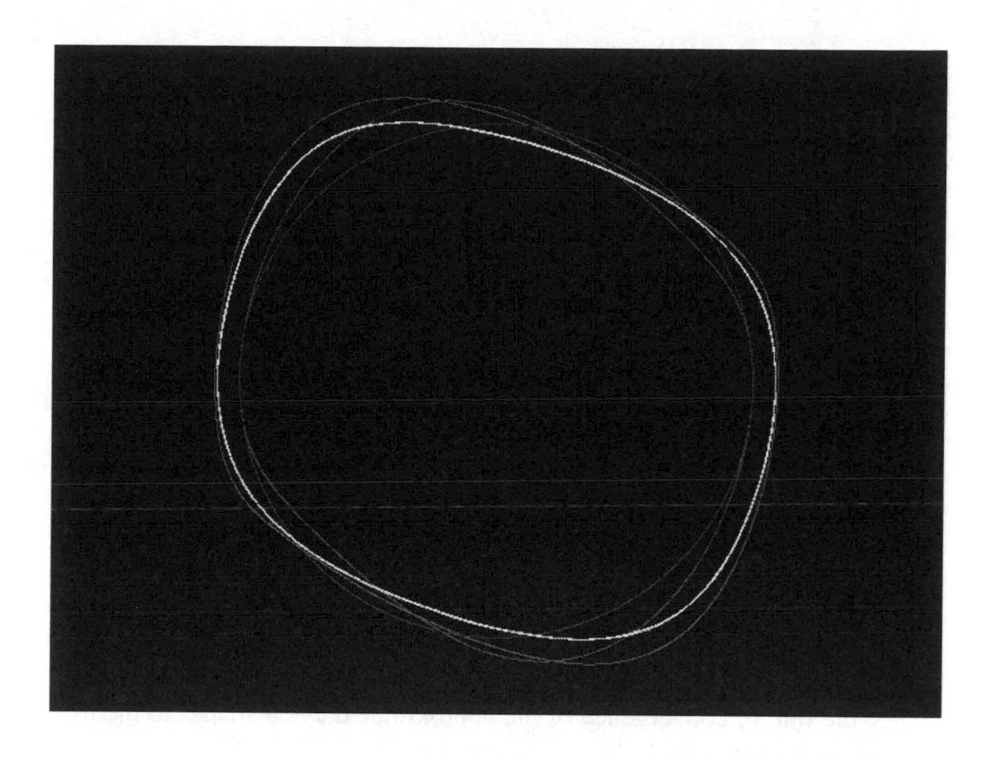

Fig. 8.2. Three successive perturbation series compared with simulated solution.

The coefficient for the sine term is again fixed from the initial conditions. We get $B_1 = -1/(4\sqrt{3})$. To determine the coefficient of the cosine term we again have to wait to the elimination of the secular terms in dealing with the next equation (8.20). Completing this we get $A_1 = 0$ and $\omega_2 = -1/16$.

The solution now follows the same procedure. From previous solutions we calculate the derivatives in the right hand sides of the differential equations, using the standard trigonometric formulas to express the powers of sines and cosines as sums of sines and cosines of the basic frequency and its harmonics. We eliminate all secular terms, thus fixing coefficients and obtaining new approximations to the period. The solving procedure always concerns linear differential equations, and the initial conditions serve to determine the remaining coefficients.

To get an impression of the rate of convergence to the true solution we state the first three approximations with numerical coefficients:

$$Y_0 = 1.155\cos\tau \tag{8.31}$$

$$Y_1 = -0.144\sin\tau + 0.048\sin 3\tau \tag{8.32}$$

$$Y_2 = 0.081\cos\tau + 0.036\cos 3\tau - 0.005\cos 5\tau \tag{8.33}$$

To this approximation the period is adjusted by $\tau = (15/16)t$. We see that the successive terms have rapidly decreasing amplitudes, so that the series converges when we just add these together, as we have to do with $\varepsilon = 1$. Of course, the convergence would be much faster with powers of a really small ε as weights.

In Fig. 8.2 we display the three first approximate solutions as compared with the solution obtained by simulation with the four point Runge-Kutta method. We see that, despite the apparently rapid convergence of the coefficients, the rate of convergence of the approximated cycle shapes to the true one is not so impressively fast.

8.6 Limit Cycles: Transients and Stability

The Poincaré-Lindstedt method can serve to calculate the limit cycle, but it tells us nothing about the rate of approach to it. To know something about stability it is interesting to find the way of approach to the orbits. For this reason another standard perturbation method, the so called two-timing, can be used.

We recall that, ε being small, the product $T = \varepsilon t$ is a new time-like variable proceeding slower than the normal time. The method proposes to treat fast time t and slow time T as different independent variables. This may seem somewhat absurd in our case where in reality $\varepsilon = 1$, and so the time scales are the same. But we still pretend that the smallness parameter is small. The only harm that can arise is that the approximations fail to converge as fast as they would otherwise. Again we try a power series solution:

$$Y(t,T) = Y_0(t,T) + \varepsilon Y_1(t,T) + \varepsilon^2 Y_2(t,T) + \ldots \tag{8.34}$$

and insert it in the differential equation:

$$\ddot{Y} + Y = \varepsilon\left(\dot{Y} - \dot{Y}^3\right) \tag{8.35}$$

Differentiating with respect to time, as required in the equation we want to solve, results in derivatives both with respect to fast and slow time. In the last case we get the factor ε from the definition by the chain rule. Formally:

$$\frac{d}{dt} = \frac{\partial}{\partial t} + \varepsilon \frac{\partial}{\partial T} \tag{8.36}$$

Inserting the attempted series solution (8.34) along with the composite differential operator (8.36) into the differential equation (8.35) and assembling the coefficients of different powers of ε , we again realize that the differential equation must vanish for each power separately. We thus obtain a new sequence of differential equations, of which we record the first two:

$$\ddot{Y}_0 + Y_0 = 0 \tag{8.37}$$

and

$$\ddot{Y}_1 + Y_1 = -2\frac{\partial^2 Y_0}{\partial t \partial T} + \frac{\partial Y_0}{\partial t} - \left(\frac{\partial Y_0}{\partial t}\right)^3 \tag{8.38}$$

The solution of the equation (8.37) is:

$$Y_0 = A(T)\cos t + B(T)\sin t \tag{8.39}$$

where we have taken into consideration that the amplitudes may not be constant but vary with slow time. This solution is inserted into the second equation which becomes:

$$\ddot{Y}_1 + Y_1 = \tag{8.40}$$

$$+\left(2\frac{dA}{dT} + \frac{3}{4}A(A+B) - A\right)\sin t - \frac{1}{4}\left(A^3 - 3AB^2\right)\sin 3t$$

$$-\left(2\frac{dB}{dT} + \frac{3}{4}B(A+B) - B\right)\cos t - \frac{1}{4}\left(B^3 - 3A^2B\right)\cos 3t$$

Again we must equate the coefficients of the sine and cosine expressions to zero in order to avoid secular terms in the solution. We note that this now results in two differential, not algebraic, equations:

$$\frac{dA}{dT} = \frac{1}{2}A - \frac{3}{8}A(A+B) \tag{8.41}$$

and

$$\frac{dB}{dT} = \frac{1}{2}B - \frac{3}{8}B(A+B) \tag{8.42}$$

To solve these differential equations (8.41)-(8.42) it is most convenient to recast the problem in polar coordinates, defined by:

$$A = \rho\cos\phi \tag{8.43}$$

and

$$B = \rho\sin\phi \tag{8.44}$$

which transforms the differential equations (8.41)-(8.42) for the amplitudes A and B into a pair of uncoupled equations in the polar coordinates that is easy to solve:

$$\frac{d\rho}{dT} = \frac{1}{2}\rho - \frac{3}{8}\rho^3 \tag{8.45}$$

and

$$\frac{d\phi}{dT} = 0 \tag{8.46}$$

The last equation just renders a constant phase angle, so in order to simplify we can assume it to be zero. The first equation is of the Bernoulli type and can be readily solved in closed form. As a result we get the first approximation:

$$Y_0 = \frac{2}{\sqrt{3}} \frac{\cos t}{\sqrt{1-(1-K)e^{-t}}} \tag{8.47}$$

where K is an integration constant determined by the initial conditions, and where we have used the fact that $t = T$.

We see that as T goes to infinity the derived solution goes to the first approximation:

$$Y_0 = \frac{2}{\sqrt{3}} \cos t \tag{8.48}$$

obtained by the previously tried Poincaré-Lindstedt method. It is worthwhile to note that the new two-timing method provides a check on the stability of the limit cycle as well. The transient motions approaching the limit cycle are shown in Fig. 8.3.

We have actually gone less far than by the first method, as we only used the second equation to eliminate the secular terms in the first one, but we have not yet attempted solving itself. If we try we find that the computations become increasingly messy much faster than in the previous method. This is the price we have to pay for obtaining a first estimate of how the system approaches the limit cycle. If we continue we, however, again get ever better estimates of the approaching process at the same time as we get better estimates of the final cyclic movement. Examples of the computation of several terms by the two-timing method can be found in Kevorkian and Cole.

Fig. 8.3. Stability and approach of trajectories to limit cycle.

The perturbation methods, both the Poincaré-Lindstedt and two-timing, as well as other closely related types, such as averaging, were mostly useful for computational purposes before the advent of the computer. But they also were important in getting systematic information of systems behaving close to the chaotic state. Stoker and Hayashi in fact demonstrate which enormous volume of information can be collected through these methods. Routines for solving differential equations both by Poincaré-Lindstedt and by multiple time scales (even more than two) can be handled by the symbolic mathematics software Maple V. In particular the detailed worksheets prepared by Heck to the second edition of his excellent book, and available from his website, can be recommended.

It is curious to note that, to my knowledge, perturbation methods were never used by economists for the study of dynamical systems, though they were developed in the haydays of economic business cycle modelling in the 50es and 60es.

We will continue to use these methods in the context of coupled and forced systems throughout this Chapter.

8.7 The Two-Region Model

In second order processes in continuous time limit cycles, apart from equilibria, are the most exciting thing we can get across. For this reason, in order to see something more of interesting dynamical phenomena, we have to raise the order of the process. This end is very easily achieved by coupling several regions by interregional trade. Two coupled regions already, with a second order process in each, are sufficient to raise the total system to order four, which already, at least in principle, admits chaotic outcomes.

The complete coupled system becoming very complex, we also study a third order system where the influence through trade is in one direction only. This approximates the case of the small open economy versus the world market, and mathcmatically it renders a forced oscillator, rather than coupled oscillators. More is known about the forced systems as they have been studied for a long time, some even very closely resemblant of the Hicks business cycle model.

The coupling can be achieved through a linear trade multiplier, the nonlinearity needed being there already in terms of the investment function.

To identify the regions we obviously have to index all the variables. We will use i or j as identification index, ranging through 1, 2. Whenever both indices are used in one single equation it is implied that they refer to different regions.

Let us so restate equations (8.1) and (8.2) for convenience with the variables properly indexed.

$$\dot{Y}_i = I_i - s_i Y_i \tag{8.49}$$

and

$$\dot{I}_i = v_i \dot{Y}_i - I_i \tag{8.50}$$

with i ranging through 1,2.

Suppose now that we have a linear propensity m_i to import. Thus export surplus, i.e., exports minus imports, for the i:th region would be:

$$X_i - M_i = m_j Y_j - m_i Y_i \tag{8.51}$$

where we note that index j denotes the other region. Exports thus are proportional to income in the other region than the region considered, imports to income in itself. This is the simple idea.

But as we consider an adaptive process, we have to reformulate the equation (8.51) for export surplus appropriately in dynamic form:

$$\dot{X}_i - \dot{M}_i = m_j Y_j - m_i Y_i - X_i + M_i \tag{8.52}$$

Now, for the open economy it is no longer true that savings equal investments, they equal investments plus export surplus. Formally,

$$I_i + X_i - M_i = S_i = s_i Y_i \tag{8.53}$$

This implies that we have to reformulate the corresponding dynamic equation (8.49) to take account of interregional trade:

$$\dot{Y}_i = I_i + X_i - M_i - s_i Y_i \tag{8.54}$$

We now, as in the one region case, differentiate (8.54) once more, use (8.50) and (8.52) in the new second order equations to eliminate the time derivatives of investments, exports and imports, and (8.54) to eliminate these variables themselves. In this way we arrive at the following pair of differential equations in the income variables alone:

$$\ddot{Y}_i + (s_i + m_i) Y_i - m_j Y_j = (v_i - 1 - s_i)\dot{Y}_i - \frac{v_i}{3}\dot{Y}_i^3 \tag{8.55}$$

where i and j range through 1,2 and $i \neq j$.

8.8 The Persistence of Cycles

To get a first rough idea of how this pair of differential equations behaves we multiply them by $m_i \dot{Y}_i$ and add. In the resulting left hand side we identify the products $Y_i \dot{Y}_i$ and $\dot{Y}_i \ddot{Y}_i$ as half the derivatives of Y_i^2 and $\dot{Y}_i^2$ respectively.

Accordingly

$$\frac{1}{2}\frac{d}{dt}\left(m_1\left(\dot{Y}_1^2 + s_1 Y_1^2\right) + m_2\left(\dot{Y}_2^2 + s_2 Y_2^2\right) + \left(m_1 Y_1 - m_2 Y_2\right)^2\right) = \tag{8.56}$$

$$= m_1\left(v_1 - 1 - s_1\right)\dot{Y}_1^2 - \frac{m_1 v_1}{3}\dot{Y}_1^4$$

$$+ m_2\left(v_2 - 1 - s_2\right)\dot{Y}_2^2 - \frac{m_2 v_2}{3}\dot{Y}_2^4$$

We first note that the left hand parenthesis can never become negative, and that it is zero in only one case, when $Y_i = \dot{Y}_i = 0$.

In passing we note that the stationary zero incomes refer to the homogeneous model without autonomous expenditures. For linear models, as mentioned above, it is well known that even when there are autonomous expenditures the homogeneous model makes sense, as referring to the deviations from stationary equilibrium. The present model is nonlinear, but, the nonlinearity being confined to the $\dot{Y}_i$ terms, the superposition principle still holds.

By analogy to mechanical systems the left hand parenthesis represents the "energy" of the system. The rate of change squared $\dot{Y}_i^2$ refers to kinetic energy, whereas potential energy has two components. One is the deviation from equilibrium squared Y_i^2, the other is the measure of spatial inhomogeneity, net export surplus squared $\left(m_1 Y_1 - m_2 Y_2\right)^2$.

Accordingly, the right hand side tells us when the system is gaining or losing energy. Suppose that

$$v_i > 1 + s_i \tag{8.57}$$

is true for both regions. Also suppose that the system is approaching the zero equilibrium state. As then the quadratic terms in the right hand side of (8.56) dominate over the quartic, and as their coefficients are positive according to (8.57) the system is gaining energy.

We conclude that motion once present in the system can never die out whenever (8.57) holds.

On the other hand, the quartic terms in the right hand side of (8.56) certainly dominate for large deviations from equilibrium, and, their coefficients being always negative, there is damping.

The suspicion is that the system preserves a limited cyclic motion forever. If not, the only remaining possibility would be that (8.56) transforms kinetic energy into potential energy. This however does not seem to be a plausible option, because, as we will see, there is no stable equilibrium state for any constant Y_i at all.

As mentioned, Goodwin in 1951 was able to demonstrate that a one region model of this kind possessed a limit cycle. Our present task is to analyse the coupled model. As we will see there still remains the possibility of cyclic motion under frequency locking in the two coupled regions. Another possibility is quasiperiodic motion where the basic frequency is constantly changing and where the trajectory in each state space fills a set of nonzero area measure. These possibilities we can analyse by classical perturbation methods, and this will be done in the sequel.

Below we illustrate cases of frequency locking and quasiperiodicity in Figs. 8.4 and 8.5, where the phase diagrams and so called Lissajou diagrams for the variables and their derivatives have been arranged in the different quadrants. It is amusing to note that already Huygens noted frequency locking experimentally by observing synchronization in two pendulum clocks hanging on the same wall and ascribing this phenomenon to the weak coupling due to echoes in the wall.

A final possibility is chaotic motion, but the treatment of it violates the limits of elementary analysis, and we will have to simplify further in order to deal with that case.

8.9 Perturbation Analysis of the Coupled Model

For the purpose of further analysis we pretend that the nonlinearity is small and introduce a smallness parameter ε in the right hand sides of (8.55). This makes it possible to study the nonlinear equations as a departure from the corresponding linear ones. It has been demonstrated that the perturbation methods from a qualitative point of view yield good results even when it is not true that $\varepsilon << 1$. For details the reader is referred to Stoker and Hayashi, the classics of perturbation methods. Hayashi in particular checked the cases with medium and large nonlinearities by extensive simulation on analog computers.

With the smallness parameter included (8.55) reads:

$$\ddot{Y}_i + (s_i + m_i)Y_i - m_j Y_j = \varepsilon\left((v_i - 1 - s_i)\dot{Y}_i - \frac{v_i}{3}\dot{Y}_i^3 \right) \tag{8.58}$$

To this pair of differential equations we try to find a periodic solution of a common period ω:

$$Y_i = A_i(t)\cos\omega t + B_i(t)\sin\omega t \tag{8.59}$$

The coefficients A_i and B_i are assumed to be slowly varying functions of time. This assumption of slow variation along with the smallness of the parameter ε is the basis of a set of approximations. These imply that all second time derivatives of the amplitudes A_i and B_i are assumed to vanish, as are all powers and products of their first time derivatives (with each other and also with the small parameter ε).

We differentiate the attempted solution (8.59) to obtain $\dot{Y}_i$ and $\ddot{Y}_i$ and substitute into (8.58) while observing which terms vanish due to the approximations just made.

In order that (8.59) should make (8.58) hold as an identity the coefficients of the terms $\cos\omega t$ and $\sin\omega t$ must vanish individually. The trigonometric functions of period 3ω are at present of no interest, because, unlike the case of the one region model, we are presently only concerned with determining the basic frequency of oscillation.

Equating all the coefficients mentioned to zero, however, renders a pair of differential equations for the slowly varying amplitudes

$$2\dot{A}_i = \left((v_i - 1 - s_i) - \frac{v_i}{4}(A_i^2 + B_i^2)\omega^2 \right) A_i \tag{8.60}$$

$$+ \frac{s_i + m_i - \omega^2}{\omega} B_i - \frac{m_j}{\omega} B_j$$

and

$$2\dot{B}_i = \left((v_i - 1 - s_i) - \frac{v_i}{4}(A_i^2 + B_i^2)\omega^2 \right) B_i \tag{8.61}$$

$$- \frac{s_i + m_i - \omega^2}{\omega} A_i + \frac{m_j}{\omega} A_j$$

We note that whenever the system (8.60)-(8.61) has a fixed point then the solution to (8.58) has a limit cycle with a common frequency for both regions. When (8.60)-(8.61) itself has a limit cycle then (8.58) has a quasiperiodic solution where the trajectory in each phase space covers a region of nonzero area (of course assumed that the new frequency is not a rational multiple of ω which would be an unlikely occurrence).

The system (8.60)-(8.61) is best studied after some variable transformations that are standard in perturbation theory. First we set a new time scale $\tau = t/2$ to get rid of the multipliers in the left hand sides. Further we define:

$$\kappa_i = v_i - 1 - s_i \tag{8.62}$$

$$\alpha_i = \frac{1}{2}\sqrt{v_i}\,\omega A_i \tag{8.63}$$

$$\beta_i = \frac{1}{2}\sqrt{v_i}\,\omega B_i \tag{8.64}$$

$$\rho_i = \alpha_i^2 + \beta_i^2 \tag{8.65}$$

$$\sigma_i = \frac{s_i + m_i - \omega^2}{\omega} \tag{8.66}$$

$$\mu_i = \sqrt{\frac{v_j}{v_i} \frac{m_i}{\omega}} \tag{8.67}$$

Note that the κ_i are structurally determined constants. Later we shall see that they are the squared natural amplitudes of the uncoupled system. This by the way differs basically from the linear multiplier-accelerator model where the amplitudes are not determined by the system. The μ_i are directly proportionate to the import propensities and hence represent the degree of coupling in the system. The σ_i, usually called "detuning coefficients", depend on how much the interlocking frequency ω deviates from the natural frequency $\sqrt{s_i + m_i}$ of each uncoupled oscillator.

Finally the α_i and β_i are simple linear transformations of the A_i and B_i respectively. Accordingly ρ_i are the squared amplitudes of these new pairs of variables. In the uncoupled case we will see that $\rho_i = \kappa_i$. With these definitions (8.60)-(8.61) become much more handy:

$$\dot{\alpha}_i = (\kappa_i - \rho_i)\alpha_i + \sigma_i \beta_i - \mu_j \beta_j \tag{8.68}$$

$$\dot{\beta}_i = (\kappa_i - \rho_i)\beta_i - \sigma_i \alpha_i + \mu_j \alpha_j \tag{8.69}$$

8.10 The Unstable Zero Equilibrium

A study of this system naturally starts with its fixed points. As already mentioned the fixed points correspond to various cases of frequency locking and purely cyclic motion in both regions. One fixed point is trivial. As the system (8.68)-(8.69) is homogeneous, $\alpha_i = \beta_i = 0$ is an obvious fixed point. But is it stable? To answer this question we linearise (8.68)-(8.69) around any assumed equilibrium, defining new variables ξ_i and η_i as deviations from the equilibrium values of α_i and β_i respectively, thus obtaining:

$$\dot{\xi}_i = \left(\kappa_i - 3\alpha_i^2 - \beta_i^2\right)\xi_i - \left(2\alpha_i\beta_i - \sigma_i\right)\eta_i - \mu_j\eta_j \tag{8.70}$$

$$\dot{\eta}_i = \left(\kappa_i - 3\beta_i^2 - \alpha_i^2\right)\eta_i - \left(2\alpha_i\beta_i + \sigma_i\right)\xi_i + \mu_j\xi_j \tag{8.71}$$

These general expressions will be useful later on as they refer to any linearisation around a fixed point. For the present case where the fixed point is at the origin we deal with a much simpler particular case where (8.70)-(8.71) become

$$\dot{\xi}_i = \kappa_i\xi_i + \sigma_i\eta_i - \mu_j\eta_j \tag{8.72}$$

$$\dot{\eta}_i = \kappa_i\eta_i - \sigma_i\xi_i + \mu_j\xi_j \tag{8.73}$$

We could calculate the eigenvalues of these linear equations to decide on stability, but we can do better than that. It is easy to find a Lyapunov function by multiplying (8.72) by $-(-1)^i\mu_i\xi_i$ and (8.73) by $(-1)^i\mu_i\eta_i$ and adding. Thus

$$\frac{1}{2}\frac{d}{dt}L = L \tag{8.74}$$

where

$$L = \mu_1\left(\xi_1^2 + \eta_1^2\right) - \mu_2\left(\xi_2^2 + \eta_2^2\right) \tag{8.75}$$

This saddle dynamic obviously displays instability. Any weighted amplitude that dominates initially will be amplified at an exponential rate. The result is, of course, only local around the origin where the linearisation holds with good approximation.

It should be noted at this stage that the unstable zero solution $\alpha_i = \beta_i = 0$ refers to all fixed points of the original system (8.58), not just the case $Y_i = 0$.

According to (8.59) any fixed point requires $\omega = 0$. Thus, from (8.63)-(8.64), $\alpha_i = \beta_i = 0$.

What happens at very large distances from the origin in phase space can be seen easily. From (8.68)-(8.69) we see directly that at large amplitudes the third order terms dominate. Thus, using (8.65), we find that with large α_i, β_i:

$$\dot{\alpha}_i = -\rho_i \alpha_i \tag{8.76}$$

$$\dot{\beta}_i = -\rho_i \beta_i \tag{8.77}$$

At large amplitudes the system is damped almost radially inward to the origin, whereas the equilibrium there is itself unstable.

8.11 Other Fixed Points

This again arouses a suspicion that there is a limit cycle, this time for the slowly varying amplitudes in the solution (8.59). This would result in a quasiperiodic solution for (8.58), where the trajectories cover an area of nonzero area. In terms of economics the solution would be "cyclic", but no cycle would be repeated exactly. Rather phase and amplitude would be slowly changing all the time. Fig. 8.4 illustrates the case of quasiperiodic motion for two coupled oscillators. The phase diagrams for both regions are displayed in the second and fourth quadrants with Lissajou figures in the first and third showing the covariation of the two oscillators.

This, however, is only one possibility. The discussion above departed from the tacit assumption that there were no other fixed points for (8.68)-(8.69) than the unstable one at the origin. This is, of course, not necessary, and we are going to find out which other fixed points there may be in the system. We are also going to discuss the stability of such fixed points, because they can put the system into frequency locked cyclic motion only when the corresponding fixed points are stable.

Let us thus see what conditions we have for the fixed points. First we obviously have to put (8.68)-(8.69) equal to zero. Next, we see from (8.65) how the squared amplitude ρ_i is related to the variables α_i, β_i. Finally equations (8.66) tell us how the detuning constants σ_i are related to the unknown

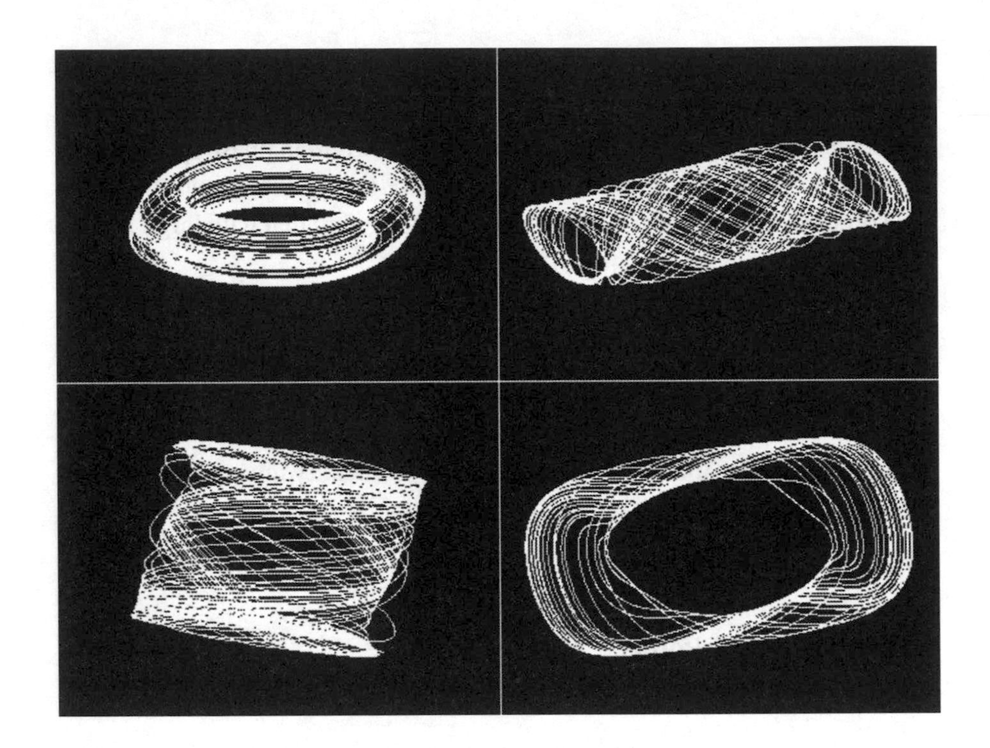

Fig.8.4. Quasiperiodic motion in two coupled oscillators.

interlocking period ω. Since the region index ranges through two values each equation counts twice. So we have in all discussed eight equations. The unknowns are $\alpha_i, \beta_i, \rho_i, \sigma_i$, and ω, four of them counted twice plus one, i.e. nine in all. Accordingly the unknowns exceed the equations by one.

What else do we have? The κ_i are completely determined by the structural coefficients of the model, as we can see from (8.62) and so they can be treated as constants. The only remaining variables are the μ_i. As we see from (8.67) they are completely determined by the structural coefficients, except for the reciprocal dependence on the unknown interlocking period. We can accordingly regard them as variables, but then we add just as many variables as we add equations. Still there is one equation missing.

We can, however, derive the missing equation simply by considering some elementary algebra. The set of algebraic equations, (8.68)-(8.69) put equal to zero, can be regarded as a linear system in α_i, β_i, provided we for the moment regard the ρ_i as constants along with the κ_i and the σ_i. We know they are not. In fact (8.65) shows that the system (8.68)-(8.69) is third degree in α_i, β_i. But the rules of algebra for a homogeneous linear system to possess a nonzero solution still hold for our purely formal set of four "linear" equations.

We already dealt with the zero solution, and now we are interested in the nonzero solutions. Accordingly we have to equate the determinant of the linear system to zero, and this furnishes the missing equation. Defining

$$M = \begin{bmatrix} \kappa_1 - \rho_1 & \sigma_1 & 0 & -\mu_2 \\ -\sigma_1 & \kappa_1 - \rho_1 & \mu_2 & 0 \\ 0 & -\mu_1 & \kappa_2 - \rho_2 & \sigma_2 \\ \mu_1 & 0 & -\sigma_2 & \kappa_2 - \rho_2 \end{bmatrix} \tag{8.78}$$

we have Det(M)=0, or, expanded

$$\left((\kappa_1 - \rho_1)^2 + \sigma_1^2\right)\left((\kappa_2 - \rho_2)^2 + \sigma_2^2\right) + \mu_1^2\mu_2^2 \tag{8.79}$$

$$+ 2\mu_1\mu_2(\kappa_1 - \rho_1)(\kappa_2 - \rho_2) - 2\mu_1\mu_2\sigma_1\sigma_2 = 0$$

We note that in this equation α_i, β_i are no longer present. We are going to manipulate the rest of the equations as well to such a form where these variables only enter by their squared amplitude ρ_i.

In equations (8.68)-(8.69) put equal to zero we move the j-indexed variables to the other side of the equality sign, square, and add, thus obtaining:

$$\left((\kappa_i - \rho_i)^2 + \sigma_i^2\right)\rho_i = \mu_j^2\rho_j \tag{8.80}$$

Multiplying together the two equations (8.80) and cancelling ρ_1, ρ_2, which we now want to be nonzero, in both sides we get:

$$\left((\kappa_1 - \rho_1)^2 + \sigma_1^2\right)\left((\kappa_2 - \rho_2)^2 + \sigma_2^2\right) = \mu_1^2\mu_2^2 \tag{8.81}$$

Subtracting (8.81) from (8.79) and dividing by $2\mu_1\mu_2\sigma_1\sigma_2$ next yields

$$\left(\frac{\kappa_1-\rho_1}{\sigma_1}\right)\left(\frac{\kappa_2-\rho_2}{\sigma_2}\right)=1-\frac{\mu_1\mu_2}{\sigma_1\sigma_2} \tag{8.82}$$

which substituted in (8.81) yields

$$\frac{(\kappa_1-\rho_1)^2}{\sigma_1^2}+\frac{(\kappa_2-\rho_2)^2}{\sigma_2^2}=2\left(\frac{\mu_1\mu_2}{\sigma_1\sigma_2}-1\right) \tag{8.83}$$

From (8.82)-(8.83) we now readily obtain

$$\frac{\kappa_1-\rho_1}{\sigma_1}+\frac{\kappa_2-\rho_2}{\sigma_2}=0 \tag{8.84}$$

and

$$\frac{\kappa_1-\rho_1}{\sigma_1}-\frac{\kappa_2-\rho_2}{\sigma_2}=\pm\sqrt{\frac{\mu_1\mu_2}{\sigma_1\sigma_2}-1} \tag{8.85}$$

which shows how the ρ_i are determined once we know μ_i, σ_i. Even more explicitly we have from (8.84)-(8.85)

$$\rho_i=\kappa_i\pm(-1)^i\sigma_i\sqrt{\frac{\mu_1\mu_2}{\sigma_1\sigma_2}-1} \tag{8.86}$$

But, how do we determine the μ_i, σ_i, and the period ω on which they ultimately depend? To accomplish this we have to realize that in deriving (8.81) from (8.80) we did not exhaust all the information contained in this latter pair of equations. Let us move the terms of (8.84) to either side of the equality sign, take squares, and substitute into (8.80) which we have first divided through by $\sigma_i^2\rho_i$. Thus we obviously get:

$$\frac{\mu_2^2\rho_2}{\sigma_1^2\rho_1}=\frac{\mu_1^2\rho_1}{\sigma_2^2\rho_2} \tag{8.87}$$

8.12 Properties of Fixed Points

Let us stop here for a while to discuss some properties of fixed points. The left hand side of (8.83), being a sum of squares, is certainly nonnegative. Thus, we must have

$$\mu_1\mu_2 \geq \sigma_1\sigma_2 > 0 \tag{8.88}$$

and, according to (8.82),

$$(\kappa_1 - \rho_1)(\kappa_2 - \rho_2) < 0 \tag{8.89}$$

We conclude that the detuning coefficients have equal sign. According to the definitions (8.66) this means that the common frequency of the coupled oscillators is either smaller than the smallest or larger than the largest of the natural frequencies $\sqrt{s_i + m_i}$. The interlocked system will thus never oscillate at an intermediate frequency between the natural ones for the free oscillators.

Moreover, we see from (8.89) that one of the oscillators has an amplitude larger than the natural one in the uncoupled case, whereas the other has an amplitude smaller than the natural one. The case of frequency locking is illustrated in Fig. 8.5, drawn along the same principles as Fig. 8.4.

In terms of economics, coupling of two regions by trade speeds up or slows down the business cycle in both regions. At the same time it increases the size of the movements in one region but decreases it in the other. What we presently need is the fact that $\sigma_1\sigma_2$ have the same sign. This translates (8.87) into

$$\mu_1\sigma_1\rho_1 = \mu_2\sigma_2\rho_2 \tag{8.90}$$

The solution strategy is now clear. We substitute from (8.66)-(8.67) for σ_i, μ_i and from (8.86) for ρ_i into (8.90). The resulting equation, which is too complicated to give any direct information and therefore is not reproduced, involves the variable ω^2 alone. After squaring in order to get rid of the square roots originating in (8.86) we get a polynomial equation for the interlocking frequency. The coefficients of this polynomial are determined by the structural coefficients of the model, the accelerators and propensities

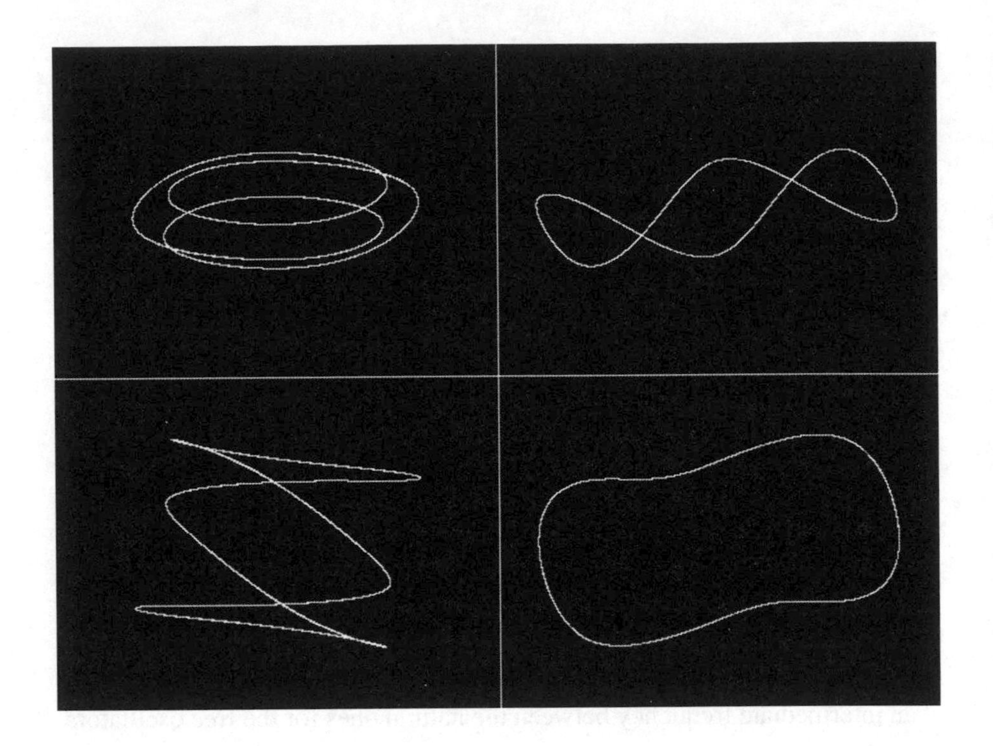

Fig. 8.5. Frequency locking in two coupled oscillators.

to save and to import. The polynomial equation can be solved on any microcomputer. The frequency can then be substituted in (8.66)-(8.67) and (8.86) to get the rest of the variables. To be quite exact we only get the squared amplitudes ρ_i. To obtain the α_i, β_i we have to return to (8.68)-(8.69) equated to zero and solve the resulting linear system. We will then find that there remains an arbitrary phase angle to be determined.

8.13 The Arbitrary Phase Angle

To see this let us define the phase angles

$$\theta_i = \arctan \frac{\beta_i}{\alpha_i} \tag{8.91}$$

With this definition along with (8.65) above we can readily transform the system to polar coordinates. Accordingly

$$\alpha_i = \sqrt{\rho_i}\cos\theta_i \tag{8.92}$$

$$\beta_i = \sqrt{\rho_i}\sin\theta_i \tag{8.93}$$

The root sign is due to the fact that we found it convenient to work with the squared amplitudes. Defining the phase difference:

$$\theta = \theta_1 - \theta_2 \tag{8.94}$$

we find, using some elementary trigonometric identities:

$$\alpha_1\alpha_2 + \beta_1\beta_2 = \sqrt{\rho_1\rho_2}\cos\theta \tag{8.95}$$

$$\alpha_1\beta_2 - \beta_1\alpha_2 = -\sqrt{\rho_1\rho_2}\sin\theta \tag{8.96}$$

The polar coordinates can now be used to put the system (8.68)-(8.69) in a different form. As only the phase difference is involved we find that the number of equations is reduced from four to three, i.e.

$$\frac{1}{2}\dot{\rho}_1 = (\kappa_1 - \rho_1)\rho_1 - \mu_2\sqrt{\rho_1\rho_2}\sin\theta \tag{8.97}$$

$$\frac{1}{2}\dot{\rho}_2 = (\kappa_2 - \rho_2)\rho_2 + \mu_1\sqrt{\rho_1\rho_2}\sin\theta \tag{8.98}$$

$$\dot{\theta} = -(\sigma_1 - \sigma_2) + \left(\mu_1\sqrt{\frac{\rho_1}{\rho_2}} - \mu_2\sqrt{\frac{\rho_2}{\rho_1}}\right)\cos\theta \tag{8.99}$$

The dynamic system in polar coordinates (8.97)-(8.99) is handy in particular to analyse drifting oscillations, but again it is difficult to get any closer information without considerable work at the computer.

We also see that (8.97)-(8.99) equated to zero yield a new solution strategy for the fixed points. From (8.66)-(8.67) we see that multiplication by ω transforms the difference $(\sigma_1 - \sigma_2)$ in (8.99) into $(s_1 + m_1 - s_2 - m_2)$, which is a constant. Likewise the μ_i are transformed to constants. Accordingly, (8.99) equated to zero is by this multiplication changed into an equation in ρ_1, ρ_2, θ only. It is true that (8.97)-(8.98) involve the variables μ_i, but these variables have a very simple reciprocal relation to the interlocking frequency. So, we can pretend that the μ_i are constants, solve the system for the three variables ρ_1, ρ_2, θ, calculate the interlocking frequency, and thus see what import propensities we assumed.

8.14 Stability of the Coupled Oscillators

Once we have found a fixed point by the solution procedure indicated above we have to check the stability of the system. To this end let us write down the stability matrix of the linearised system (8.70)-(8.71) as it is not so easy in the general case to proceed directly with Lyapunov functions. The matrix, denoted N, is:

$$\begin{bmatrix} \kappa_1 - \rho_1 - 2\alpha_1^2 & -2\alpha_1\beta_1 + \sigma_1 & 0 & -\mu_2 \\ -2\alpha_1\beta_1 - \sigma_1 & \kappa_1 - \rho_1 - 2\beta_1^2 & \mu_2 & 0 \\ 0 & -\mu_1 & \kappa_2 - \rho_2 - 2\alpha_2^2 & -2\alpha_2\beta_2 + \sigma_2 \\ \mu_1 & 0 & -2\alpha_2\beta_2 - \sigma_2 & \kappa_2 - \rho_2 - 2\beta_2^2 \end{bmatrix} \tag{8.100}$$

For stability we must compute the eigenvalues of

$$\operatorname{Det}(N - \lambda I) = 0 \tag{8.101}$$

and check that their real parts are negative. There is no simple rule in terms of the signs of the trace and determinant as there is for the two by two matrix. For the convenience of the reader we, however, give the polynomial expansion of (8.101) mainly in terms of the traces and determinants of the upper left and lower right two by two submatrices to (8.100) which is almost decomposed when the coupling by import propensities, the μ_i, is weak.

The polynomial equation is

$$\lambda^4 - (T_1 + T_2)\lambda^3 + (D_1 + D_2 + T_1 T_2 + 2\mu_1\mu_2)\lambda^2 \tag{8.102}$$

$$- (T_1 D_2 + T_2 D_1 + \mu_1\mu_2(T_1 + T_2))\lambda$$

$$+ (D_1 D_2 + \mu_1\mu_2 H + \mu_1^2\mu_2^2) = 0$$

For convenience let us also write down the traces T_i and determinants D_i of the two main submatrices from (8.100) and of what has been denoted H in (8.102):

$$T_i = 2(\kappa_i - 2\rho_i) \tag{8.103}$$

$$D_i = (\kappa_i - \rho_i)(\kappa_i - 3\rho_i) + \sigma_i^2 \tag{8.104}$$

$$H = 2((\kappa_1 - \rho_1)(\kappa_2 - \rho_2) - \rho_1\rho_2 \cos 2\theta - \sigma_1\sigma_2) \tag{8.105}$$

where θ is the angular difference in polar coordinates for the two oscillators, as defined in (8.94).

We see immediately that (8.102) factorizes into

$$(\lambda^2 - T_1\lambda + D_1)(\lambda^2 - T_2\lambda + D_2) = 0 \tag{8.106}$$

whenever one of the μ_i is zero. In those cases a negative trace and a positive determinant for each of the regions according to (8.103)-(8.104) gives necessary and sufficient conditions for stability. We see from (8.102) that if the coupling of the two regions is sufficiently weak then the conditions on traces and determinants still hold. It is well known from Stoker that in σ_i, ρ_i-space the trace is negative whenever the squared amplitude exceeds a certain value, whereas the determinant is positive whenever the value combination for the squared amplitude and the detuning coefficient lies outside an elliptic area. In any case we see from (8.103)-(8.104) that for the ρ_i sufficiently large both the traces and determinants have correct signs for stability, and with weak coupling this carries over to the general case.

8.15 The Forced Oscillator

As we have seen the procedure of actually finding a fixed point and checking its stability may be a formidable computational task. In order to get some insight in these matters we can choose a much simplified case where one of the import propensities is zero. In terms of economics it means that we consider the "small open economy" that is influenced by "the rest of the world" but itself has practically no influence on it.

Mathematically this is the case of the forced oscillator that has been studied since the early work by van der Pol, and that has been summed up by Stoker. Although the material is well known it has not dissipated to the literature in mathematical economics, and therefore it may be instructive to see how this case fits in the present more general framework.

We know that the results obtained when one import propensity is zero still hold qualitatively due to basic mathematical principles provided this propensity is sufficiently small.

8.16 The World Market

For definiteness let 1 denote the small open economy and 2 denote the rest of the world. Accordingly $\mu_1 = 0$. Then in equilibrium from (8.98) either $\rho_2 = 0$ or $\rho_2 = \kappa_2$. The first case is of no interest as it represents the unstable zero equilibrium. The other possibility is that the rest of the world, independently of the small open economy, oscillates at it own natural frequency and amplitude. To see the latter fact we substitute $\mu_1 = 0$ and $\rho_2 = \kappa_2$ into (8.80). Thus $\sigma_2 = 0$, and, according to (8.66) $\omega = \sqrt{s_2 + m_2}$.

As the amplitude is fixed, the coefficients α_2, β_2 according to (8.65) depend on the phase only, i.e., $\alpha_2 = \sqrt{\rho_2}\cos\theta_2$ and $\beta_2 = \sqrt{\rho_2}\sin\theta_2$.

Above we however saw that the phase angle was arbitrary and only the difference of phase angles mattered. We can accordingly choose any value for θ_2 we wish.

From (8.59) we have the solution:

$$Y_2 = A_2 \cos\omega t + B_2 \sin\omega t \tag{8.107}$$

where the coefficients from (8.63)-(8.64) are the constants:

$$A_2 = \frac{2}{\omega}\frac{v_2 - 1 - s_2}{\sqrt{v_2}}\cos\theta_2 \tag{8.108}$$

$$B_2 = \frac{2}{\omega}\frac{v_2 - 1 - s_2}{\sqrt{v_2}}\sin\theta_2 \tag{8.109}$$

and the period is fixed by

$$\omega = \sqrt{s_2 + m_2} \tag{8.110}$$

In a form where we can easily recognize how everything, except the arbitrary phase, is determined by the structural coefficients we write:

$$Y_2 = \frac{2}{\sqrt{s_2 + m_2}}\frac{v_2 - 1 - s_2}{\sqrt{v_2}}\cos\left(\sqrt{s_2 + m_2}\,t - \theta_2\right) \tag{8.111}$$

This gives us all information we need concerning the world market. According to our (approximate) solution it describes a simple periodic motion. If we pursue the perturbation approach by deriving more terms of the series we, as always, get a better approximation to the periodic motion of the true nonlinear oscillator, and see that it is by no means simple harmonic. But we leave these complications.

8.17 The Small Open Economy

What happens in the small open economy? Consider the remaining equation in the set (8.80), i.e. for $i = 1$. From (8.67) $\mu_2^2 = (v_1 / v_2) \cdot m_2 / (s_2 + m_2)$. Further we had $\rho_2 = \kappa_2$, and from (8.62) $\kappa_2 = v_2 - 1 - s_2$. Using $F^2 = \mu_2^2 \kappa_2$ as a new symbol for a constant that is completely determined by the structural coefficients, and deleting the index 1 for the small open economy we get:

$$\left((\kappa - \rho)^2 + \sigma^2\right)\rho = F^2 \tag{8.112}$$

We note that $\sigma^2 = \sigma_1^2 = (s_1 + m_1 - s_2 - m_2) / \sqrt{s_2 + m_2}$ according to (8.66). Thus σ^2 represents the "detuning", the difference of the natural and the driving frequencies, whereas F represents the driving amplitude. The only variable in equation (8.112) is ρ, with implicit index 1. This variable represents the amplitude of the driven oscillation. It is implicit that, after a transient period, the frequency is entrained at the driving frequency, if it settles at all.

8.18 Stability of the Forced Oscillator

We are now in a position to investigate the phenomenon of resonance according to (8.112) more closely. In this we can rely on the work by Stoker.

In Fig. 8.6 the amplitude ρ is shown as a function of the detuning coefficient σ. The family of response curves is obtained for different values of the parameter F^2. For high values of the parameter, i.e., the forcing amplitude there is just one amplitude for the driven oscillator to each detuning. With low forcing amplitudes the response curves are narrowed at the base so that there are three different response amplitudes to each detuning, and finally the curves become disjoint. When there are three different response amplitudes we can expect the middle one to be unstable. We should remember that dealing with equilibria for the amplitudes and their stability means dealing with limit cycles for the driven oscillator and their stability.

Stability of the present case can be fully analysed by simple means. The linearised equations around a supposed equilibrium point corresponding to (8.70)-(8.71) are now:

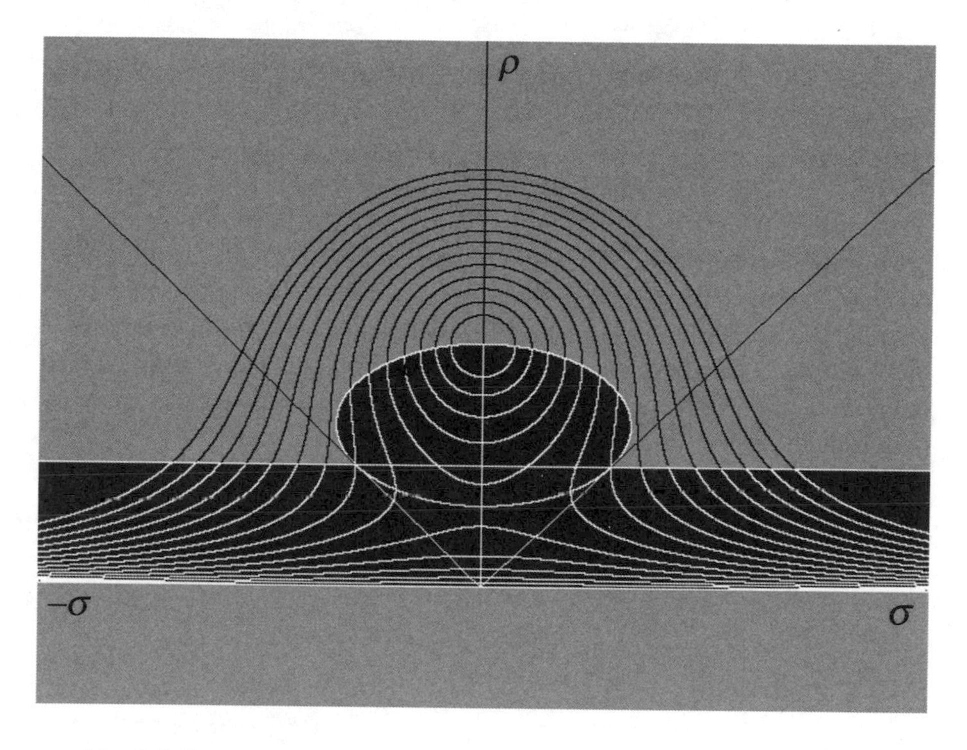

Fig. 8.6. Response curves and regions of stability for the forced oscillator.

$$\dot{\xi} = (\kappa - 3\alpha^2 - \beta^2)\xi - (2\alpha\beta - \sigma)\eta \tag{8.113}$$

$$\dot{\eta} = (\kappa - \alpha^2 - 3\beta^2)\eta - (2\alpha\beta + \sigma)\xi \tag{8.114}$$

where we have again deleted the indices and where the terms in (8.70)-(8.71) due to the driving world economy do not enter the linearisation as the world economy is on an equilibrium limit cycle. Due to Poincaré the equilibrium points can be completely characterized by the trace, the determinant and the discriminant of this linear system.

We have:

$$\mathrm{Tr} = 2(\kappa - 2\rho) \tag{8.115}$$

$$\mathrm{Det} = (\kappa - \rho)(\kappa - 3\rho) + \sigma^2 \tag{8.116}$$

and

$$\Delta = \mathrm{Tr}^2 - 4\,\mathrm{Det} = 4\left(\rho^2 - \sigma^2\right) \tag{8.117}$$

If the determinant is negative then the equilibrium is a saddle, otherwise it is a node or a spiral. A node or spiral is unstable if the trace is positive, stable if it is negative. Putting the determinant equal to zero we obviously get a second order equation which describes an ellipse drawn in Fig. 8.6. Inside this ellipse there are saddles, outside it nodes and spirals. From the formula for the trace we readily see that the demarcation line between stability and instability is the value 1/2. This is displayed as a horizontal straight line above which there is stability, below which there is instability. To be more exact, stability is outside the ellipse and above the line.

The discriminant put equal to zero defines two straight lines through the origin at a slope of + or - 45 degrees. Above the wedge defined by these lines there are nodes (or saddles), below it there are spirals.

8.19 Catastrophe

An interesting feature of the ellipse is that it passes through the points of the response curves having vertical tangents. Formally a vertical tangent means a boundary case between one and three response amplitudes. In fact (8.112) furnishes a third degree algebraic equation for the squared amplitude ρ. We compute its discriminant D:

$$27D = \sigma^2\left(1+\sigma^2\right)^2 - \left(1-9\sigma^2\right)F^2 + \frac{27}{4}F^4 \tag{8.118}$$

If the sign is negative there is only one real root to (8.112), if it is positive there are three. When (8.118) is zero the case is degenerate with three coincident roots. This is a signal of bifurcation. Eliminating ρ between the equilibrium condition (8.112) and the equation obtained by putting the determinant (8.116) of the linearised dynamical system equal to zero actually yields $D = 0$ in (8.118). As a zero determinant implies that one eigenvalue of the linearised system becomes zero the case is indeed associated with loss of stability.

Actually the whole set of response curves seen as level curves of a surface has the look of a catastrophe surface of the double cusp variety, and the relation between σ^2, F^2 furnished by putting (8.118) equal to zero is a typical bifurcation set in control space. We can thus expect that changes of forcing frequency (detuning) and/or amplitude cause changes bearing all the marks of catastrophe: sudden jumps, irreversibility, hysteresis and divergence. It should be recalled that all these phenomena apply to the limit cycles.

Even if the control parameters are fixed the trajectory may be quite complex. For sufficiently small detuning there are, as we saw, three different roots to (8.112), corresponding to three different equilibrium amplitudes of the driven limit cycle. The discussion on stability tells us that the smallest root is an unstable node or spiral, the intermediate one a saddle, and the largest root a stable node. If we start close to the unstable node, pass close to the saddle and finally approach the stable node, the system may seem to settle at limit cycles twice (as the amplitudes change slowly close to equilibrium points, stable or not) only to leave them and finally settle at the stable limit cycle. Computer simulations sometimes indicate this deceptive behaviour.

8.20 Period Doubling and Chaos

Let us now look back at the equations (8.55), more precisely the one for $i = 1$, representing the small open economy being treated as forced by the world market.

$$\ddot{Y}_1 + (s_1 + m_1)Y_i = (v_1 - 1 - s_1)\dot{Y}_1 - \frac{v_1}{3}\dot{Y}_1^3 + m_2 Y_2 \tag{8.119}$$

In (8.111) we however found a handy expression for Y_2. We just need to multiply through by m_2. Moreover, we put the phase lead equal to zero, use (8.110) for the constant driving frequency, and introduce K/ω as an abbreviation for the somewhat messy structural constant, the amplitude of the driving force. The reason for division by the frequency will very soon become apparent. This results in:

$$\ddot{Y}_1 + (s_1 + m_1)Y_i = (v_1 - 1 - s_1)\dot{Y}_1 - \frac{v_1}{3}\dot{Y}_1^3 + \frac{K}{\omega}\cos\omega t \tag{8.120}$$

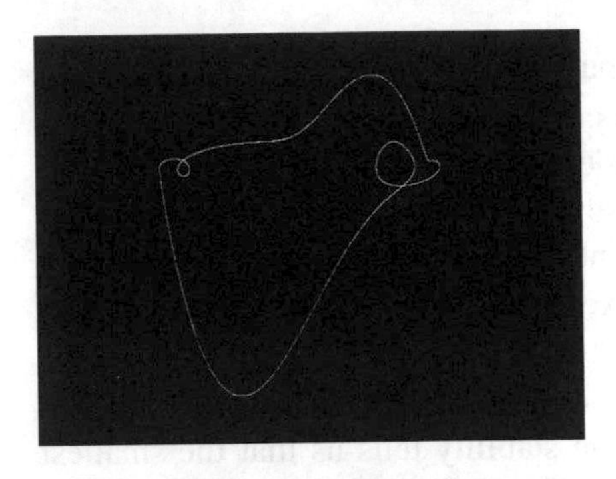

Fig. 8.7. Period doubling cascade to chaos. $\omega = 2.457$

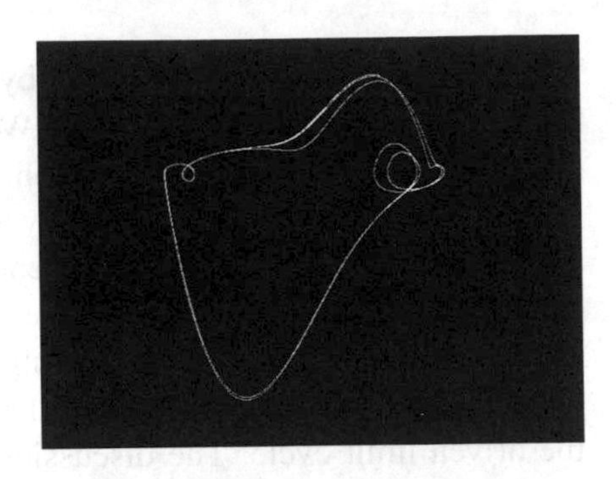

Fig. 8.8. Period doubling cascade to chaos. $\omega = 2.460$

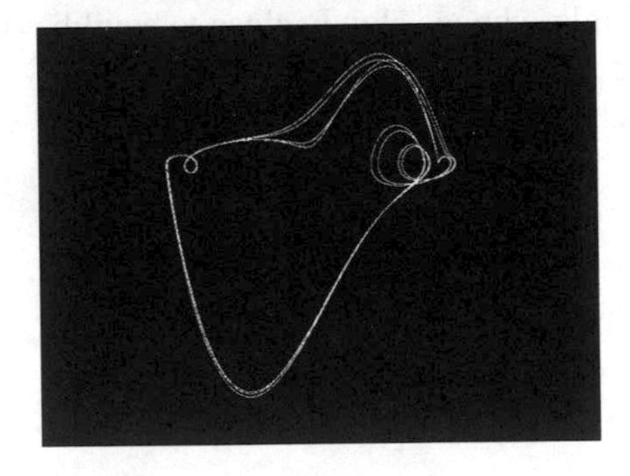

Fig. 8.9. Period doubling cascade to chaos. $\omega = 2.462$

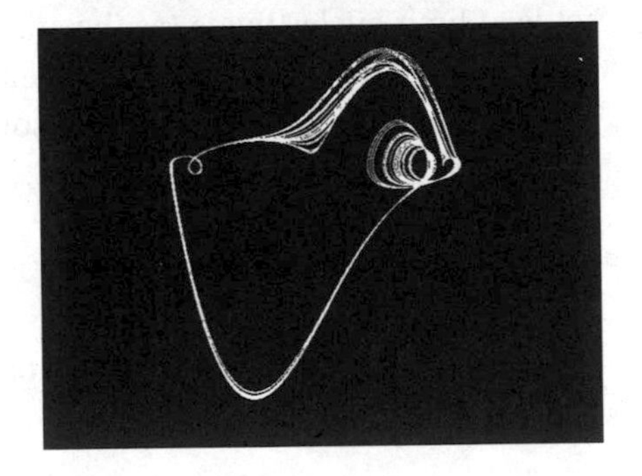

Fig. 8.10. Period doubling cascade to chaos. $\omega = 2.463$

As we are studying just one region, the small open economy, there is no point in keeping the index. Moreover, suppose the structural coefficients are such that we after rescaling can write:

$$\ddot{Y} + Y = 5\dot{Y} - \frac{5}{3}\dot{Y}^3 + \frac{K}{\omega}\cos\omega t \tag{8.121}$$

It is now better to pass to the original van der Pol form, because, though equivalent to our system, it has been much more studied. To this end we differentiate (8.121):

$$\dddot{Y} + \dot{Y} = 5\left(1 - \dot{Y}^2\right)\ddot{Y} - K\sin\omega t \tag{8.122}$$

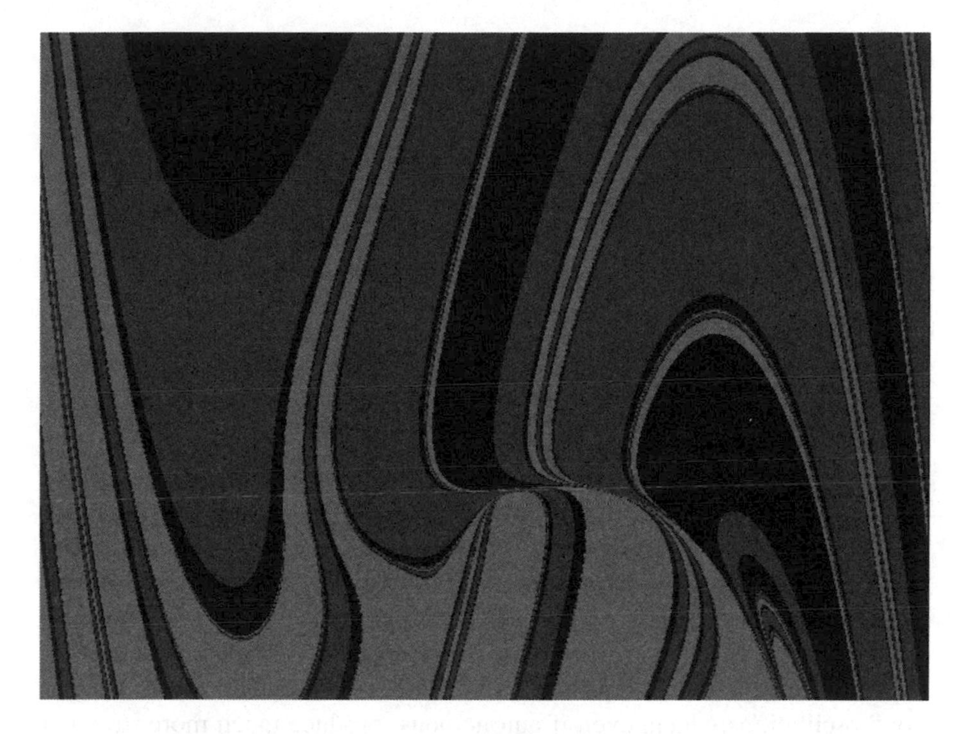

Fig. 8.11. Basins for competing attracors. $\omega = 2.457$

To see the complete analogy define $X = \dot{Y}$, and we have:

$$\ddot{X} + X = 5\left(1 - X^2\right)\dot{X} - K \sin \omega t \tag{8.123}$$

and we are there.

Kapitaniak, following Parlitz and Lauterborn, has demonstrated that equation (8.123) has solutions that go through a period doubling cascade to chaos when $K = 5$ and ω takes values in the interval from 2.457 to 2.463. In Figs. 8.7-8.10 we display a period doubling cascade of bifurcations starting from a four period cycle, via period eight, sixteen, and thirtytwo to fully developed chaos.

As the attractors in Figs. 8.7-8.10 are asymmetric, but the system itself symmetric, we may expect coexistence of multiple attractors. See Ueda. For this reason we have several basins of attraction with fractal boundaries. In Fig. 8.11 we show the attraction basins for the case of the simple periodic cycle shown in Fig. 8.7. There are two shades of gray for the bounded attractors, whereas black indicates escape to infinity.

All this means that in practice simulations may become unpredictable even in the case of cycles, because a slight computation error may bring the process into the basin of the competing attractor.

8.21 Relaxation Cycles

We will now deal with another case of the forced van der Pol oscillator which was studied much earlier, and which is of great interest for our study, i.e. the case of so called relaxation cycles.

In the beginning of this Chapter we used the perturbation approach based on the assumption that the nonlinearity was small. The perturbation approach can, however, also be used for the contrary case, i.e., when the nonlinearity is very large. Then the system is forced to move along the nonlinear "characteristic" most of the time, except for sudden transitions from one branch to another. In our case the characteristic in phase space would be the nonlinear expenditure function of equation (8.6). Stoker investigated these "relaxation" oscillations which, even if autonomous, produce much more irregular cycles than in the case of small nonlinearities.

Cartwright and Littlewood in 1945 studied the case of driven van der Pol oscillations with such large nonlinearities, and discovered a series of astonishing phenomena. All these studies were applied to the van der Pol oscillator, but, as we just saw, this is closely related to our system, as we only need to differentiate the latter once more to arrive at the former.

These studies revealed periodic solutions of very long periods, several hundred times that of the driving force, and multiplicity of such periodic solutions. Moreover, it was discovered that, for certain parameter values, what seemed to be a limit cycle in phase space, was actually a very thin orbit of high complexity that would nowadays be called a "strange attractor" as it is neither a curve nor an area. It was also detected that the successive time periods at which the phase variables attained certain values could be arranged as a random sequence defying predictability.

This lack of predictability together with a strange attractor characterize what is now called chaos in an ever expanding literature on this fascinating subject. The case of the forced van der Pol oscillator was neatly summed up by Levi. The conclusion of highest interest for us is that the business cycle model with a nonlinear accelerator, exactly of the form Hicks proposed, just extended to the case of interregional trade triggered by the simplest and most accepted mechanism, the linear import multiplier, was capable of producing chaotic motion.

The interesting fact is that one of our most basic business cycle models coincides with one of the earliest chaotic models produced.

The practical importance is to call the whole process of forecasting into doubt. Maybe economic forecasts are poor because economic systems, like meteorological ones, are unpredictable, although simple and determinate. To be more exact, short run forecasting can still be used with confidence, because the exponential separation of nearby trajectories does not work when the period is short enough. And in the very long run the system just traces out the strange attractor, which may even be so simple that it can be mistaken for a limit cycle. It is the medium run forecasting that is really tricky, the business of telling where on such a strange attractor the system will be after a couple of rounds.

8.22 Relaxation: The Autonomous Case

As mentioned, the above use of perturbation methods required that the nonlinearity was small. Let us now consider the opposite case, when it instead is very large. Van der Pol termed those oscillations by the attribute "relaxation" as the oscillations consist in two very distinct stages - storing and discharging energy.

Consider again the equation (8.121). We now assume somewhat different coefficients, so that we have:

$$\ddot{Y} + Y = \left(\dot{Y} - \dot{Y}^3\right) + K\cos\omega t \tag{8.124}$$

We also delete the frequency in the denominator of the forcing terms, because we now have no intention to reformulate the equation into the van der Pol style.

To study the relaxation oscillations themselves we delete the forcing term for a start, thus arriving at equation (8.14) again. In (8.15) we introduced a smallness parameter for the nonlinear term of the autonomous equation. Now we do the reverse, applying it to the rest of the equation, thus obtaining:

$$\varepsilon\left(\ddot{Y} + Y\right) = \left(\dot{Y} - \dot{Y}^3\right) \tag{8.125}$$

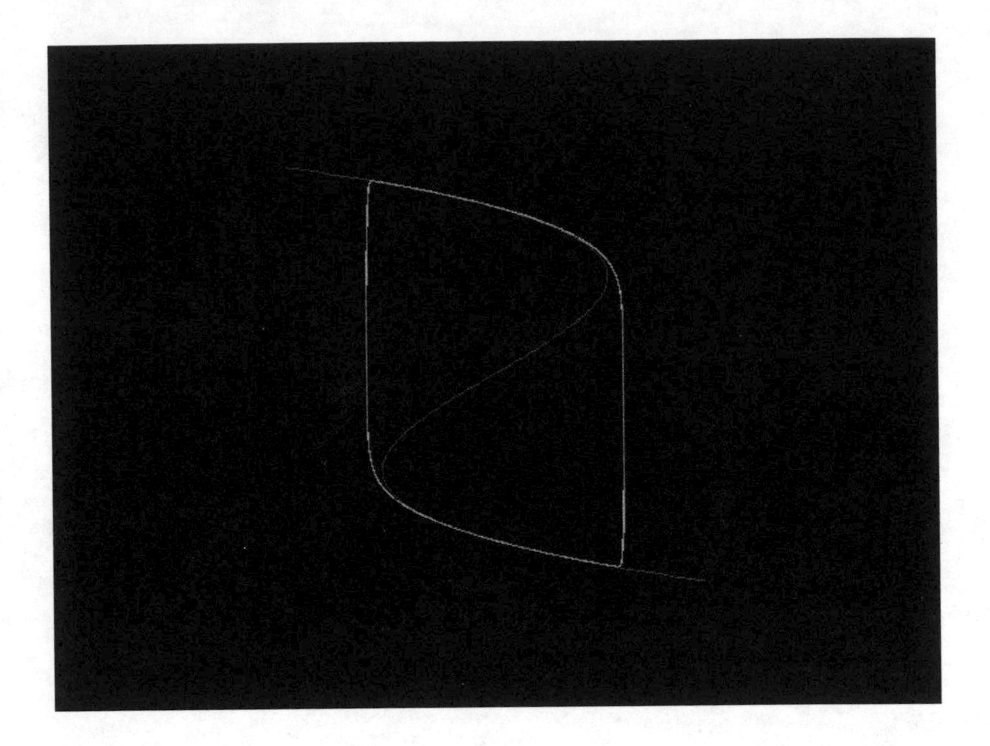

Fig. 8.12. Relaxation cycle for the autonomous case.

To study the system in the phase plane it is most convenient to split it into two first order differential equations. Thus:

$$\frac{dY}{dt} = Z \tag{8.126}$$

and

$$\frac{dZ}{dt} = \frac{1}{\varepsilon}\left(Z - Z^3\right) - Y \tag{8.127}$$

We want to study the field directions in the phase plane and therefore form the quotient (8.127) to (8.126), but first we make a change of variable appropriate in the limiting process when we let ε go to zero. We make the substitution $X = \varepsilon Y$. Thus:

$$\frac{dZ}{dX} = \frac{1}{\varepsilon^2} \frac{Z - Z^3 - X}{Z} \tag{8.128}$$

When $\varepsilon \rightarrow 0$ we see that the field directions become vertical, except when:

$$X = Z - Z^3 \tag{8.129}$$

which defines the so called characteristic. As the field direction is vertical except on the characteristic we conclude that the system either follows the characteristic or moves vertically to it. To be more exact we see in Fig. 8.12 that only the falling parts of it can be portions of a cycle, the rising middle section implying a rate of change inconsistent with the sign of Z.

We can thus easily conclude what the relaxation oscillation looks like: It moves clockwise as always, and follows the upper branch of the characteristic rightwards until its rightmost projection, then moves vertically down to the lower branch of the characteristic, following it leftwards again to the leftmost projection, and then jumps vertically back to the upper section. From then the process is repeated over and over again. This behaviour is corroborated by simulation experiments as shown in Fig. 8.12. All this is discussed in dctail in Stoker. The result for relaxation oscillations is not very drastic. All that happens is that the almost elliptic orbits for small nonlinearities are deformed into lcss smooth shapes, but this we know from simulations. Much more interesting things happen when we reintroduce the forcing term.

8.23 Relaxation: The Forced Case

For the discussion of the forced system we have to rely on Levi. For the proofs Levi assumes that the third order nonlinearity can be represented by a curve close to a kinked linear one: this is not too bad as the case exactly represents the Hicksian case with two linear constraints. The forcing term is likewise assumed to be approximately sectionally linear, more like the signum of a cosine, rather than the cosine itself. This is not too bad either, as the driving world economy can be seen to oscillate with a profile somewhat inbetween a smooth sinoid function and its signum when sufficiently many terms in the asymptotic approximation are retained. It should also be recalled that the proofs need coarse nearness to these functions only.

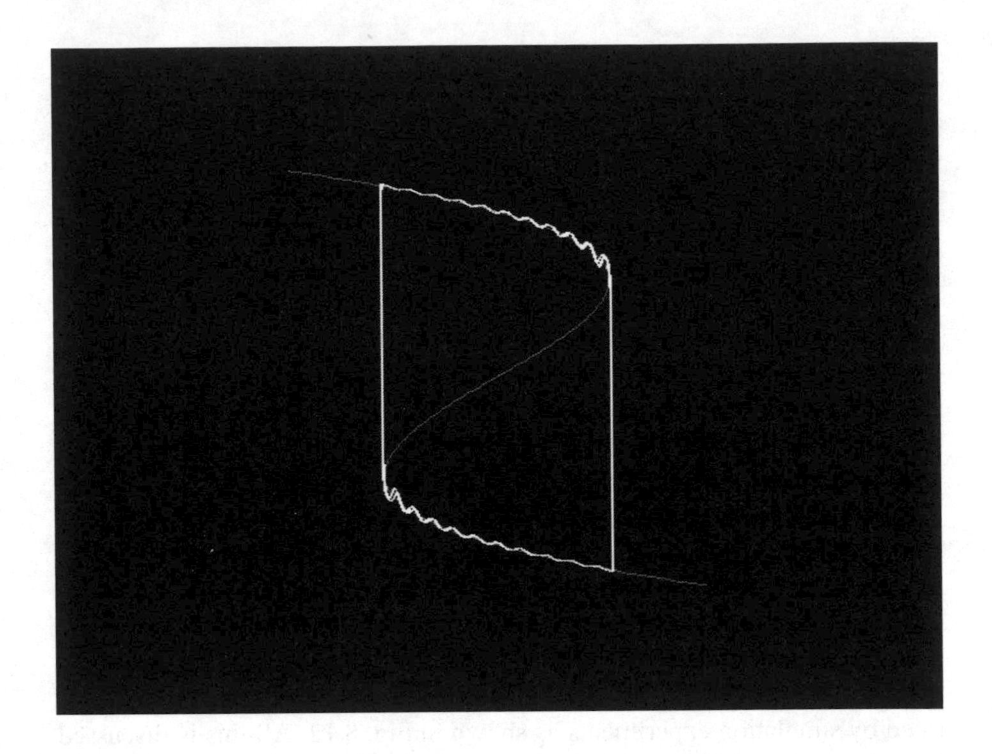

Fig. 8.13. Relaxation cycle for the forced case.

It is shown that even for the driven system the movement is composed by motion along the characteristic and sudden transitions between its branches. The trace of the system even resembles a limit cycle of the relaxation type. The system is strongly contractive, rapidly squeezing any region of phase space into a narrow boundary layer around the pure relaxation cycle as illustrated in Fig. 8.13.

What is intriguing is, however, that the trace is not a limit cycle. It is not even a curve but a tangle of curves compressed within a very thin annulus - its thickness has the order of magnitude of $\exp(-1/\varepsilon^2)$. This tangle is a strange attractor having a fractal dimension higher than unity.

When the system moves along this attractor its pace varies, which would be impossible with a smooth dynamical system along a curve. It is also shown that successive passages of a certain section of phase space occur at time intervals which on average differ in length by $\pm T/2$, where $T = 2\pi/\omega$ is the period of the driving force. For certain parameter combinations any sequence of such variable periods can provide a solution. This is the very definition of unpredictability, as the expected error amounts to half a cycle.

8.24 Three Identical Regions

Let us now consider the case of three coupled oscillators, i.e. three regions linked by interregional trade. This, of course, increases the complexity of the model. With two oscillators we had a fourth order system, which was reduced to the third order in the forced one way influence model. With three oscillators we obviously have the order of the system multiplied to the sixth, so we have to make some simplifications in order to maintain tractability.

The simplifications proposed consist in making the oscillators identical, i.e dropping the region indices on the structural constants in equations (8.55). Accordingly we have:

$$\ddot{Y}_i + (s+3m)Y_i - m\sum_{j=1}^{j=3} Y_j = (v-1-s)\dot{Y}_i - \frac{v}{3}\dot{Y}_i^3 \tag{8.130}$$

Except for the obvious fact that the indices now run from 1 through 3, there is a slight change in the style of presentation as compared to (8.55). We have added and subtracted fictitious imports and exports to the region from itself in order to make the formula look nicer.

The equilibria of system (8.130) are easily located. Putting the time derivatives equal to zero in (8.130) we obtain:

$$(s+3m)Y_i = m\sum_{j=1}^{j=3} Y_j \tag{8.131}$$

the right hand sides being equal we conclude that $Y_1 = Y_2 = Y_3$. Substituting in (8.131) we thus find that $(s+3m)Y_i = 3mY_i$ must hold. This is possible only if all three income variables are zero or if the rate of saving is zero. The last possibility being an uninteresting special case, we conclude that the only equilibrium is the zero equilibrium for all oscillators.

The next move is to introduce a formal smallness parameter in the right hand side exactly as it was done in (8.58) and to try to substitute a periodic solution of the same type as in equation (8.59):

$$Y_i = A_i(t)\cos\omega t + B_i(t)\sin\omega t \tag{8.132}$$

In order that the solution be periodic the coefficients of secular terms again have to vanish. Dealing with the fundamental frequency, differential equations for the coefficients A_i, B_i are obtained exactly as in the case of (8.60)-(8.61) above.

$$2\dot{A}_i = \left((v-1-s) - \frac{v}{4}\left(A_i^2 + B_i^2\right)\omega^2\right)A_i \tag{8.133}$$

$$+\frac{s+3m-\omega^2}{\omega}B_i - \frac{m}{\omega}\sum_{j=1}^{j=3} B_j$$

$$2\dot{B}_i = \left((v-1-s) - \frac{v}{4}\left(A_i^2 + B_i^2\right)\omega^2\right)B_i \tag{8.134}$$

$$-\frac{s+3m-\omega^2}{\omega}A_i + \frac{m}{\omega}\sum_{j=1}^{j=3} A_j$$

We recall that fixed points of the system (8.133)-(8.134) correspond to periodic solutions of the original system (8.130). To make the system more tractable we will again use rescaling of the coefficients and variables. Thus we again halve the time scale and define the following:

$$\kappa = \nu - 1 - s \tag{8.135}$$

$$\alpha_i = \frac{1}{2}\sqrt{\nu}\omega A_i \tag{8.136}$$

$$\beta_i = \frac{1}{2}\sqrt{\nu}\omega B_i \tag{8.137}$$

$$\rho_i = \alpha_i^2 + \beta_i^2 \tag{8.138}$$

$$\sigma = \frac{s + 3m - \omega^2}{\omega} \tag{8.139}$$

$$\mu = \frac{m}{\omega} \tag{8.140}$$

which yields

$$\dot{\alpha}_i = (\kappa - \rho_i)\alpha_i + \sigma\beta_i - \mu\sum_{j=1}^{j=3}\beta_j \tag{8.141}$$

$$\dot{\beta}_i = (\kappa - \rho_i)\beta_i - \sigma\alpha_i + \mu\sum_{j=1}^{j=3}\alpha_j \tag{8.142}$$

These equations correspond to (8.68)-(8.69) above.

8.25 On the Existence of Periodic Solutions

To find out which periodic solutions the three-region model has we have to investigate the fixed points of (8.141)-(8.142). One is obvious, viz. the zero solution $\alpha_i = \beta_i = 0$ but as before it is unstable. We will return to the stability matters later on. Presently we consider nonzero solutions only. Moreover, we consider distinct solutions for the regions, because whenever at least two solutions are identical then we virtually deal with the two-region case already discussed. Let us now equate (8.141)-(8.142) to zero thus obtaining

$$\sigma\beta_i + (\kappa - \rho_i)\alpha_i = \mu\sum_{j=1}^{j=3}\beta_j \tag{8.143}$$

$$\sigma\alpha_i - (\kappa - \rho_i)\beta_i = \mu\sum_{j=1}^{j=3}\alpha_j \tag{8.144}$$

Next multiply (8.143) through by α_i and (8.144) by β_i then subtract the results and sum over index *i*. The terms containing σ and μ cancel, and, using (8.138), we obtain

$$\sum_{i=1}^{i=3}(\kappa - \rho_i)\rho_i = 0 \tag{8.145}$$

Similarly we multiply (8.143) by β_i and (8.144) by α_i and add in stead, again summing over *i*. Now different terms cancel and we get

$$\sigma\sum_{i=1}^{i=3}\rho_i = \mu\left(\left(\sum_{i=1}^{i=3}\alpha_i\right)^2 + \left(\sum_{i=1}^{i=3}\beta_i\right)^2\right) \tag{8.146}$$

As κ is a structural constant equation (8.145) tells us that the fixed points for (8.143)-(8.144) must be located on a sphere in ρ_1, ρ_2, ρ_3-space with centre in $(\kappa/2, \kappa/2, \kappa/2)$ and radius $\sqrt{3}\kappa/2$. Equation (8.146) represents a plane of equal intercepts in the same space, so the intersection defined by (8.145)-(8.146) is a circle. To determine the solution more exactly we need to extract some more information from (8.143)-(8.144).

This is obtained by taking squares of both sides of (8.143) and (8.144) and adding. Again using (8.138) we get

$$\left((\kappa-\rho_i)^2+\sigma^2\right)\rho_i=\mu^2\left(\left(\sum_{j=1}^{j=3}\alpha_j\right)^2+\left(\sum_{j=1}^{j=3}\beta_j\right)^2\right) \tag{8.147}$$

This provides us with three new equations. We can use (8.146) to eliminate the awkward terms in the right hand sides, thus obtaining

$$\left((\kappa-\rho_i)^2+\sigma^2\right)\rho_i=\mu\sigma\sum_{j=1}^{j=3}\rho_j \tag{8.148}$$

The set of four equations (8.145) and (8.148) determine the ρ_i and σ and by (8.136)-(8.140) together with (8.143)-(8.144) everything else as well.

Let us now take any two of the equations (8.148) and subtract. Rearranging terms we get:

$$\left(\rho_i^3-\rho_j^3\right)-2\kappa\left(\rho_i^2-\rho_j^2\right)+\left(\kappa^2+\sigma^2\right)\left(\rho_i-\rho_j\right)=0 \tag{8.149}$$

But, as we are presently interested in distinct solutions only, we may divide through by $\left(\rho_i-\rho_j\right)$ thus obtaining:

$$\left(\rho_i^2+\rho_i\rho_j+\rho_j^2\right)-2\kappa\left(\rho_i+\rho_j\right)+\left(\kappa^2+\sigma^2\right)=0 \tag{8.150}$$

There are three different combinations of i, j among the three regions. Adding expressions (8.150) for those three we get:

$$2\sum_{i=1}^{i=3}\rho_i^2+(\rho_1\rho_2+\rho_1\rho_3+\rho_2\rho_3)-4\kappa\sum_{i=1}^{i=3}\rho_i+3\left(\kappa^2+\sigma^2\right)=0 \tag{8.151}$$

Using (8.145) we can eliminate the squares so that

$$(\rho_1\rho_2+\rho_1\rho_3+\rho_2\rho_3)-2\kappa\sum_{i=1}^{i=3}\rho_i+3\left(\kappa^2+\sigma^2\right)=0 \tag{8.152}$$

Let us now define the sum

$$P = \sum_{i=1}^{i=3} \rho_i \tag{8.153}$$

of the squared amplitudes of our three coupled oscillators. Taking squares of both sides of (8.153) we next get

$$P^2 = \sum_{i=1}^{i=3} \rho_i^2 + 2(\rho_1\rho_2 + \rho_1\rho_3 + \rho_2\rho_3) \tag{8.154}$$

But, again using (8.145) to eliminate the squares and substituting from (8.153), we get

$$P^2 = \kappa P + 2(\rho_1\rho_2 + \rho_1\rho_3 + \rho_2\rho_3) \tag{8.155}$$

This last expression can be used to eliminate the product terms in (8.152). Doing this and using the definition (8.153)

$$P^2 - 5\kappa P + 6(\kappa^2 + \sigma^2) = 0 \tag{8.156}$$

is obtained. This is a simple second order equation with solutions

$$P = \frac{5\kappa}{2} \pm \frac{1}{2}\sqrt{\kappa^2 - 24\sigma^2} \tag{8.157}$$

It is obvious that solutions may exist only if the detuning coefficient σ is small in relation to the common natural amplitude κ of the three identical oscillators. Once we have solved for the sum P of the squared amplitudes, we can substitute the result in the right hand sides of equations (8.148), thus obtaining third order equations for each of the three.

Let us so take a look at one of those three identical equations:

$$\rho_i^3 - 2\kappa\rho_i^2 + (\kappa^2 + \sigma^2)\rho_i - \mu\sigma P = 0 \tag{8.158}$$

where P is any of the roots from (8.157). From the assumption that the three solutions be all distinct (8.158) must have three real and different roots, each taken on for one of the regions. Equation (8.158) must accordingly be equivalent to:

$$(\rho_i - \rho_1)(\rho_i - \rho_2)(\rho_i - \rho_3) = 0 \tag{8.159}$$

We could inspect the discriminant of (8.158) for real and distinct solutions, but it is easier to just consider the relation between roots and coefficients. The coefficient of the quadratic term, sign reversed, thus equals the sum of the roots, i.e.

$$\mathrm{P} = \sum_{i=1}^{i=3} \rho_i = 2\kappa \tag{8.160}$$

Equating (8.160) to (8.157), rearranging and taking squares we find that

$$\kappa^2 = \kappa^2 - 24\sigma^2 \tag{8.161}$$

i.e.

$$\sigma = 0 \tag{8.162}$$

According to (8.139) this equates the frequency of the periodic solution to the square root of the propensity to save plus the total propensity to import (including the fictitious imports from the region to itself), i.e.

$$\omega = \sqrt{s + 3m} \tag{8.163}$$

We see from equation (8.146) that if (8.162) holds, then the nonnegative right hand side of the former must equal zero, i.e.

$$\mu\left(\left(\sum_{i=1}^{i=3} \alpha_i\right)^2 + \left(\sum_{i=1}^{i=3} \beta_i\right)^2\right) = 0 \tag{8.164}$$

This, however, is possible only provided that either all the amplitudes α_i, β_i

or the propensity to import μ are zero. Both alternatives are contrary to assumption.

The conclusion is that we cannot expect to find any solutions as simple as periodic in the case of three coupled identical oscillators.

8.26 Stability of Three Oscillators

How does the system then behave, if there are neither any stable stationary solutions, nor any periodic solutions? It does not just explode to infinity, as we see from the following argument.

Take equations (8.141)-(8.142), multiply them with α_i and β_i respectively, add and sum over index i. According to the definition (8.138) we get

$$\frac{1}{2}\sum_{i=1}^{i=3}\dot{\rho}_i = \sum_{i=1}^{i=3}(\kappa - \rho_i)\rho_i \tag{8.165}$$

As mentioned above the right hand side is zero on a sphere centred in the point $\rho_1 = \rho_2 = \rho_3 = \kappa / 2$ and having radius $\sqrt{3}\kappa / 2$. Inside this sphere trajectories cross any plane $\rho_1 + \rho_2 + \rho_3 = \text{constant}$ outwards, outside the sphere the direction is reversed. The plane touches the sphere in the point $\rho_1 = \rho_2 = \rho_3 = \kappa$. Provided we thus take the constant larger than 3κ the plane does not meet the sphere at all. It cuts off a tetrahedron in the positive orthant of ρ_1, ρ_2, ρ_3-space. All trajectories starting sufficiently far from the origin in the positive orthant tend to cross this plane, thus entering the tetrahedron, and none can ever leave it this way. On the other hand these squared amplitudes ρ_i according to (8.138) are nonnegative, so the other planes of the tetrahedron cannot be crossed either. Any motion is thus bound to enter the tetrahedron defined by

$$\rho_i \geq 0 \tag{8.166}$$

$$\sum_{i=1}^{i=3}\rho_i \leq \text{constant} > 3\kappa \tag{8.167}$$

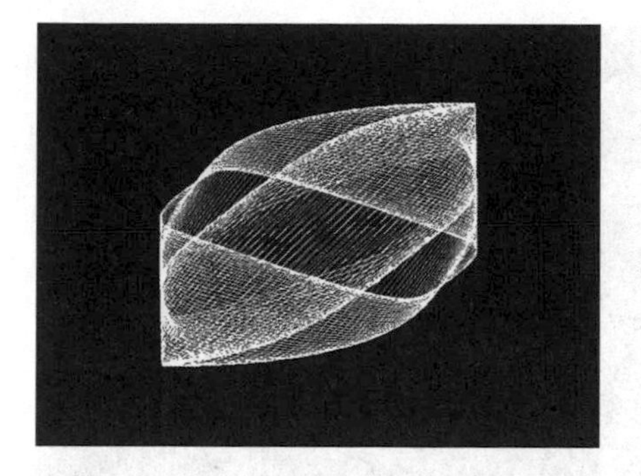

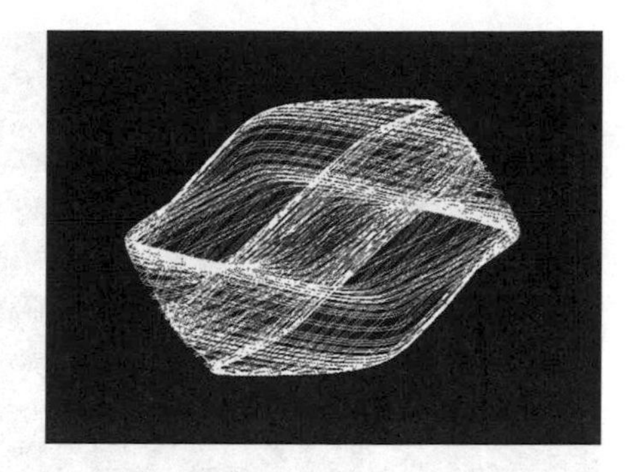

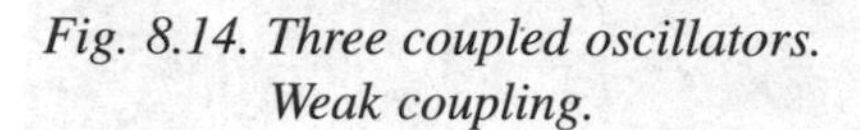

Fig. 8.14. Three coupled oscillators. Weak coupling.

Fig. 8.15. Three coupled oscillators. Strong coupling.

and be confined there eternally. We can even restrict the region further by truncating the tetrahedron through the condition

$$\sum_{i=1}^{i=3} \rho_i \geq \text{constant} < \kappa \tag{8.168}$$

The part of such a truncating plane contained in the positive orthant is also entirely contained in the sphere defining the zero of (8.165), and so all trajectories starting closer to the origin are bound to cross it outwards, never to leave through it in the reverse direction again. By isolating the origin this construction also shows that the single stationary solution that exists is an unstable equilibrium.

8.27 Simulations

We have to rely on computer simulation to find out more about the case of three oscillators. Figs. 8.14 and 8.15 show trajectories of the original system (8.130) in three-dimensional Y_1, Y_2, Y_3-space. In Fig. 8.14, obtained for a weak coupling of the regions by a very low import propensity, the solution is quasiperiodic, nearby trajectories remaining nearby. In Fig. 8.15 the coupling is stronger and the travel on the attractor has a more turbulent character.

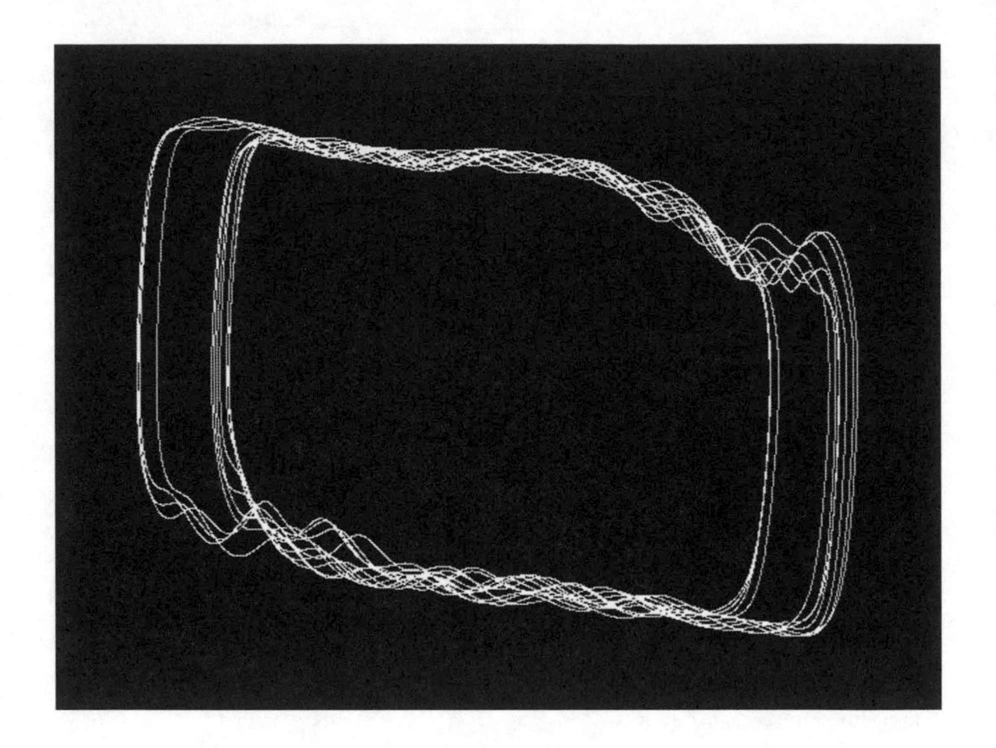

Fig. 8.16. Qusiperiodically forced oscillator. Chaotic outcome.

Whether its is chaotic or not is not very easy to see by casual inspection, but a closer look at a live simulation displays certain unexpected divergence in seemingly very nearby trajectories. Looking at a three-dimensional picture it is, however, always hard to say whether two lines looking close are in fact close or are just one behind the other.

We also know that three coupled oscillators quite easily become chaotic. A special case studied is the quasiperiodically forced van der Pol oscillator, or our equivalent one. A quasiperiodically forced oscillator is of the type:

$$\ddot{Y} + Y = \left(\dot{Y} - \dot{Y}^3\right) + K \sin \omega t \sin \Omega t \tag{8.169}$$

It arises when we skip the simplifying assumption that the regions are identical with respect to the characteristic coefficients, but assume that two of the regions are of considerable size, whereas the third is very small. The two big regions influence each other, and they can easily establish a quasiperiodic motion as we have seen. Provided the effects of the third small economy on the two giants can be neglected, the result will be a one way forcing of that small economy, as formulated in equation (8.169). It is well known that such quasiperiodic forcing easily results in chaotic motion. Fig. 8.16 displays a simulated case with a manifestly chaotic outcome.

The case is somewhat similar to the classical so called restricted three body problem where the gravity fields of two large planets influence each other and a speck of dust, whose influence back on the planets, however, can be ignored. The speck of dust is precisely what the small open economy is.

9 Business Cycles: Continuous Space

9.1 Introduction

In economics there has been very scant modelling by means of partial differential equations. In business cycle modelling they, however, arise in a natural way if we put the problem studied in a spatial setting by introducing interregional trade, considering a dynamical process that evolves both in continuous space and time. This will be done in the simplest and most obvious way, i.e., by a linear import-export multiplier, as is in line with the multiplier for local expenditures already present and with the general Keynesian macroeconomic outlook.

It will be seen that this adds surprisingly much even to the linear model. The limit cycle obtained with a nonlinear investment function, no matter how far it is in shape from the simple sinoid solution to the original Samuelson model, is still strictly periodic. In the setting of two-dimensional geographical space this is no longer true, not even for the linear model. Periodicity is lost. And even with one spatial dimension the solution to the linear model can be as irregular as any nonlinear limit cycle.

To understand the second point we note that when the effects spread through space they are reflected back, sooner from close locations and later from remote locations. Space thus functions as a distributed lag system. We know from the careful studies by Hicks how complex the solutions for distributed temporal lags become by superposing waves of different periods.

The difference between one and two dimensions of space can be made clear by an analogy to a well known physical system - the acoustics of strings and sound-boards in musical instruments. The one dimensional string has a specific pitch related to density, thickness, length, and tension. Depending on how it is set in motion it can produce a variety of colourings of that basic pitch, but the sounding note always is a combination of the basic frequency and its natural harmonics. Resonance in the string only occurs when we sound a pitchfork with the frequency of one of the natural harmonics. Mathematically, the eigenvalues of the system are in a rational relationship.

For a two-dimensional soundboard this is no longer true. Although it has its own basic pitch, it is capable of being resonant with almost any note sounded on a string mounted on it. The reason is that the eigenvalues, corresponding to the natural harmonics, no longer are in rational relation to each other, and that they come so close to each other that there is near resonance to almost any note. For instance the irrational series of natural frequencies for a square membrane is 1, 1.58, 2, 2.24, 2.55, 2.92, 3, 3.16, 3.54, 3.61, 3.81, 4 to be compared with the series of the natural numbers 1, 2, 3, 4 for the string. This density of natural harmonics increases with ascending eigenvalues. It is also interesting to note that most of the eigenvalues can arise from vibration of the square in several modes. The eigenvalue 2.24 can, for instance, arise from the square vibrating in two, three, or four parts.

The richness of two dimensions is easy to understand if we realize that subdivision of space in 1D only contains intervals, whereas subdivision of space in 2D also involves the concept of shape. The superposition two irrational frequencies results in quasiperiodic motion, i.e., one where no cycle is ever repeated again. For a soundboard this means that it can produce mere noise, for a model of the space economy that we run into trouble when we try to use standard statistical methods for analysing periodicity.

To, summarize: One spatial dimension transforms the business cycle model so that it can produce any periodic motion, however irregular, two spatial dimensions destroy even periodicity.

9.2 Interregional Trade

We assume that imports to any location depend on local income Y, which is now a function of time and two space coordinates t, x_1, x_2. As stated above this dependence is assumed to be linear, the constant propensity to import being denoted m. Assuming local action as a first approximation, imports will be generated by the import propensity from income in the point studied, whereas exports will be generated by the action of the same import propensity from incomes in the surroundings of the point. Net export surplus would then be the constant import propensity multiplied by the income difference between the surrounding points and the point itself.

The question is how to measure this difference. Let us start with one dimensional space. In a continuous framework we know how to measure differences in a variable over time: By the derivative. Now, space is different from time in the respect that there is no obvious forward direction prescribed. When we ask about spatial differences of conditions at a point and its surroundings we have to consider that there is a difference to the right and another to the left. What then is of interest is the difference of right and left differences, which by the limiting process in the continuous case becomes the second derivative. This is also the lowest derivative that remains invariant upon reversing space directions.

In two dimensions we deal with two coordinate directions, and so the sum of the second direct spatial derivatives becomes the measure of income differences between the surroundings of a point and the point itself, all possible directions of departure taken into consideration. This sum of derivatives is called the Laplacian, and is denoted:

$$\nabla^2 Y = \frac{\partial^2 Y}{\partial x_1^2} + \frac{\partial^2 Y}{\partial x_2^2} \tag{9.1}$$

It is no more strange than a common derivative and has been the basic concept in all spatial physical models where there is linear diffusion (of heat or matter). Hotelling also applied it in his model of the growth and dispersal of populations. That this measure really is the general measure of spatial differences is formally ensured by Gauss's Integral Theorem, derived in Chapter 3, stating that:

$$\iint_R \nabla^2 Y dx_1 dx_2 = \iint_R \nabla \cdot \nabla Y dx_1 dx_2 = \oint_{\partial R} (\nabla Y)_n ds \tag{9.2}$$

where the Laplacian has been decomposed in two differential operators used in sequence: innermost the gradient $\nabla Y = (\partial Y / \partial x_1, \partial Y / \partial x_2) = (\phi_1, \phi_2) = \phi$, producing an intermediate result, a vector which we provisionally denote ϕ, and then the divergence $\nabla \cdot \phi = \partial \phi_1 / \partial x_1 + \partial \phi_2 / \partial x_2$, again producing a scalar from the vector ϕ. It is easy to check that the divergence of the gradient in fact results in the Laplacian. We refer to Chapter 3 for more detail.

The right hand side integrand of (9.2) is the projection, normal to the boundary, of the gradient vector ∇Y, and the curve integral is taken all the way round on the boundary of the region. The integrand thus at any point measures the rate of change of income, and the curve integral then becomes the net change of income as we leave the enclosed region. The size or shape of the boundary do not matter. As we shrink the region of integration around a point to a shapeless nothing, we just retain the Laplacian in that point on the left, and the net rate of change, all directions of departure considered, on the right.

We can thus define export surplus in the following way:

$$X - M = m\nabla^2 Y \tag{9.3}$$

9.3 The Linear Model

The only restrictiveness is that we assume local action between spatially contiguous locations only. This is a natural first approximation, similar to our treatment of time where we only deal with immediate action between temporally contiguous moments, once we formulate a process in terms of derivatives. This, of course, does not prevent effects to spread through space by chains of influences. As we already stated space acts in a manner equivalent to a distributed lag structure. What the assumption of local action amounts to is that remote places are affected later than close places.

The causation of trade by the Laplacian operator also ensures that there is always a global balance, what is exported from one location is always imported to another, and vice versa. What is stated above is, however, only an equilibrium condition, and what we need is a relation in adaptive form. To simplify notation we again assume that the adjustment speed is the same as in the adjustments of income and of investments (an assumption equivalent to the choice of a commensurable lag structure in discrete models). Thus the coefficient of adjustment becomes unitary and we may write:

$$\dot{X} - \dot{M} = m\nabla^2 Y - X + M \tag{9.4}$$

We should remark at this stage that we keep the dot to denote time derivatives although it from now on is a partial derivative. The adjustment equation for income (8.1) must be modified, as exports act like investment, imports like savings. Thus:

$$\dot{Y} = I + X - M - sY \tag{9.5}$$

whereas the linear investment equation $\dot{I} = v\dot{Y} - I$ (8.2) is not changed.

We now proceed as before, differentiate the income adjustment equation once more, so raising its degree, and eliminate investments, exports and imports along with their derivatives by the three original equations. Thus:

$$\ddot{Y} + sY - m\nabla^2 Y = (v - 1 - s)\dot{Y} \tag{9.6}$$

To find out the general behaviour of this model we multiply (9.6) through by $\dot{Y}$, and integrate over space. Thus:

$$\iint_R \left(\dot{Y}\ddot{Y} + s\dot{Y}Y - m\dot{Y}\nabla^2 Y\right) dx_1 dx_2 = (v - 1 - s)\iint_R \dot{Y}^2 dx_1 dx_2 \tag{9.7}$$

From Gauss's Integral Theorem (9.2) applied to the following identity from vector analysis $\nabla \cdot \left(\dot{Y}\nabla Y\right) = \nabla\dot{Y} \cdot \nabla Y + \dot{Y}\nabla^2 Y$ we however find:

$$\iint_R \nabla\dot{Y}\nabla Y dx_1 dx_2 + \iint_R \dot{Y}\nabla^2 Y dx_1 dx_2 = \oint_{\partial R} \dot{Y}(\nabla Y)_n ds = 0 \tag{9.8}$$

where the right hand side is zero provided we study change within a region where income always is in equilibrium on the boundary. We need some kind of boundary condition and this is the most obvious. In dealing with a closed region it is merely natural to choose the boundary in such a way that the equilibrium condition is satisfied. And if we deal with a closed surface (like the surface of the earth) the boundary condition is automatically satisfied.

Using this result, and identifying in (9.7) on the left as half the sum of squares of income, its first time derivative, and its gradient, we obtain:

$$\frac{1}{2}\frac{\partial}{\partial t}\iint_R \left(\dot{Y}^2 + sY^2 + m(\nabla Y)^2\right) dx_1 dx_2 = (v - 1 - s)\iint_R \dot{Y}^2 dx_1 dx_2 \tag{9.9}$$

We see that depending on the sign of $(v - 1 - s)$ the system is damped towards stationary equilibrium, or explodes away from it.

9.4 Coordinate Separation

This exercise was carried out for future purposes where nonlinearities prevent us from finding closed form solutions. For the linear system (9.6) we can find a nice closed form solution by the method of coordinate separation. For this reason we try a solution:

$$Y = T(t)S(x_1, x_2) \tag{9.10}$$

i.e., factorizing the solution in a time-dependent and a space-dependent factor. Substituting into the differential equation (9.6), dividing through by $Y = ST$, and rearranging the terms we get:

$$\frac{T''}{T} - (v - 1 - s)\frac{T'}{T} + s = m\frac{\nabla^2 S}{S} \tag{9.11}$$

As the right hand side is independent of time, whereas the left hand side is independent of space we conclude that this equation can hold only if both sides are constant. Denote this constant by $-m\lambda^2$. We thus get two equations out of one:

$$\nabla^2 S + \lambda^2 S = 0 \tag{9.12}$$

and:

$$T'' - (v - 1 - s)T' + (s + m\lambda^2)T = 0 \tag{9.13}$$

The first equation is well known from the physical application in the wave equation. The values for λ for which it can be solved are the eigenvalues, and the corresponding solutions are the eigenfunctions associated with the kind of boundary conditions: With constant equilibrium on the boundary, as assumed above, they are sine and cosine functions for a square, Bessel functions for a circular disc, and Legendre polynomials for a sphere. The eigenvalues can be proved to be real by the use of Gauss's Integral Theorem. To this end we multiply equation (9.12) through by *S*, and integrate over space. The derivation, which we do not repeat here, was given in Section 3.12. The result is:

$$\iint_R (\nabla S)^2 dx_1 dx_2 = \lambda^2 \iint_R S^2 dx_1 dx_2 \tag{9.14}$$

Unless we are dealing with the system in spatially homogeneous equilibrium, with $S^2 \equiv (\nabla S)^2 \equiv 0$, we have $\lambda^2 > 0$.

The temporal equation is ordinary, and its solution is:

$$T = \exp(-\alpha t)(A\cos\omega t + B\sin\omega t) \tag{9.15}$$

where:

$$\alpha = \frac{v - 1 - s}{2} \tag{9.16}$$

and:

$$\omega = \sqrt{s + m\lambda^2 - \alpha^2} \tag{9.17}$$

We see that if the eigenvalues are not too small, when the solution can be overdamped, the solution is always oscillatory. As always with linear systems these oscillations can be damped or explosive, which depends on whether the accelerator exceeds the rate of saving by more or less than unity. We also note that damping is uniform for all spatial modes, as it only depends on the structural coefficients, not on the eigenvalue.

The solution process is now clear. We first find out any eigenvalue and eigenfunction to the spatial equation, then solve the temporal one, whose solution depends on the eigenvalue, and combine the product. All such products are solutions, and so are any weighted sums of such products. This is true even for infinite sums capable of producing any periodic or quasiperiodic solutions with profiles far from the simple sinoid, as we know from Fourier's theorem.

More about the mathematics can be found in Courant and Hilbert and in Duff and Naylor, which are both highly recommended. So is Lord Rayleigh's classic "Theory of Sound". To make some introductory points clear we will give three examples of solutions for two-dimensional regions: the square, the circular disc, and the surface of the sphere.

9.5 The Square Region

The procedure followed here works for a rectangle, but we follow the general strategy of avoiding to complicate the formulas for the mere sake of generality. As a rectangle differs from a square by a proportional change of scale only, we deal with the square and assume it to have side π.

As the spatial equation still is a partial differential equation we try the method of coordinate separation again. Putting:

$$S(x_1, x_2) = X_1(x_1)X_2(x_2) \tag{9.18}$$

the equation splits again in two:

$$X_1'' + i^2 X_1 = 0 \tag{9.19}$$

$$X_2'' + j^2 X_2 = 0 \tag{9.20}$$

We will not repeat the discussion, which involves substituting the attempted solution, dividing through by it, rearranging terms, and recognizing two sides that only depend on each one of the space coordinates, so that the equality can hold only when both sides are constant. For convenience two new constants, i^2 and j^2, are introduced, but they must obey the following relation:

$$\left(i^2 + j^2\right) = \lambda^2 \tag{9.21}$$

Through the right hand side of (9.21) the solution modes for the spatial equation are linked to the solutions for the temporal equation. The equations for the separated coordinate functions are readily obtained. We have:

$$X_1 = \sin ix_1 \tag{9.22}$$

$$X_2 = \sin jx_2 \tag{9.23}$$

for any integers *i* and *j*. The solutions are pure sines without phase lead or lag because our square was assumed to have the interval $(0, \pi)$ as sides, and because the boundary condition stated that income be in equilibrium on the boundary. We obtain the complete solution:

$$Y = \exp(-\alpha t)\sum_{i=1}^{\infty}\sum_{j=1}^{\infty} \sin ix_1 \sin jx_2 \left(A_{ij} \cos\omega_{ij}t + B_{ij} \sin\omega_{ij}t\right) \tag{9.24}$$

where:

$$m\left(i^2 + j^2\right) + s = \alpha^2 + \omega_{ij}^2 \tag{9.25}$$

determines the frequency ω_{ij} of each solution term. Again we note that damping is uniform whereas the oscillatory speed in time depends on the spatial wavelength, so that long waves move slowly, short ones fast. For the future we note that a nonlinearity would be needed to avoid damping (or explosion).

Each mode, except the lowest one, involves a subdivision of the square region by nodal lines where the points on the nodal lines are constantly at rest whereas adjacent regions on either side of a nodal line move in opposite direction. In terms of our context the whole region is divided in subregions where prosperity and depression alternate. The square may be subdivided by several networks of coordinate lines, and the total movement is a compound of these. The speed of motion is also different for the different modes, the finer the subdivision the faster the cyclic variation is.

By conclusion let us note that the coefficients A_{ij}, B_{ij} are determined from the relations:

$$A_{ij} = \frac{4}{\pi^2}\iint_R Y_0 \sin ix_1 \sin jx_2 dx_1 dx_2 \tag{9.26}$$

$$B_{ij} = \frac{4}{\omega_{ij}\pi^2}\iint_R Y_0' \sin ix_1 \sin jx_2 dx_1 dx_2 \tag{9.27}$$

where Y_0, Y_0' are the initial income distribution and its rate of change distribution over the region. This displays the full generality of the solution as any shapes of these initial profiles may be part of an oscillatory solution to the model.

9.6 The Circular Region

Another case easily solvable in closed form is that of the circular region. For this case it is convenient to change the Cartesian coordinates to polar, i.e., to define:

$$x_1 = \rho \cos\theta \tag{9.28}$$

$$x_2 = \rho \sin\theta \tag{9.29}$$

The Laplacian in polar coordinates can be easily calculated as:

$$\nabla^2 S = \frac{\partial^2 S}{\partial \rho^2} + \frac{1}{\rho}\frac{\partial S}{\partial \rho} + \frac{1}{\rho^2}\frac{\partial^2 S}{\partial \theta^2} \tag{9.30}$$

The choice of coordinates suggests the separation to be attempted:

$$S = \mathrm{P}(\rho)\Theta(\theta) \tag{9.31}$$

Substituting this into the spatial equation (9.12) yields:

$$\Theta'' + i^2\Theta = 0 \tag{9.32}$$

and

$$\rho^2\mathrm{P}'' + \rho\mathrm{P}' + \left(\lambda^2\rho^2 - i^2\right)\mathrm{P} = 0 \tag{9.33}$$

Equation (9.32) has the obvious solution:

$$\Theta = \cos i\theta \tag{9.34}$$

where we delete any phase lead or lag, as it does not mean anything more than a rotation of the whole set of node lines which are equally spaced radials $2i$ in number.

The second equation (9.33) is Bessel's differential equation with the Bessel functions of the first and second kind as solutions. Only those of the first kind make sense as they stay finite at the origin. In the conventional symbols we write the solution:

$$P = J_i(\lambda\rho) \tag{9.35}$$

and hence from (9.31) and (9.34)-(9.35)

$$S = J_i(\lambda\rho)\cos i\theta \tag{9.36}$$

The Bessel functions like sines and cosines undulate around zero and thus define a number of nodal lines. As they refer to the radial coordinate they define sets of concentric circles as nodal lines, dividing the disk into rings, the number of which depends on λ. Of course, the boundary of the region must be among these nodal lines.

Like the nodal lines for the square divided it in small rectangles by a network of horizontal and vertical lines, the circular disk is divided in sectors and rings by radials and circles. Again we may expect the frequency of a certain mode to depend on the mesh, the finer it is the faster the vibration. The formal relation is:

$$m\lambda^2 + s = \alpha^2 + \omega^2 \tag{9.37}$$

Speaking about the nets of nodal lines, we of course refer to the elementary modes. The superposition principle holds again, and combinations of the basic modes may lead to more complex sets of nodal lines. The general solutions and the way of determining the arbitrary coefficients are exactly parallel to the previous case, so we will not dwell on these details again

9.7 The Spherical Region

The final case where we can obtain a solution in closed form is that of the closed two-dimensional surface of a sphere embedded in three-dimensional space. It is different from the previous cases as the surface is curved. The circular case could have been dealt with in terms of the original Cartesian coordinates, even if the expressions would have become unnecessarily awkward. Presently, as the surface is curved, we must revert to curvilinear coordinates.

Suppose we deal with a sphere of unit radius and identify its points by the angles of colatitude θ and longitude ϕ. If the sphere is embedded in three-dimensional space with Cartesian coordinates ξ, η, ζ we have the coordinate transformation:

$$\xi = \sin\theta\cos\phi \tag{9.38}$$

$$\eta = \sin\theta\sin\phi \tag{9.39}$$

$$\zeta = \cos\theta \tag{9.40}$$

With some labour we can work out the Laplacian in these coordinates:

$$\nabla^2 S = \cos\theta\frac{\partial S}{\partial\theta} + \frac{\partial^2 S}{\partial\theta^2} + \frac{1}{\sin^2\theta}\frac{\partial^2 S}{\partial\phi^2} \tag{9.41}$$

Again the choice of coordinates determines the choice of separation to be attempted. We try $S = \Theta(\theta)\Phi(\phi)$, and this again splits the partial spatial equation in two ordinary ones:

$$\Phi'' + i^2\Phi = 0 \tag{9.42}$$

$$\Theta'' + \cos\theta\Theta' + \left(j(j+1) - \frac{i^2}{\sin^2\theta} \right)\Theta = 0 \tag{9.43}$$

where

$$j(j+1) = \lambda^2 \tag{9.44}$$

The first equation again has a solution in terms of simple trigonometric functions, so dispensing with arbitrary rotations, we can write:

$$\Phi = \cos i\phi \tag{9.45}$$

The node lines defined by this solution are parallel circles. The second equation can be expected to yield great circles through the poles. It is the associated Legendre equation and its solutions are the Legendre functions:

$$\Theta = \mathrm{P}_j^i(\cos\theta) \tag{9.46}$$

where the most handy definition is:

$$\mathrm{P}_j^i(\cos\theta) = \frac{\sin^j\theta}{2^j j!}\frac{d^{i+j}(\sin^{2j}\theta)}{d(\cos\theta)^{i+j}} \tag{9.47}$$

This expression makes it easy to calculate the Legendre functions in terms of sine and cosine expressions, at least for low i, j. For more details see Duff and Naylor. As already indicated the nodal lines form a network corresponding to the common geographical coordinates of longitude and latitude. In Table 9.1 we list the lowest modes.

Table 9.1. Spatial vibration modes for the sphere.

j	i	$\mathrm{P}_j^i(\cos\theta)\cos i\phi$
0	0	1
1	0	$\cos\theta$
1	1	$\sin\theta\cos\phi$
2	0	$3\cos^2\theta-1$
2	1	$\cos\theta\sin\theta\sin\phi$
2	2	$\sin^2\theta\cos 2\phi$
3	0	$5\cos^3\theta-3\cos\theta$
3	1	$\sin\theta(5\cos^2\theta-1)\cos\phi$
3	2	$\cos\theta\sin^2\theta\cos 2\phi$
3	3	$\sin^3\theta\cos 3\phi$

Note that there are two pure cases, corresponding to $i=0$ and $i=j$ respectively. In the first case ϕ is not involved. So the nodal lines are polar great circles, and the vibration mode is pure sectorial. In the second θ is only involved by a power of its sine, which defines the poles as two degenerate circles. The remaining coordinate defines parallel circles and the vibration mode is zonal. The remaining modes are mixed. The frequency of vibration is again related to mesh size, the formal relation being:

$$mj(j+1)+s=\alpha^2+\omega^2 \tag{9.48}$$

The solutions are superposed, so that the general solution reads:

$$Y=\exp(-\alpha t)\sum_{i=1}^{\infty}\sum_{j=1}^{\infty}\mathrm{P}_j^i(\cos\theta)\cos i\phi\left(A_{ij}\cos\omega_{ij}t+B_{ij}\sin\omega_{ij}t\right) \tag{9.49}$$

We note that there is again an overall damping (or explosion) of all oscillatory modes of motion.

The arbitrary coefficients can be determined from:

$$A_{ij}=\frac{(j-i)!}{(j+i)!}\frac{2n+1}{\pi}\iint_R Y_0\,\mathrm{P}_j^i(\cos\theta)\cos i\phi d\theta d\phi \tag{9.50}$$

$$B_{ij}=\frac{(j-i)!}{(j+i)!}\frac{2n+1}{\pi}\iint_R Y_0'\mathrm{P}_j^i(\cos\theta)\cos i\phi d\theta d\phi \tag{9.51}$$

This again makes the solution compatible with all imaginable initial conditions.

9.8 The Nonlinear Spatial Model

There is nothing more we can learn from the linear spatial model. It is interesting to note that the mere introduction of space generalizes the model to such an extent that any periodic or quasiperiodic motion, however irregular its profile is, can be produced. In all cases, however, linearity still implies uniform damping or explosion at an exponential rate, and so all our initial objections to linear modelling still apply.

The natural next step is to reintroduce the nonlinearity while keeping the spatial setting with interregional trade. Trouble can be expected, as the field of nonlinear partial differential equations is little exploited. So after an introductory discussion, we will have to simplify the model again.

Consider the spatial variant of the model (8.5) with the nonlinear investment function:

$$\ddot{Y} + sY - m\nabla^2 Y = (v - 1 - s)\dot{Y} - \frac{v}{3}\dot{Y}^3 \tag{9.52}$$

To find out its general behaviour let us multiply through by $\dot{Y}$ and integrate over space, as we did in the linear case, and use Gauss's integral theorem, given the boundary condition of constant rest on the boundary. In this way we get:

$$\frac{1}{2}\frac{\partial}{\partial t}\iint_R \left(\dot{Y}^2 + sY^2 + m(\nabla Y)^2\right)dx_1 dx_2 = \tag{9.53}$$

$$(v - 1 - s)\iint_R \dot{Y}^2 dx_1 dx_2 - \frac{v}{3}\iint_R \dot{Y}^4 dx_1 dx_2$$

Provided the accelerator exceeds the rate of saving by more than one we note that the right hand side has a positive term and a negative one. For low rates of change the positive quadratic term dominates, whereas for high rates of change the negative quartic one dominates. Thus, far from stationary equilibrium the system is damped, whereas close to it the system is explosive. Accordingly, the stationary equilibrium is unstable, but the system does not explode. Damping at high rates of change keeps the system within a bounded distance from stationary equilibrium. So, the stage is set for never ending bounded motion defined by the system itself.

9.9 Dispersive Waves

The easiest type of solution is that of dispersive waves, applicable when there are no boundary conditions to be satisfied. To find out something more about possible motions we realize that the left hand side of (9.52) is a linear Klein-Gordon equation, known to produce dispersive waves. See Nayfeh. So, define

$$\theta = \kappa\rho - \omega t \tag{9.54}$$

where

$$\rho = \sqrt{x_1^2 + x_2^2} \tag{9.55}$$

and try solutions of the form $Y(\theta)$.

With the left hand side of (9.52) equal to zero the solution would fit provided it fulfils the dispersion relation:

$$\omega^2 - m\kappa^2 = s \tag{9.56}$$

Let us simplify by choosing $v = 2$ and $m = s = 1/2$. Then the dispersion relation admits a wave number $\kappa = 1$ along with a frequency $\omega = 1$. The way space was introduced through radius vector suggests that we again revert to polar coordinates, and only consider change in the radial direction.

Now, the equation we want to solve is not the linear Klein-Gordon equation, but suppose the right hand side of our equation is small. We indicate this by reintroducing the smallness parameter. With our simplifications:

$$Y'' + Y = \varepsilon\left(\left(1 - \frac{1}{\rho}\right)Y' - \frac{4}{3}Y'^3\right) \tag{9.57}$$

where the dashes denote differentiations with respect to the composite argument $\theta = \rho - t$.

We can solve this system numerically by a four point Runge-Kutta method used in other simulations in this book. Keeping distance from the origin ρ

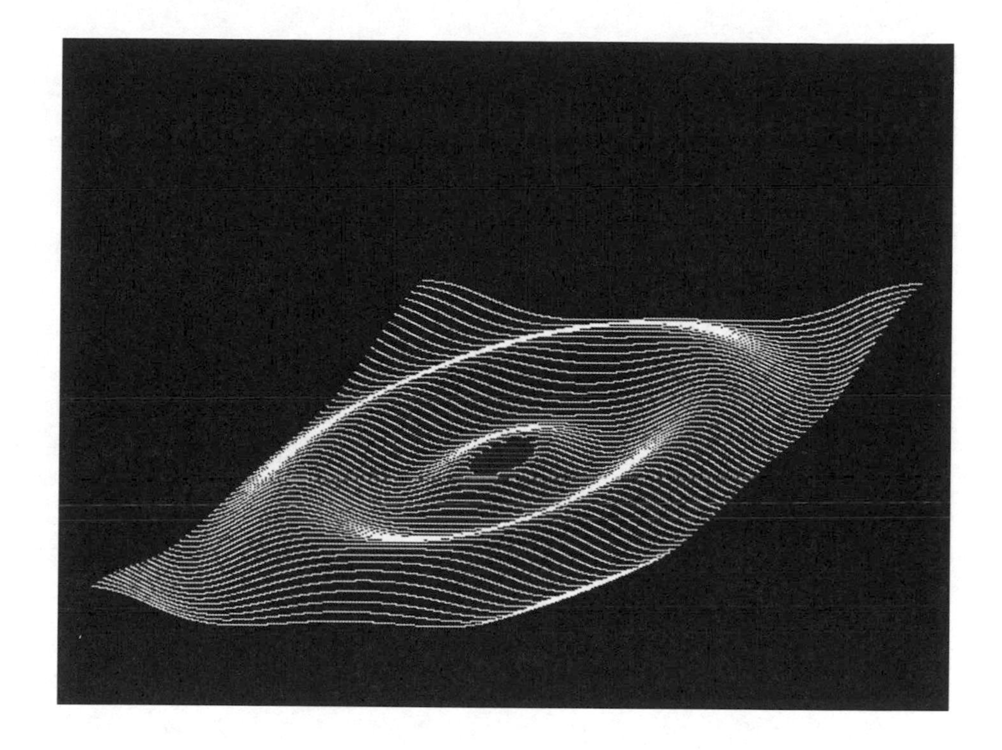

Fig. 9.1. Dispersive wave.

fixed we get limit cycles for each distance. We find that for distances less than unity from the origin there are no limit cycles. At unit distance a bifurcation occurs and a limit cycle appears. For increasing distance this amplitude is gradually increased going asymptotically towards a limiting value of finite amplitude. This is shown in Fig. 9.1.

The same results are obtained if we use the two-timing perturbation method, to which end we introduced the smallness parameter. Without repeating the details from Chapter 2, we record the final result for the first approximation:

$$Y_0 = \left(1-\frac{1}{\rho}\right)\frac{\cos(\rho-t)}{\sqrt{1-K\left(1-\frac{1}{\rho}\right)^2 \exp\left(\left(1-\frac{1}{\rho}\right)(\rho-t)\right)}} \tag{9.58}$$

The asymptotic solution for radii larger than unity is:

$$Y_0 = \left(1 - \frac{1}{\rho}\right)\cos(\rho - t) \tag{9.59}$$

At any moment this gives a spatial pattern of radial waves, and for any given distance we get an approximation to the limit cycle. As we saw the solution is not relevant for distances less than unity, where the motion is damped out to eternal equilibrium. It is interesting to note that the space coordinate acts as a parameter causing bifurcation.

It must be emphasized that the present conclusions must be modified by all reservations due to the very crude first approximation, and we also have to remember that only dispersive wave phenomena have been dealt with.

9.10 Standing Waves

Another possibility to exploit is that of standing waves. Suppose that the rate of saving is small. We can find out the behaviour of the system for rates of saving that are sufficiently small if we can say something about the case where it is exactly zero, i.e., $s = 0$. The general differential equation then becomes:

$$\ddot{Y} - m\nabla^2 Y = (v-1)\dot{Y} - \frac{v}{3}\dot{Y}^3 \tag{9.60}$$

Suppose that the spatial waves are made up from plane facets alone. Then $\nabla^2 Y = 0$ everywhere. Suppose moreover that the rate of change over time jumps between three values:

$$\dot{Y} = 0 \tag{9.61}$$

$$\dot{Y} = \pm\sqrt{\frac{3(v-1)}{v}} \tag{9.62}$$

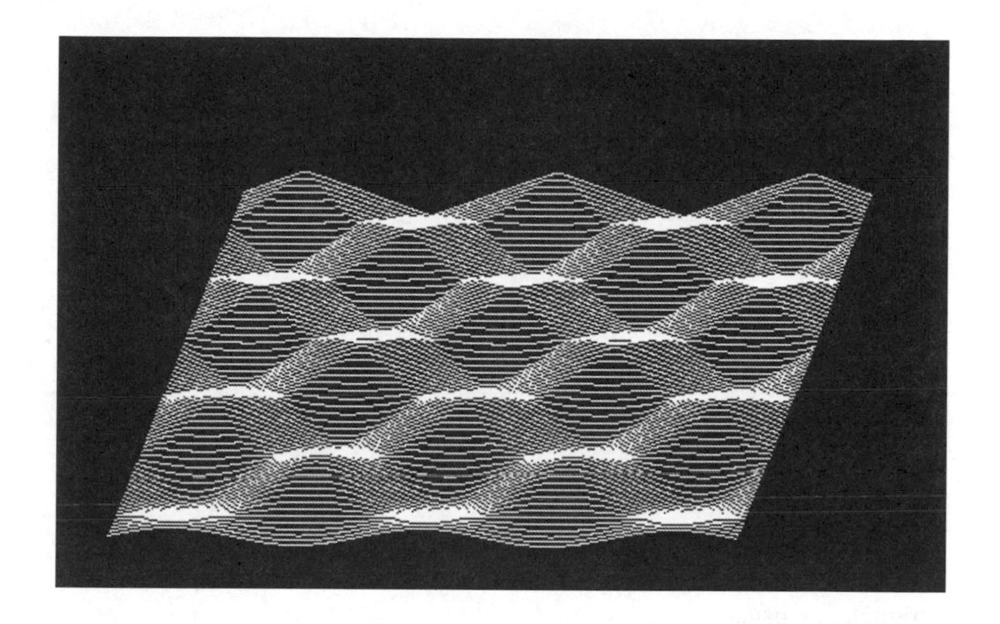

Fig. 9.2. Transformation of sine wave into triangular.

Then $\ddot{Y}=0$ except at isolated moments when the rate of change makes a jump. Such jumps occur in the spatial wave pattern too where the plane facets have common edges, but suppose we can disregard these discontinuities that occur at isolated spots in time and space.

When the second time derivative and the Laplacian are zero the left hand of the differential equation is zero. And when the first time derivative takes on any of the three values listed the right hand too is zero. Thus the differential equation is satisfied when the various plane facets are travelling at constant speed up or down or are unchanged.

For the one-dimensional case, where $\nabla^2 Y = \partial^2 Y / \partial x^2$. Myerscough, studying wind induced oscillations in overhead lines, did show that standing triangular waves were attractive solutions to this equation.

That this holds for small but nonzero s is shown in Fig. 9.2 where the results of simulations of the partial differential equation in time and one space dimension by the finite element method are shown. Initially regular sinoid waves of different wave numbers were introduced. Simulation invariably shows the transformation of the smooth sine waves into almost triangular ones, but in each case keeping the wave number from the initial conditions. We also see how temporal speed of the waves is related to their spatial wave number.

9.11 Perturbation Analysis

We can, however, also analyse the nonlinear model (9.52) with $s \neq 0$ using perturbation methods. Both Poincaré-Lindstedt, and two-timing are applicable. Let us restate (9.52) as:

$$\ddot{Y} + sY - m\nabla^2 Y = \varepsilon\left(\alpha\dot{Y} - \beta\dot{Y}^3\right) \tag{9.63}$$

where we defined

$$\alpha = (v - 1 - s) \qquad \beta = \tfrac{1}{3}v \tag{9.64}$$

in order to shorten the coming formulas, and also introduced the perturbation parameter as usual.

Next we try a power series solution:

$$Y = Y_0(\tau, x) + \varepsilon Y_1(\tau, x) + \ldots \tag{9.65}$$

where we introduced a new time scale:

$$\tau = (\omega_0 + \varepsilon\omega_1 + \ldots)t \tag{9.66}$$

Substituting (9.66) in (9.65), we obtain:

$$\dot{Y} = \omega_0 Y_0' + \ldots \tag{9.67}$$

$$\dot{Y}^3 = \omega_0^3 Y_0'^3 + \ldots \tag{9.68}$$

Further:

$$\ddot{Y} = \omega_0^2 Y_0'' + \varepsilon\left(\omega_0^2 Y_1'' - 2\omega_0\omega_1 Y_0''\right) + \ldots \tag{9.69}$$

$$\nabla^2 Y = \nabla^2 Y_0 + \varepsilon\nabla^2 Y_1 + \ldots \tag{9.70}$$

We included expansions up to the first power of the perturbation parameter, except for the first derivative and its third power where we stopped with the zero power terms because they are preceded by the perturbation parameter already.

As always, we denote derivatives with respect to the new time scale variable by dashes, whereas those with respect to real time are denoted by dots.

Substituting from (9.67)-(9.70) in (9.63) and requiring that the differential equation be fulfilled for each power of the perturbation parameter separately, we obtain the first two linear differential equations in an infinite sequence:

$$\omega_0^2 Y_0'' + sY_0 - m\nabla^2 Y_0 = 0 \tag{9.71}$$

and

$$\omega_0^2 Y_1'' + sY_1 - m\nabla^2 Y_1 = \alpha\omega_0 Y_0' - \beta(\omega_0 Y_0')^3 - 2\omega_0\omega_1 Y_0'' \tag{9.72}$$

Equation (9.71) is a linear Klein-Gordon equation. We can separate temporal and spatial coordinates to solve it, quite as in Section 9.4, though the formulas are not identical, because we presently do not have any linear friction term present in (9.71). Friction, of third order, along with the negative linear friction, was included in the perturbation term, and it turns first up in the following equation (9.72).

Supposing for specificity that our region is one dimensional, the interval $(0,\pi)$, with boundary conditions prescribing rest at the endpoints, we get the solution:

$$Y_0 = A_i \cos\tau \sin ix \tag{9.73}$$

Choosing a suitable origin for the time scale, we can make the coefficient of the temporal sine term zero, which shortens the formulas considerably. From (9.73) we find $Y_0'' = -A_i \cos\tau \sin ix$ and $\nabla^2 Y_0 = -i^2 A_i \cos\tau \sin ix$, so the solution (9.73) indeed fits the differential equation (9.71) provided the dispersion relation

$$\omega_0^2 = s + mi^2 \tag{9.74}$$

is fulfilled. Hence we see how the basic frequency component is determined by the model parameters, the propensities to save and to import, and by the spatial solution mode *i*. The higher the eigenvalue, the faster the vibrations are.

Note that (9.73) works for any integral *i*. Which mode it goes to is determined by the initial conditions, as Keller and Kogelman demonstrated.

Given the solution (9.73) we can proceed with next equation (9.72). We have $\omega_0 Y_0' = -\omega_0 A_i \sin\tau \sin ix$, so expanding the cubes of sines we obtain:

$$\alpha\omega_0 Y_0' - \beta(\omega_0 Y')^3 = \left(\tfrac{9}{16}\beta\omega_0^2 A_i^2 - \alpha\right)\omega_0 A_i \sin\tau\sin ix \tag{9.75}$$

$$-\tfrac{3}{16}\beta\omega_0^3 A_i^3 \sin\tau\sin 3ix - \tfrac{3}{16}\beta\omega_0^3 A_i^3 \sin 3\tau\sin ix$$

$$+\tfrac{1}{16}\beta\omega_0^3 A_i^3 \sin 3\tau\sin 3ix$$

as a substitution for the right hand side of (9.72). Likewise:

$$-2\omega_0\omega_1 Y_0'' = 2\omega_0\omega_1 A_i \cos\tau\sin ix \tag{9.76}$$

Considering a particular solution for equation (9.72) we note that the term $\sin\tau\sin ix$ in (9.75) and the term $\cos\tau\sin ix$ would lead to secular terms of the forms $\tau\cos\tau\sin ix$ and $\tau\sin\tau\sin ix$ respectively. To avoid this we put their coefficients equal to zero which is obtained by putting:

$$A_i = \frac{4}{3}\sqrt{\frac{\alpha}{\beta}}\frac{1}{\omega_0} \tag{9.77}$$

and

$$\omega_1 = 0 \tag{9.78}$$

Hence the frequency is not changed to the next approximation, and the amplitude of the basic vibration mode is determined by the structural coefficients of the model $\alpha = (v - 1 - s)$, $\beta = v/3$, and $\omega_0 = \sqrt{s + i^2 m}$.

Substituting from (9.75)-(9.76) in the right hand side of (9.72), and taking note of (9.77)-(9.78) we have:

$$\omega_0^2 Y_1'' + sY_1 - m\nabla^2 Y_1 = \tfrac{4}{27}\sqrt{\alpha^3\beta}\sin 3\tau \sin 3ix \tag{9.79}$$

$$-\tfrac{4}{9}\sqrt{\alpha^3\beta}\sin 3\tau \sin ix - \tfrac{4}{9}\sqrt{\alpha^3\beta}\sin \tau \sin 3ix$$

We can try a particular solution to (9.79) as a composite with undetermined coefficients of the three spatio-temporal vibration modes present in the right hand side. Substituting and matching coefficients we find:

$$Y_1 = -\frac{1}{54}\frac{\sqrt{\alpha^3\beta}}{s}\sin 3\tau \sin 3ix \tag{9.80}$$

$$+\frac{1}{18}\frac{\sqrt{\alpha^3\beta}}{s+i^2m}\sin 3\tau \sin ix - \frac{1}{18}\frac{\sqrt{\alpha^3\beta}}{i^2m}\sin \tau \sin 3ix$$

To the particular solution we add a complementary function which is a solution to the homogeneous for (with zero right hand side) of (9.72). The resulting differential equation is of the same form as (9.71), so the complementary function is of the same type as (9.73), but with coefficients determined by initial conditions.

We have now found the first two solutions to the series (9.65). The rest of the procedure is similar, though each step is computationally more messy. A third step would add a fifth harmonic, a fourth step a seventh harmonic and so forth. The superposition of odd harmonics only tends to produce a somewhat saw-toothed displacement curve, just as we saw in Section 9.10 as a result of simulation. However, we could not consider the perturbation solution for the case where $s \to 0$, because the first term in (9.80) then goes to infinity.

We should also recall that the basic vibration mode in terms of the number i was arbitrary. Even though we cannot now compound the solutions for different i in our present case where the differential equation (9.63) is nonlinear, we get a series of odd harmonics to each i once we have chosen it.

Each such series solution (9.65) is an attractor with its proper basin in the space of initial conditions.

By an advanced use of two-timing, which is beyond the scope of the present text, Keller and Kogelman demonstrated how the initial conditions pick out a given basic frequency, and how each such solution is locally stable. In principle, if we feed in one, three, or five sinoid waves on the given interval of length π as initial condition, the system picks this wave number, keeps it, and just adds the further odd harmonics to produce a sharper profile as time goes on.

10 Business Cycles: Discrete Time

10.1 Introduction

Although we for reasons given prefer to work with continuous time models it must be admitted that there are certain advantages in displaying the details of chaos for discrete time models. This is so because, before the tools of analysis, such as symbolic dynamics, can be applied to the continuous models we need to construct the return map on the Poincaré section for the orbit investigated. This, however, means that we first have to integrate the system over a complete cycle. The details of such an integration can easily become just much too complex.

It is easier to display the chaotic regime for a discrete mapping where the recurrence map exists from the outset and we thus do not need this first step of constructing it. In economics, in particular the modelling of business cycles, discrete time always was the main framework. In what follows the Samuelson-Hicks multiplier-accelerator model will be discussed, though changed to a nonlinear format.

The original model as proposed by Alvin Hansen and Paul Samuelson in 1939 could generate cyclical change, vaguely resemblant of real business cycles, and the model remained the basis for business cycle modelling for many years. When it was solved in closed form the weakness of the linear model became apparent. Due to the linearity, it could only produce exponentially explosive or damped amplitudes.

The reader may also find it absurd that the model had a zero equilibrium, and that income would oscillate around this value, becoming negative in the depression phases of the cycle. As a matter of fact the complete model, however, also includes various "autonomous" expenditures, such as government investments not being induced by the business cycle. Those expenditures result in a positive equilibrium income, a stationary particular solution, from which the homogeneous system presented produces the deviations. This superposition principle also works in all the nonlinear variations on the model dealt with in this Chapter.

10.2 Investments

The principle of acceleration claims that for technological reasons the stock of capital always is in a given proportion to production, i.e. to real income. Investments, being defined as the rate of change of capital, would therefore be in proportion to the rate of change of income. In discrete time the rate of change is just the difference between income in two successive periods. Though being crude, in the sense that a Leontief type of technology without substitution is implied, the idea has a certain appeal due to its straightforwardness.

Investments, themselves being among the determinants for income, the principle of acceleration provides for a feed back appropriate for formulating an interesting dynamics.

The underlying linear investment function was, however, questioned on factual grounds, as it implied active destruction of capital to keep the exact proportionality of capital stock to income whenever income declined at a faster rate than capital did in the complete absence of replacements.

Sir John Hicks in 1950 suggested to replace the linear investment function by a piecewise linear function with upper and lower bounds. Whenever income decreased faster than the natural rate of capital depreciation, realised when no machinery or buildings at all that wore out were replaced, disinvestment just stayed at the negative value of this natural depreciation.

Likewise, whenever income increased so fast that other factors of production, such as labour, land, or raw materials, became limiting, then there was no point in investing beyond a certain maximum amount.

As an alternative Richard Goodwin suggested that the upper and lower bounds could be approached asymptotically by a hyperbolic tangent type of investment curve. Both the piecewise linear Hicksian function, and the smooth Goodwin alternative are shown on top of Fig. 10.1.

Both types can be approximated by a linear-cubic Taylor series expansion, and this variant in particular presents possibilities for an interesting dynamics. This Taylor series function is, however, back-bending, which is a feature

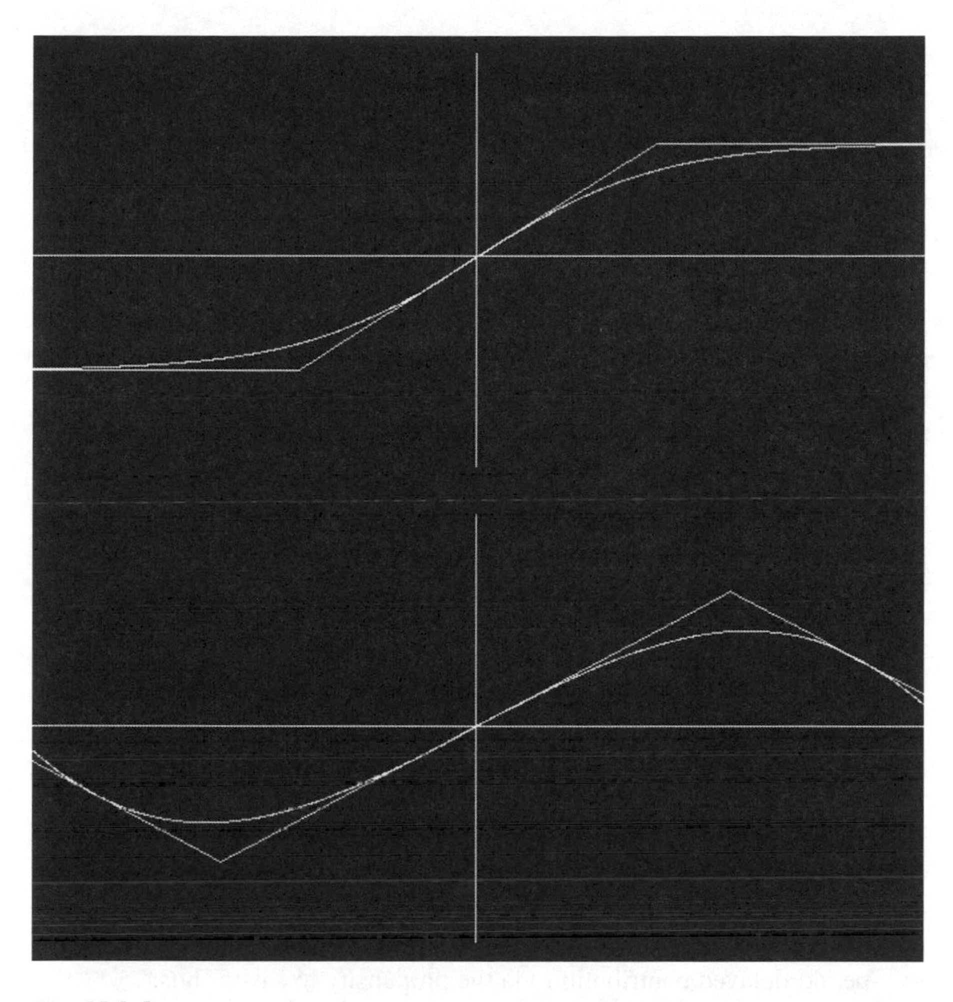

Fig. 10.1. Investments plotted versus income change. Top: Private investments alone with Hicksian floor and ceiling and smooth approximation. Bottom: Contracyclical public investments included, kinked variant and smooth cubic.

that the original functions do not share. As this back-bending is essential for some of the more spectacular phenomena produced, we should state a factual reason for it.

This is quite easy. We just include government investments which tend to be distributed contracyclically over the business cycle, partly as an attempt to counteract the cycle itself, partly in order to profit from the lower cost levels during periods of depression as compared to periods of prosperity. The way of financing these investments through a more severe taxation in periods of overheating sustains this effect. The lower part of Fig. 10.1 shows the cubic along with a piecewise linear function which takes account of the Hicksian argument and the contracyclical government policy.

Accordingly we assume total investments to be a cubic function in the rate of change of income:

$$I_t = v(Y_{t-1} - Y_{t-2}) - v(Y_{t-1} - Y_{t-2})^3 \tag{10.1}$$

In continuous time variants of the model we use the coefficient $v/3$ in the second term, which in fact appears for instance in the Taylor series for the hyperbolic tangent or for the arc tangent, but the second coefficient can be changed at will by simply rescaling the currency in which we measure income. A rescaling by the factor $\sqrt{3}$ does the job. Unlike the case of other Chapters, where we may need to differentiate the investment function, formalism is presently simplified by disposing of the denominator. Later on we will use further rescaling.

10.3 Consumption

As for consumption, suppose we have a distributed lag of exactly two periods for the disposal of income. Savings are for just one period and are completely consumed after the period has elapsed, so that, except the one-period delayed contribution to consumption via the propensity to consume there is also a two-period delayed contribution via the propensity to save. Thus:

$$C_t = (1 - s)Y_{t-1} + sY_{t-2} \tag{10.2}$$

We also have the definition of income formation by consumption and investments:

$$Y_t = C_t + I_t \tag{10.3}$$

Inserting the expressions (10.1) and (10.2) into the equation (10.3) we have, after a slight rearrangement:

$$Y_t - Y_{t-1} = (v - s)(Y_{t-1} - Y_{t-2}) - v(Y_{t-1} - Y_{t-2})^3 \tag{10.4}$$

and can now introduce the definition:

$$Y_t - Y_{t-1} = Z_{t-1} \tag{10.5}$$

so that we get the simple recurrence relation:

$$Z_t = (v - s)Z_{t-1} - vZ_{t-1}^3 \tag{10.6}$$

This was the purpose of our assumption that there was a distributed lag of exactly two periods. From a factual point of view this assumption is no worse than the usual one, according to which savings are for eternity. In view of model performance the case, however, represents an isolated specific case. We therefore later assume a general distributed lag pattern. Nevertheless, it is instructive to study the behaviour of equation (10.6) first, because we will need it for the discussion of the general case.

We now note that the second coefficient v can be given any numerical value we wish by a simple linear change of the unit of measurement of income. We can thus choose:

$$Z_t = \lambda Z_{t-1} - (\lambda + 1)Z_{t-1}^3 \tag{10.7}$$

where we recall that $\lambda = (v - s)$, being linear and thus not affected by the rescaling, is the difference of the accelerator and the propensity to save. (Past empirical measurements, whatever their value at this crude level, have evaluated this to the neighbourhood 2 to 4.)

10.4 The Cubic Iterative Map

The choice of coordinate change was made in order to ensure that the cubic function:

$$f_\lambda(Z) = \lambda Z - (\lambda + 1)Z^3 \tag{10.8}$$

should always pass the points (-1,1), (0,0), and (1,-1) in the diagrams to follow. These diagrams contain a square box with the interval [-1,1] as edges. Provided $\lambda < 3$ the cubic is contained in the box. Through the diagrams is drawn a diagonal by the help of which we can shift any outcome of an iteration (vertical coordinate) to become an initial value for the next iteration (horizontal coordinate). Thus the process of successive iterations can be traced in any number of steps just as in the cobweb model of price adjustments, or in innumerable expositions of the quadratic mapping iteration in the recent literature on chaos.

10.5 Fixed Points, Cycles, and Chaos

We note that for $\lambda > 1$ there are two fixed points defined by:

$$f_\lambda(Z) = Z \tag{10.9}$$

or

$$Z = \pm\sqrt{\frac{\lambda - 1}{\lambda + 1}} \tag{10.10}$$

These fixed points are stable if the absolute slope of the derivative does not exceed the unitary slope of the diagonal, i.e.,

$$\left|f_\lambda'(Z)\right| = \left|\lambda - 3(\lambda + 1)Z^2\right| = \left|3 - 2\lambda\right| < 1 \tag{10.11}$$

This condition obviously simplifies to:

$$\lambda < 2 \tag{10.12}$$

Any process starting in the interval [-1,1] then goes to one of the fixed points as shown in Fig. 10.2.

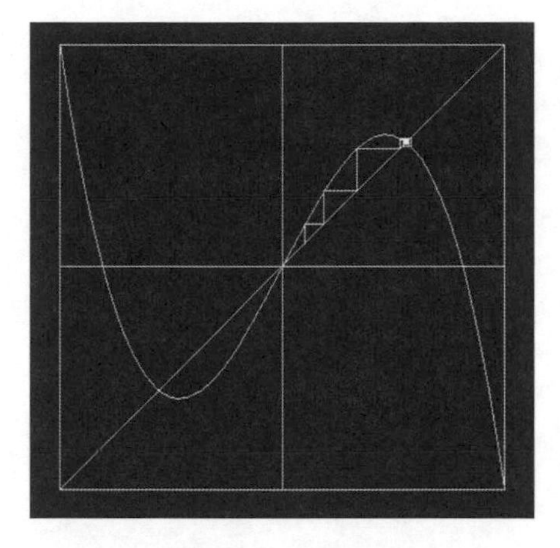

Fig. 10.2. Fixed point $\lambda = 1.9$.

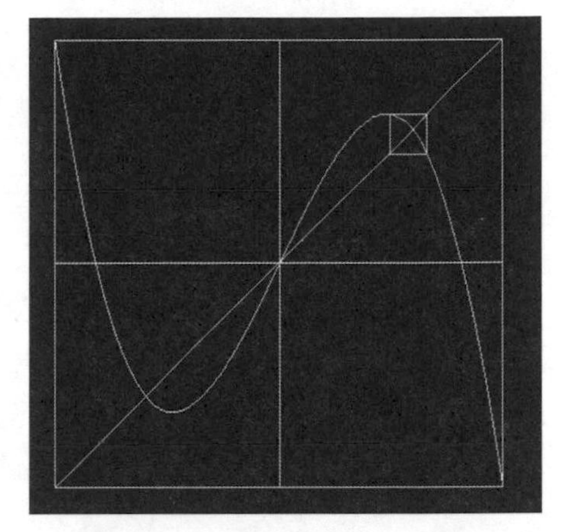

Fig. 10.3. A 2-cycle $\lambda = 2.1$.

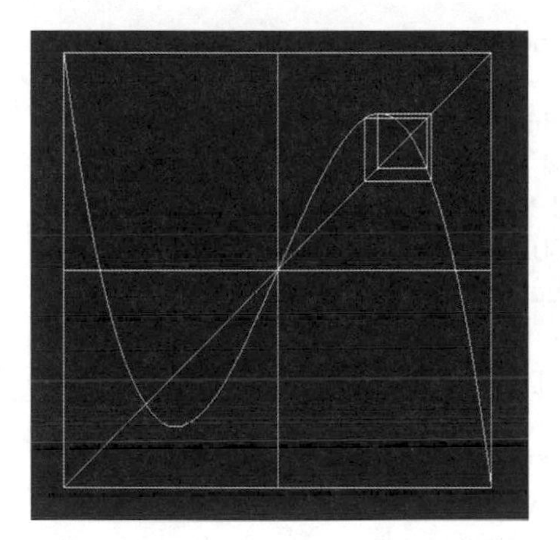

Fig. 10.4. A 4-cycle $\lambda = 2.25$.

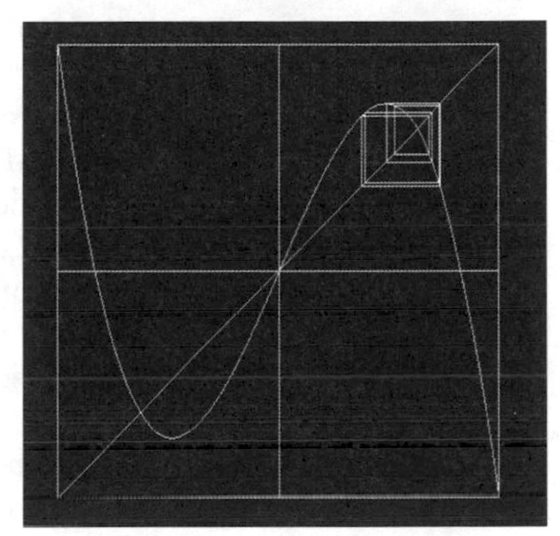

Fig. 10.5. An 8-cycle $\lambda = 2.295$.

At the critical value 2 of the parameter a bifurcation occurs, the fixed point loses stability, and in stead there emerges a stable cycle alternating between two values. Any process starting in the interval specified is attracted to the cycle. Alternatively expressed, the twice iterated map $f_\lambda(f_\lambda(Z))$, unlike $f_\lambda(Z)$ itself, retains stability, i.e. its derivative (evaluated at one of its fixed points) remains less than unity in absolute value.

In Fig. 10.3 we show the final 2-cycle for a parameter value 2.1, without obscuring the picture by the transitory process. Note that the fixed point and

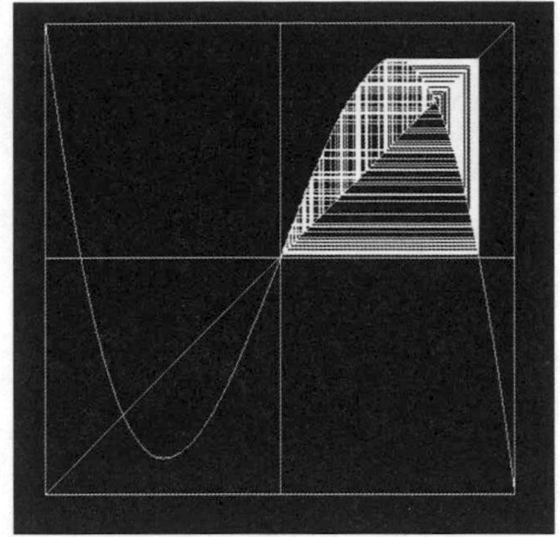

Fig. 10.6. Chaos $\lambda = 2.59$.

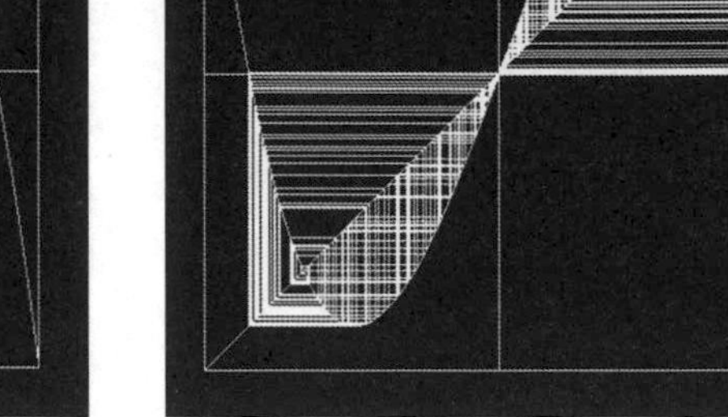

Fig. 10.7. Spillover $\lambda = 2.6$.

the cycle, as well as all the following ones, could equally well have occurred in the lower left quadrant.

Increasing the parameter value introduces further period doubling. In Figs. 10.4 and 10.5 we see cycles of period 4 and 8, occurring at parameter values 2.25 and 2.295 respectively. The period doubling points accumulate and chaos occurs around the parameter value 2.302, the so called Feigenbaum point.

The chaotic process is at first confined to that quadrant only where it starts, as we see in Fig. 10.6 where the parameter value is 2.59. In Fig. 10.7, obtained for the parameter value 2.6, the process spills over to both quadrants, no matter in which it starts.

The critical case occurs where the maximum (minimum) of the cubic equals the value at which the cubic has a zero, when the process from the extremum can be reflected by the diagonal into another quadrant.

An extremum occurs where:

$$f_{\lambda}'(Z) = \lambda - 3(\lambda+1)Z^2 = 0 \tag{10.13}$$

or

$$Z = \pm \frac{1}{\sqrt{3}} \sqrt{\frac{\lambda}{\lambda+1}} \tag{10.14}$$

The extremum value then is:

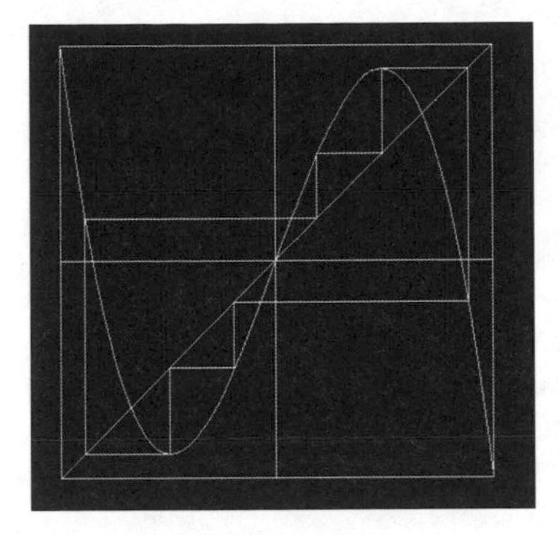

Fig. 10.8. Order regained $\lambda = 2.7$.

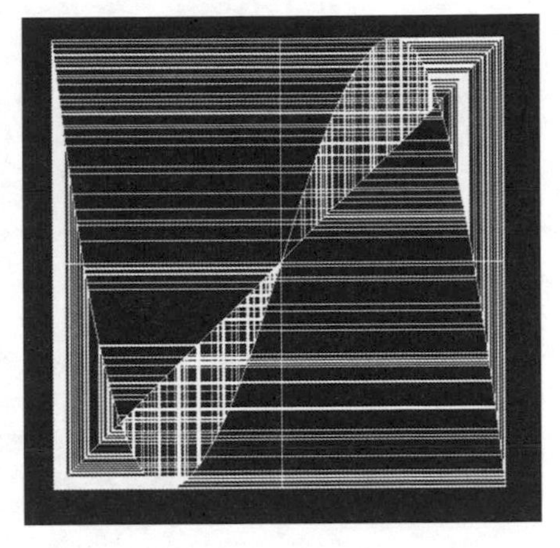

Fig 10.9. Chaos regained $\lambda = 3.0$.

$$f_\lambda(Z) = \pm \frac{2\lambda}{3\sqrt{3}} \sqrt{\frac{\lambda}{\lambda+1}} \tag{10.15}$$

The roots of the cubic, not counting the one at the origin, are:

$$Z = \pm \sqrt{\frac{\lambda}{\lambda+1}} \tag{10.16}$$

Thus the maximum (minimum) of the cubic equals its root when

$$\lambda = \frac{3\sqrt{3}}{2} \approx 2.5981 \tag{10.17}$$

The significance of this dividing point is that it separates growth processes from cyclic processes. To see this we have to recall that we after introducing the Z variable are dealing with successive income differences. As long as all values are positive we deal with growth, whenever their sign alternates the process is cyclical.

As in all chaotic processes there are opened up "windows" of ordered behaviour. This can occur in only one quadrant or in two as we show in Fig. 10.8, obtained for a parameter 2.7. This 6-cycle is attractive for all initial

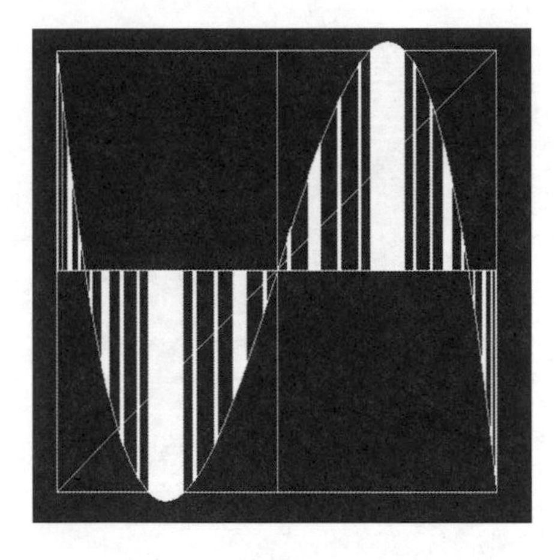

Fig. 10.10. 3-step escape $\lambda = 3.1$.

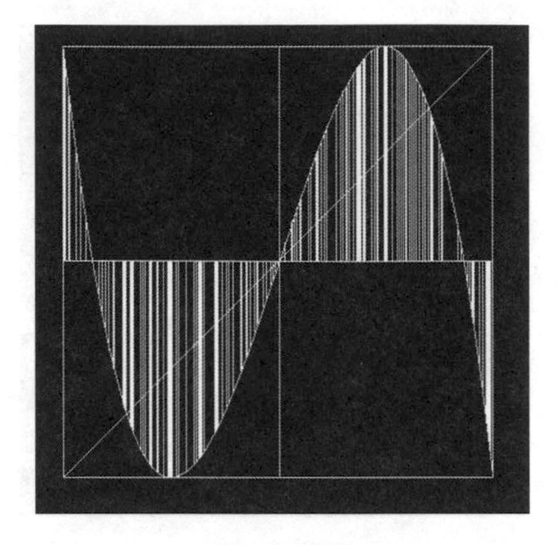

Fig. 10.11. 8-step escape $\lambda = 3.001$

conditions in the interval [-1,1]. (We should underline at this point that in chaotic regimes there exist all periodic solutions, but they are not attractive.)

When the parameter increases the ordered behaviour disappears again and we get chaos. For a parameter value of 3 chaos occurs on the whole interval (except the sparse set of rational points associated with unstable periodic solutions). See Fig. 10.9.

The case is a dividing line, because whenever the parameter exceeds 3 the model ceases to be reasonable. This is so as then the chaotic behaviour only remains on a sparse Cantor dust, whereas all other initial conditions make the system explode. The boundary case is also interesting as it enables us to explore all the details of chaos by analytical means.

We can intuitively see why the value 3 is critical. Whenever the parameter does not exceed 3 the cubic is confined within the box. If it exceeds 3, then the maximum and the minimum project outside the box, so that if we start in this part we move outside the box and the process spirals away in ever increasing steps. As long as the projecting portion is an interval, we are bound to hit it in a finite number of steps no matter where we start (except on the Cantor dust).

Figs. 10.10-10.11 show the regions of escape as white bars. For a parameter of 3.1 most of the interval [-1,1] is coloured white even when we only consider escape in three steps. With a parameter exceeding the critical value by as little as 3.001, still eight steps suffice to make the process escape from the white bars of Fig. 10.11.

In Fig. 10.12 the behaviour of the system over the whole range of parameter values is summarized. In the parameter range 1 to 2 there are two fixed

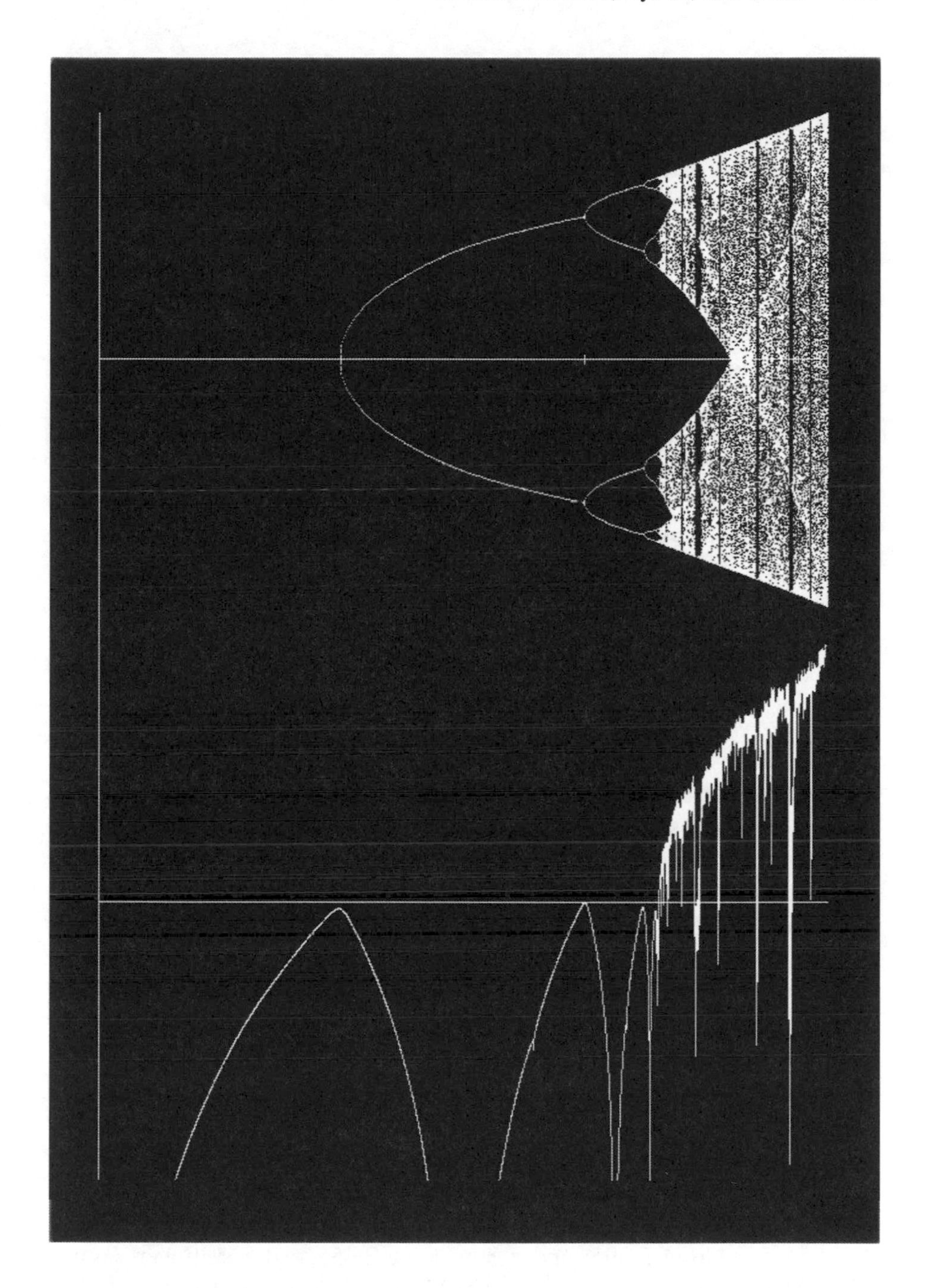

Fig. 10.12. Parameter intervals of fixed points, cycles, and chaos. Upper part: Amplitude versus parameter. Lower part: Lyapunov exponent versus parameter.

points (positive and negative), at 2 these bifurcate, and further bifurcations lead to chaos.

In reading the upper part of the diagram, where amplitude (vertical axis) is plotted against parameter (horizontal axis), we must remember that the bifurcations at parameter value 1, and at parameter value 2 are different. Thus at parameter value 1 the branching curve displays the emergence of two alternative (positive and negative) fixed points, the zero fixed point losing stability. The branching point at parameter value 2 shows the birth of oscillations between two different amplitudes in the same process. We can see the whole cascade of further period doubling bifurcations and finally the onset of chaos that is first confined to positive or negative amplitudes, but spills over around a parameter 2.598. We can also see the windows of ordered behaviour as narrow vertical strips.

The lower part of the diagram displays the Lyapunov exponent (vertical) plotted versus parameter (horizontal). For convenience we record the definition of the Lyapunov exponent:

$$L(Z_0) = \lim_{n\to\infty} \frac{1}{n} \sum_{i=1}^{i=n} \ln\left|f'_\lambda(Z_{i-1})\right| \tag{10.18}$$

Its sign tells us the stability or instability of fixed points and periodic orbits. We see that the curve displaying the Lyapunov exponent stays negative as long as there are stable fixed points and cycles, just touching the zero line each time a fixed point or cycle loses stability and is replaced by another one.

We can thus distinguish the loss of stability of the zero fixed point at parameter 1, the emergence of the 2-cycle at parameter 2, and its following replacements by the 4- and 8-cycles. The scale is too coarse to see anything of the following period doublings, accumulated at the Feigenbaum point where the curve definitely crosses the axis.

After that point the Lyapunov exponent curve dips in the negative region exactly where there are new attractive cycles. This shows up in the fact that the windows of order in the upper part of the diagram are located exactly above the dips of the Lyapunov exponent curve. For instance, in the region before the process spills over between the quadrants, there are a 6-cycle at 2.335, a 5-cycle at 2.395, and a 3-cycle at 2.455. After spillover, around 2.598 as we know, there is a 6-cycle at 2.7, a 4-cycle at 2.829, and another 6-cycle at 2.925.

Enlarging any of the window regions would display pictures of period-doubling routes to chaos like that displayed by the whole diagram. There is thus an infinite sequence of details self-similar to the complete diagram as usual in fractal sets. Accordingly we always find more details and more transitions between order and disorder the more we magnify resolution.

The occurrence of period-doubling is not surprising in view of the fact that the Schwarz derivative:

$$Sf_\lambda(Z) = \frac{f_\lambda'''(Z)}{f_\lambda'(Z)} - \frac{3}{2}\left(\frac{f_\lambda''(Z)}{f_\lambda'(Z)}\right)^2 = -\frac{6(\lambda+1)(6(\lambda+1)Z^2+\lambda)}{(3(\lambda+1)Z^2-\lambda)^2} \tag{10.19}$$

is negative for all parameter values considered.

10.6 Formal Analysis of Chaotic Dynamics

We are now able to proceed to a closer study of the chaotic process for a parameter value of 3. Then our recurrence relation reads $f_3(Z) = 3Z - 4Z^3$ or just

$$f(Z) = 3Z - 4Z^3 \tag{10.20}$$

For this case we can work out a simple coordinate transformation and so introduce the efficient tool of symbolic dynamics, which would otherwise require the advanced concept of topological conjugacy.

10.7 Coordinate Transformation

To this end we introduce the coordinate transformation:

$$Z = \sin(2\pi\theta) \tag{10.21}$$

However, from an elementary trigonometric identity:

$$f(Z) = 3\sin(2\pi\theta) - 4\sin^3(2\pi\theta) = \sin(6\pi\theta) \tag{10.22}$$

Therefore, there is a mirror image of the cubic mapping $f(Z)$:

$$g(\theta) = 3\theta \mod 1 \tag{10.23}$$

The modulus of the mapping means that we study the process on the circumference of the unit circle as is reasonable with a trigonometric transformation where we confine the argument to the main branch. In every iteration the angle (in radians) is simply multiplied by three, and any whole multiple of 2π is simply removed.

The successive multiplication by three yields a complete record of what is taking place in the iteration of the original cubic mapping. Though each step in the latter can be intuitively understood by the commonplace graphical technique used, the nature of the long run orbit is much better disclosed in terms of simple repeated multiplication.

10.8 The Three Requisites of Chaos

We can easily see that the requisites for chaotic motion are fulfilled: Sensitive dependence on initial conditions, topological transitivity, and density of periodic orbits. As for sensitive dependence take any two initial conditions differing by a tiny amount δ. After n iterations the difference has grown to: $\Delta = 3^n\delta$. This is an exponential and the iterates of any tiny initial interval eventually cover the whole circumference of the unit circle with n sufficiently large. Topological transitivity, or indecomposability of the orbit, can be shown by a similar argument. See Devaney.

As for the density of periodic points we proceed as follows. In order to have a periodic point we must start at a value θ for which

$$3^n\theta = \theta + m \tag{10.24}$$

with m an integer between zero and 3^n. Thus:

$$m = \left(3^n - 1\right)\theta \tag{10.25}$$

so that the n-periodic points are the $\left(3^n - 1\right)$:th roots of unity.

For instance if we consider 2-cycles, there are 8 different ones (not necessarily all distinct). They are obtained for

$$\theta = \frac{m}{8} \tag{10.26}$$

with m an integer between 1 and 8. By using the coordinate transformation (10.21) we see that the cycles are the fixed points $\pm 1/\sqrt{2}$ and 0, each counted twice, and the cycles -1, 1 and 1, -1 (thus also counted both ways). It may be noticed that the fixed points, of course, also are 2-cycles, as they are cycles of all periods we wish. These 2-cycles, as well as the fixed points, are unstable as demonstrated by any computer simulation. This is as it should be in a chaotic regime.

As n grows the density of periodic points grows exponentially, and we conclude that the set of periodic orbits is dense. This is the last requisite for identifying the dynamics of the model as chaotic.

10.9 Symbolic Dynamics

We can now introduce the shift map of symbolic dynamics explicitly. Suppose we write the numbers θ (defined on the unit interval) as continued triadic fractions with three digits 0, 1, 2 only. In this system a multiplication by "3", now written 10, is equivalent to shifting the "decimal" point one step to the right, and due to the modulus deleting any nonzero digits shifted to the left of the point. This, however, is nothing but the shift map of symbolic dynamics. The shift gives another clue to the concept of sensitive dependence on initial conditions. Take any two continued fractions that only differ in the n:th "decimal" place. After n iterations it is obvious that the difference is shifted to the first place.

We also see why the periodic orbits are so sparse. In terms of the shift map we understand that they correspond to periodic continued fractions. Now,

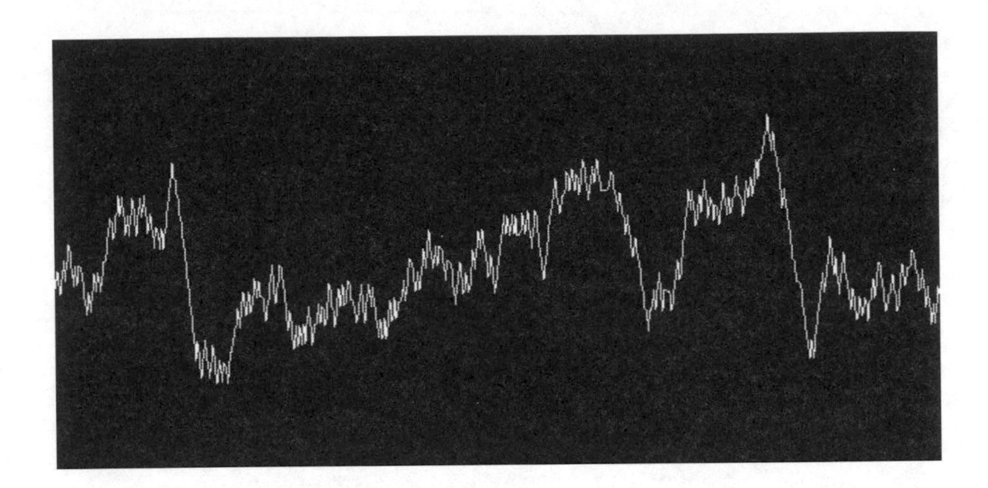

Fig. 10.13. Brownian random walk $\lambda = 2.9$ *. Iterations 1 to 1000.*

periodic sequences correspond to rational numbers in a one-to-one manner. (Binary, triadic, decimal or whatever makes no difference.) See Sondheimer and Rogerson. Thus the periodic orbits are as sparse as are the rationals among the real numbers.

We also understand that even a chaotic, i.e., non-periodic sequence is completely contained in the initial condition - but in order to predict the orbit we need to know an infinite number of digits. Only for periodic sequences is it sufficient to know a finite number of digits, i.e., so many that a complete period contains.

10.10 Brownian Random Walk

We should now consider that the resulting autonomous model in the variable Z represents successive income differences. The variable of interest to us, income Y, according to (10.5), is the cumulative sum of the outcomes of the iterative chaotic process. As the latter are like a sequence of random numbers we understand that the income variable rather moves like a Brownian particle. See Fig. 10.13, drawn for a parameter value of 2.9.

It is interesting to note that over very long periods this cumulation of random variables from a given "population" seems to create an image of cyclic

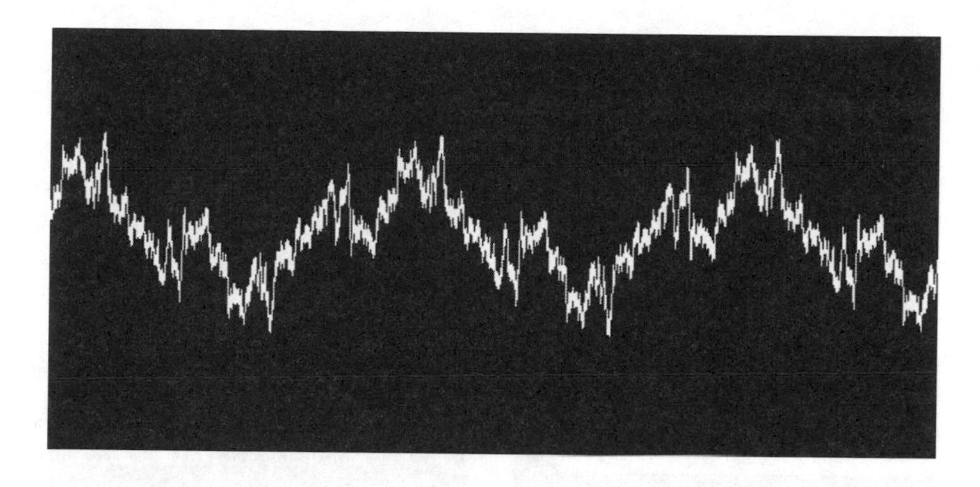

Fig. 10.14. Long-period cycles $\lambda = 2.9$ *. Iterations 5000 to 15000.*

recurrence (over thousands of iterations). This can be seen in Fig. 10.14 drawn for 10,000 iterations and the same parameter value 2.9. This is not altogether new. Chaotic movement yields a set of quasi random numbers, such as any simple random number generator on a pocket calculator. On the other hand the Gaussian probability distribution for many random variables, like the Maxwell-Bolzmann energy distribution in statistical mechanics, are instances where order is created out of disorder. Concerning the intriguing features of random walk processes we refer to Feller.

In the present case we certainly do not deal with random walk strictly speaking. The sequences of income differences generated by the chaotic mechanism are random in the sense that the outcome of next iteration cannot be predicted by any finite sequence of previous observations. This point was elaborated in Section 4.4.

On the other hand the chaotic output is far from normally distributed. Not only is there a finite range, a sub-interval of [-1,1], but the distribution tends to be bimodal with concentrations close to the extreme values. This is seen in the histogram in Fig. 10.15, computed for the same parameter value 2.9 as Figs. 10.13 and 10.14.

To get a rough understanding for why this picture of order is created we take a look at a slightly different case, for the parameter value 2.85. The corresponding time series is seen in Fig. 10.17. At a first glance it simply looks periodic.

This, however, is not true. The histogram, shown in Fig. 10.16, contains bars of a certain width which indicate whole intervals of amplitudes. We also

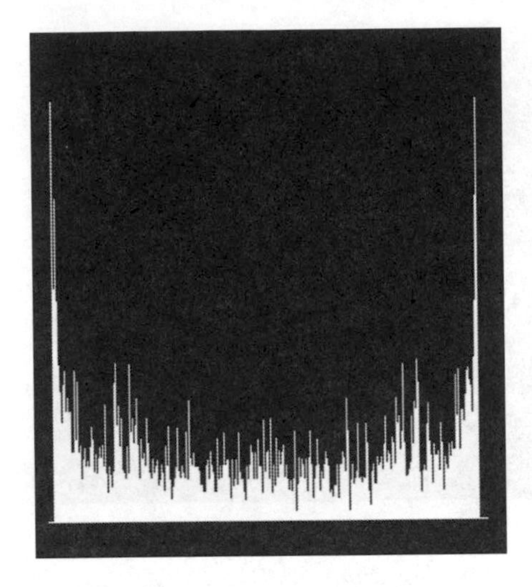

Fig. 10.15. Amplitude histogram $\lambda = 2.9$.

Fig. 10.16. Amplitude histogram $\lambda = 2.85$.

see in Figure 10.18, which is an enlargement of the lower part of Figure 10.12, the Lyapunov exponent plotted for the parameter interval [2.845,2.85]. It is definitely positive on the whole interval except two slight dips. This indicates that there is no attractive periodic solution at all for the parameter value 2.85 at the right end point.

The picture of order is created in another way. First, though the different amplitudes are not a finite set, they are sampled from a finite number of intervals with relatively narrow ranges. Second, as we see from Figure 10.19, the different amplitudes are sampled in an almost definite order as the staircase of iterations takes us through the diagram.

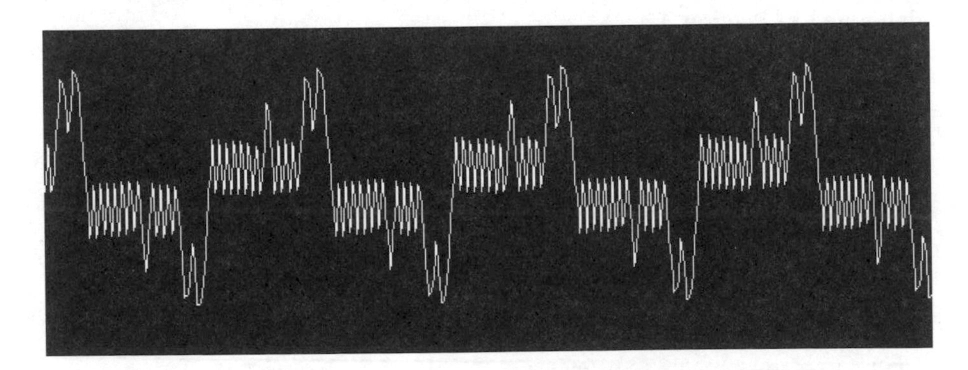

Fig. 10.17. Time series, iterations 1500 to 2000 $\lambda = 2.85$.

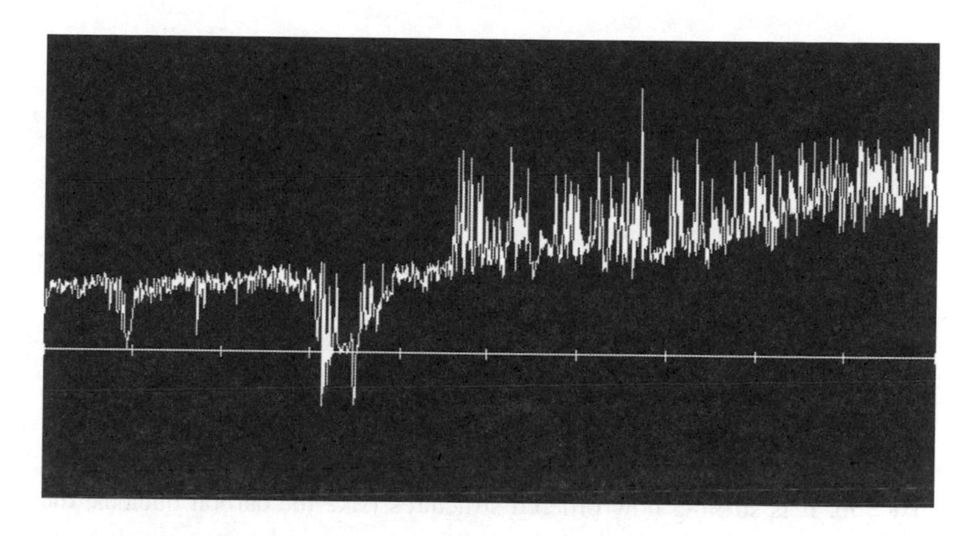

Fig. 10.18. Lyapunov exponent in the parameter interval (2.845,2.85).

From this we learn several things. Even in a chaotic process the chaotic bands may be so narrow that they are difficult to distinguish from fixed amplitudes. And, the tendency to step through the iteration diagram in a certain order in practice means sampling the various amplitudes in a certain order.

In our starting case there definitely is just one continuous band of amplitudes as can be seen from the first histogram, so the first argument is not

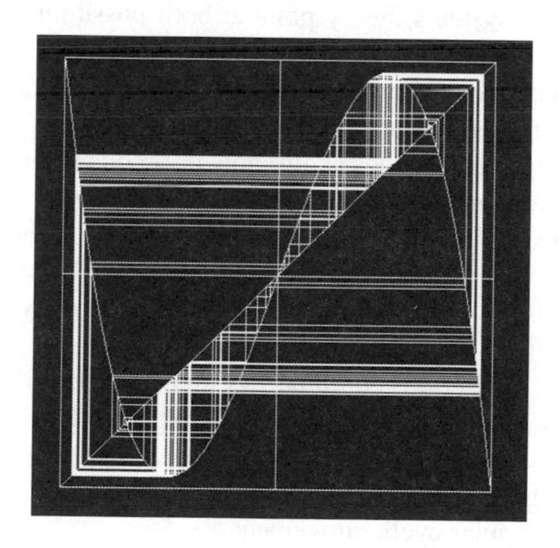

Fig. 10.19. Almost periodic process in the chaotic range $\lambda = 2.85$

relevant, but the second argument still holds. There is a tendency to step through the iteration diagram in a certain order of runs, only many more are required to create a picture of complete periodicity.

10.11 Digression on Order and Disorder

There is a beautiful film "Powers of Ten" by Morrison and Morrison taking the audience on a journey through the dimensions of the universe, from details of the carbon nucleus at the scale 10^{-16} m, to galaxy clusters at the scale 10^{24} m. It is striking how ordered structures (like the carbon nucleus, the DNA helix, the lymphocyte, our planet, and galaxy to pick but a few) alternate with disordered ones.

What Mandelbrot says about the nature of a ball of thread is applicable on the grandiose scale of the Morrisons. At distance the ball is a point. Closer it looks like a solid sphere. Still closer it becomes a mess of threads, next, each thread looks like a solid cylinder, and finally each of these is dissolved in a jumble of fibres.

As natural (or social) structures show up alternating order and disorder at different levels of resolution, it is tempting to imagine that ordered and disordered structures arise from each-other. As a matter of fact recent developments in dynamic systems theory point at both possibilities, i.e. synergetic self-organization and deterministic chaos.

Synergetics deals with complex systems organizing themselves by the action of a few order parameters and a wealth of damped (slaved) variables. See Haken. Deterministic chaos shows how unpredictable output is obtained from even the simplest recursive relations. See Devaney.

We are used to focus on one aspect at a time, gazing at the complex fractal sets that simple iterated systems can produce, or at the simple spatial patterns produced from apparently non-coordinated action by many independent agents.

In the present chapter we combine two steps in such a chain. A chaotic iterative mechanism is designed, not for the variable of interest to us, but for its successive differences, whereas the variable itself by aggregation (repeated "sampling" from the invariant population of chaotic output) can again be organized in regular cyclic movement.

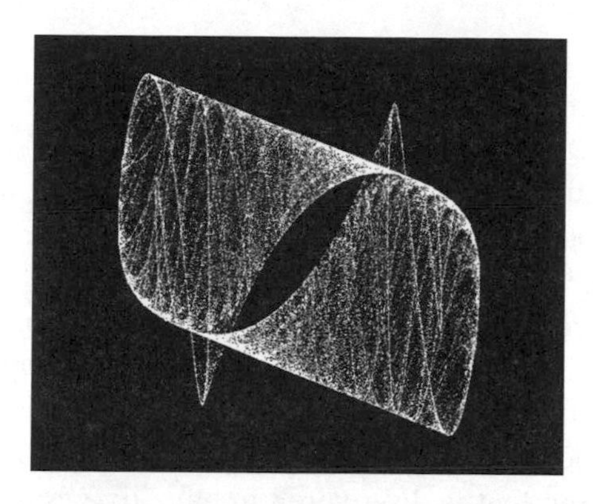

Fig. 10.20. Chaotic attractor. Savings rate = 12.5 percent.

10.12 The General Model

We will now take a look at the general case, where a fraction $0 < \varepsilon < 1$ only of savings is spent after one period. By way of introduction we dealt with the special case $\varepsilon = 1$, whereas traditionally $\varepsilon = 0$ is assumed. The special case served to uncouple equations (10.5) and (10.6). The latter could be studied as an autonomous one-equation process in income differences, and income itself could then be obtained as a cumulative sum of these differences.

This no longer works. Defining $\lambda = (v - s)$ and $\sigma = (1 - \varepsilon)s$ and rescaling as before, we now have the coupled system:

$$Y_t = Y_{t-1} + Z_{t-1} \tag{10.27}$$

$$Z_t = \lambda Z_{t-1} - (\lambda + 1)Z_{t-1}^3 - \sigma Y_{t-1} \tag{10.28}$$

The fact that we are dealing with a process in the Y_t, Z_t-space, however, implies that we can now draw phase diagrams (that of course may be expected to become fractal objects), as illustrated Fig. 10.20, drawn for parameters $\lambda = 2, \sigma = 0.125$. We note here that the exact value of ε is immaterial for the outcome as long as it is different from unity. We can treat the whole factor $\sigma = (1 - \varepsilon)s$ as an entity representing the "rate of eternal saving", i.e., the savings not spent at all according to the whole distributed lag system.

10.13 Relaxation Cycles

The following pictures show the particularly interesting metamorphosis of the chaotic attractor as the rate of saving approaches zero, the accelerator remaining unchanged. The vertical scale (income change) remains invariant, but the horizontal scale (income) is changed reciprocally to the rate of saving, more exactly to the eternal saving σ. As the "cycles" approximately retain the same size we conclude that the amplitude of the swings in the income variable are reciprocal to the rate of saving. Something that cannot be seen from the pictures, but that is discovered in any "live" simulation is that the cycle becomes slower and slower the smaller the rate of saving.

For a decreasing rate of saving the chaotic attractor becomes curve-like, as can be seen in Fig. 10.21. For even smaller rates of saving the loops become more close-wound, as can be seen in Fig. 10.22. Finally, in Fig. 10.23, at a vanishingly small savings rate, the cycle becomes a curious curve with two inserted copies of the bifurcation diagram from Fig. 10.12.

Of the two phase variables income differences move fast, but income moves slowly, especially when its total amplitude becomes large, as it does with a small savings rate. In equation (10.28) the last term is scaled down by the small rate of saving, but it remains a slowly moving item in the fast cubic iteration process, almost like a parameter. Thus income in the various phases of the cycle itself acts as a parameter in the iteration process for income differences, now causing ordered (cyclic) response, now disordered (chaotic). This gives an explanation for the occurrence of the bifurcation diagrams.

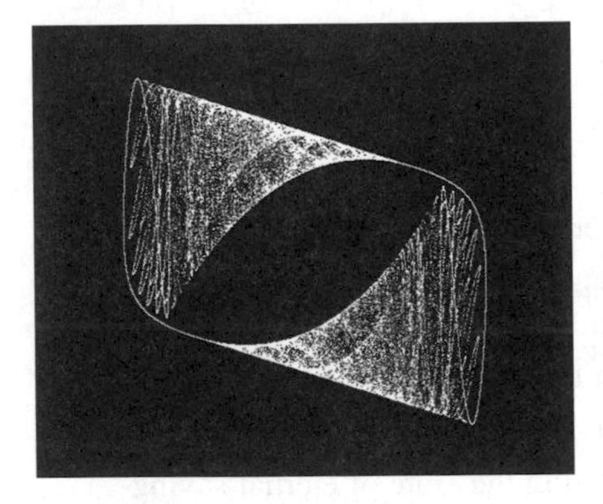

Fig. 10.21. Chaotic attractor. Savings rate = 5 percent.

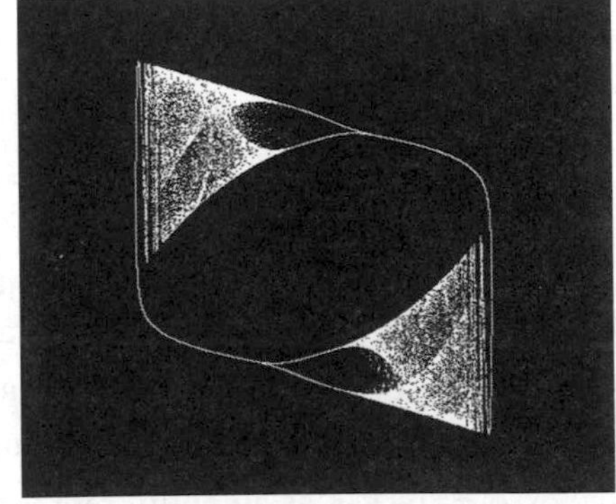

Fig. 10.22. Chaotic attractor. Savings rate = 1 percent.

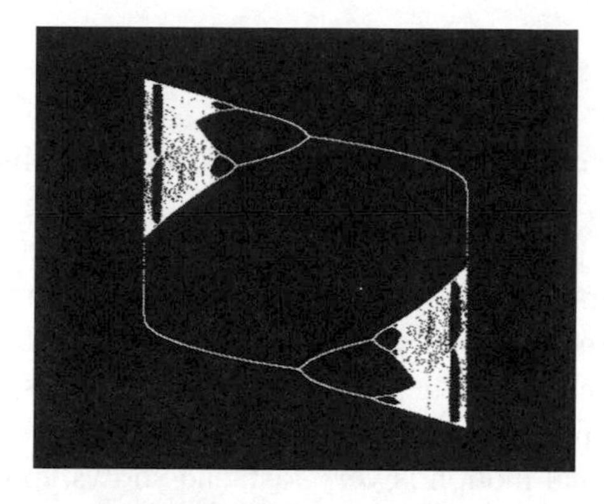

Fig. 10.23. Chaotic attractor. Savings rate vanishing.

Thus, a vanishing rate of saving has a destabilising effect on the model, causing the periods of a complete cycle to become longer and longer, and, accordingly, the amplitudes to become larger and larger.

This extreme case is interesting as it makes us see clearly the alternation of cyclic and chaotic behaviour with the slow variable acting as an internal parameter causing bifurcations. This again brings the synergetics of Haken in mind. In the next section we are going to study the feed back process by the slow variable in detail.

Except the inserted bifurcation diagrams Fig. 10.23 shows a relaxation cycle. Let us form the quotient $(Z_t - Z_{t-1}) / (Y_t - Y_{t-1})$ from equations (10.27)-(10.28), and divide by σ. Then obviously:

$$\frac{Z_t - Z_{t-1}}{X_t - X_{t-1}} = \frac{1}{\sigma} \frac{(\lambda - 1)Z_{t-1} + (\lambda + 1)Z_{t-1}^3 - X_{t-1}}{Z_{t-1}} \tag{10.29}$$

where we have re-scaled the Y variable by

$$X_t = \sigma Y_t \tag{10.30}$$

This in fact corresponds to our rescaling of the income variable in the pictures. Equation (10.29) gives the field directions in the X_t, Z_t-phase diagram. Due to the smallness of the rate of saving the first factor in the right hand side of (10.29) is very large. Thus, except when the numerator of the second factor is zero, the field directions are vertical. The numerator is zero whenever:

$$(\lambda-1)Z-(\lambda+1)Z^3-X=0 \tag{10.31}$$

is fulfilled. For simplicity we have deleted the period index, (t-1). Equation (10.31) defines a cubic in phase space.

The relaxation cycle follows the cubic, called characteristic, except its unstable rising branch. As the field directions were seen to be vertical everywhere except at the characteristic, the cycle tends to follow its stable branches, and jumps vertically to the next stable branch when it is no longer possible to follow the back bending characteristic.

Any such vertical motion is very fast, and shows up by the fact that only isolated points are marked in the vertical sections in Fig. 10.23, although several hundred thousands of iterations were computed. This is all there is in a normal relaxation cycle. To this bifurcations and chaos are presently added.

By solving for X in (10.31), differentiating with respect to Z, and equating the derivative to zero, we can locate the extrema of the cubic as:

$$Z=\pm\sqrt{\frac{\lambda-1}{3(\lambda+1)}} \tag{10.32}$$

The corresponding extremal values of X accordingly are:

$$|X|_{\max}=\frac{2}{3}\sqrt{\frac{(\lambda-1)}{3(\lambda+1)}} \tag{10.33}$$

Knowing that the relaxation cycle jumps at the extrema of the cubic characteristic we note that we have just estimated the amplitude of the cycle and thus, according to definition (10.30), confirmed the conjecture that it is reciprocal to the eternal rate of saving.

We should note that all motion is clockwise in phase diagrams. Income increases regularly, but at a decreasing rate on the upper branch of the cubic, then stops, whereas the rate of change makes a very dramatic drop entering the chaotic region, and income starts to decrease. After a while the process becomes more orderly, income drops but at a retarding pace. Then there is a sudden jump up in the phase diagram, and a new chaotic region is entered. Again, this chaotic behaviour of income differences yields to order, and we are back where we started.

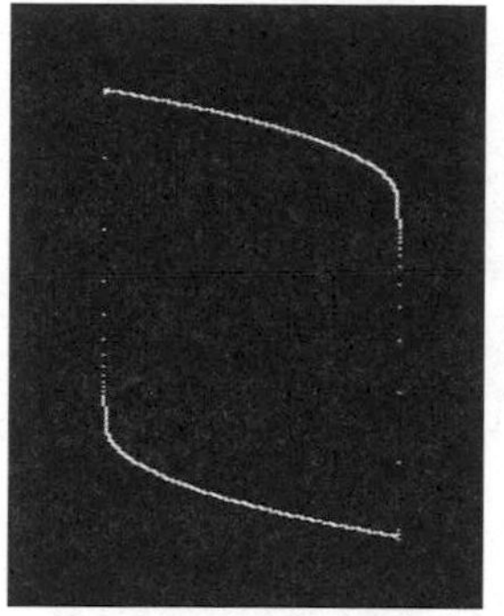

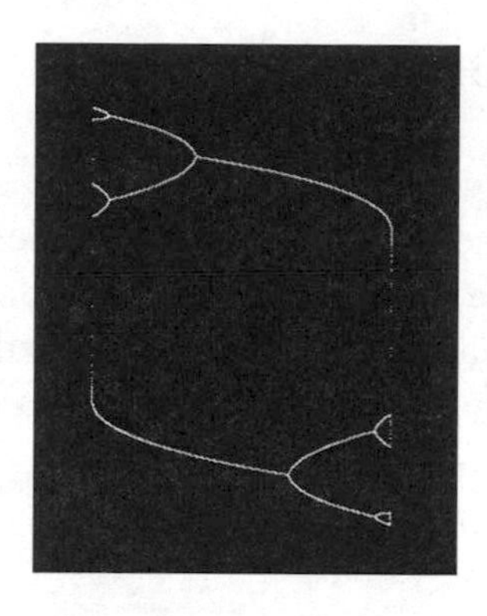

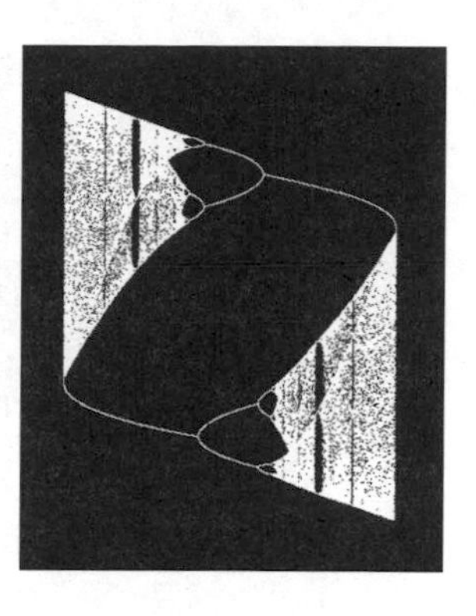

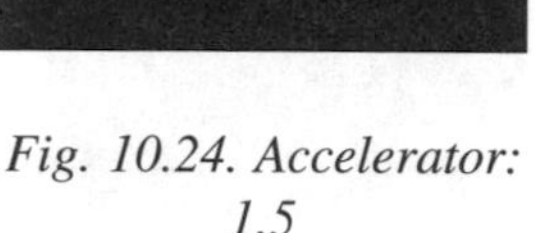

Fig. 10.24. Accelerator: 1.5

Fig. 10.25. Accelerator: 1.85

Fig. 10.26. Accelerator: 2.15

As we saw the slow variable causes bifurcations in the behaviour of the fast one, and we note that we have all the phenomena of period doubling routes to chaos, and even the windows of regained order particularly at period three. But due to the clockwise motion the process is backward, i.e. we have "period halving routes to order", whereas chaos sets in suddenly.

We observe that chaos occurs just after the turning points, at the hills of prosperity and the valleys of depression. This has the character of a testable implication. Although these phenomena are interesting, we should not push the interpretations of details too far. We must remember that the model is on the same crude level as the original Samuelson-Hicks models, and thus has a symbolic value rather than being pictorially realistic, provided there is any such thing in science.

Throughout the present discussion we kept the accelerator constant at the value 2. In order to see something more of the possibilities we take a look at some different values of this parameter, keeping the others constant. Fig. 10.24 shows the familiar case of a pure relaxation cycle. In Fig. 10.25 the period doublings are introduced, but not yet chaos. In Fig. 10.26, finally, the accelerator is higher than in Fig. 10.23, and the chaotic regime now extends over a large section of the cycle.

10.14 The Slow Feed Back

To understand more of the case of relaxation cycles when the rate of saving is small let us formalize the intuitive distinction between slow and fast variables. Y is the slow variable, whereas the Z is the fast one. Accordingly, we can treat the last term in (10.28) as if it were a constant parameter

$$\mu = \sigma Y_{t-1} \tag{10.34}$$

Accordingly the iterated map becomes:

$$Z_t = f_{\lambda\mu}(Z_{t-1}) \tag{10.35}$$

with

$$f_{\lambda\mu}(Z) = \lambda Z - (\lambda + 1)Z^3 - \mu \tag{10.36}$$

where we have again introduced the change of scale used for defining equation (10.7).

10.15 The Autonomous Term: Changes of Fixed Points

Let us see what difference the addition of a "constant" μ makes for (10.36) as compared to (10.8). In equation (10.10) we computed the fixed points of (10.8), so let us define a lower case z variable as the deviation from any of these fixed points:

$$Z = \pm\sqrt{\frac{\lambda - 1}{\lambda + 1}} + z \tag{10.37}$$

The fixed points of (10.36) are defined by:

$$(\lambda - 1)Z - (\lambda + 1)Z^3 - \mu = 0 \tag{10.38}$$

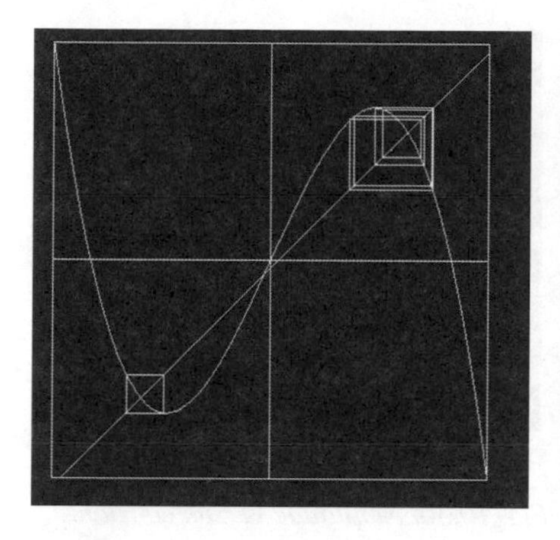

Fig. 10.27. The cubic iteration.
Parameters: $\lambda = 2.2$ $\mu = 0.04$.

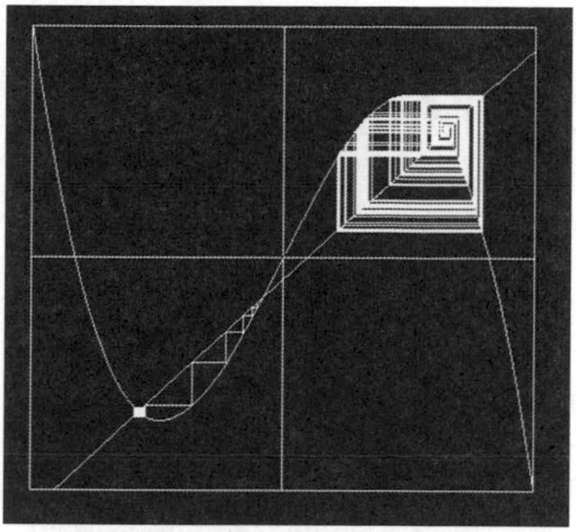

Fig. 10.28. The cubic iteration.
Parameters: $\lambda = 2.2$ $\mu = 0.1$.

Substituting from (10.37) into (10.38) and cancelling any powers of z, which is small, we obtain:

$$2(\lambda - 1)z \approx \mu \tag{10.39}$$

We note that the fixed points are changed in the same direction as the constant term whenever $\lambda > 1$ holds, which it must if we want to have any fixed points at all.

Computing the derivative for the mapping (10.36), substituting from (10.37) and again deleting all powers of z we get:

$$Df_{\lambda\mu} \approx 3 - 2\lambda \mp 6\sqrt{\lambda^2 - 1}z \tag{10.40}$$

Comparing (10.40) to (10.11) we find that the derivative is increased at one of the fixed points and decreased at the other. This stabilizing or destabilising effect does not only refer to the fixed points, but to the cycles centred around them as well. Thus, the autonomous term works as a change of the λ parameter, except that it introduces an asymmetry between the processes in the positive and negative quadrants of Figs. 10.2 through 10.9.

This is confirmed by thinking of the added constant as working a corresponding shift of the diagonal line in the Figs. 10.2 through 10.9, and using

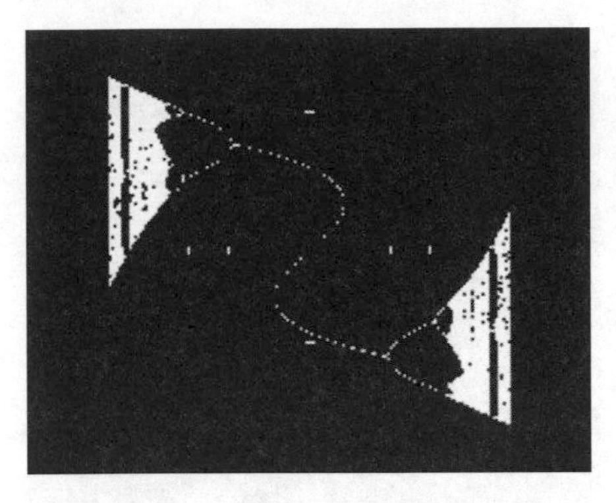

Fig. 10.29. Amplitude versus constant. $\lambda = 1.50$

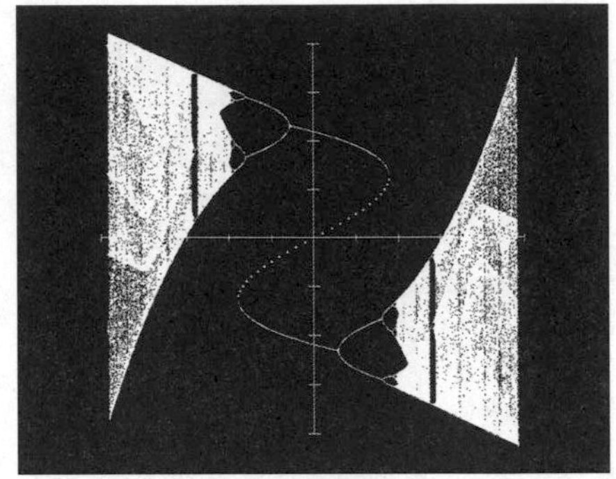

Fig. 10.30. Amplitude versus constant. $\lambda = 1.85$

the shifted diagonal for constructing the iteration staircase exactly as it was used in the original model. This is illustrated in Figs. 10.27-10.28. In Fig. 10.27 we find cycles of different periodicity simultaneously in the upper right and lower left quadrant. In Fig. 10.28 a fixed point even coexists with chaos. This is a consequence of the asymmetry caused by adding the constant term. The reader should be warned that in Figs. 10.2-10.6 we chose to illustrate the process in only one quadrant. There was no point in drawing two identical copies of the process, and doing this would have obscured the phenomenon of spillover. Presently the processes are different in the different quadrants, and so there is a point in drawing the attractors both for negative and positive initial conditions.

We can, however, again visualize the individual steps only in the iteration process, but the long run fate of the orbit with one fixed point stabilized, the other destabilised, is difficult to predict.

10.16 The Autonomous Term: Response of the Chaotic Process

More can be found out if we think in terms of the graphics of Fig. 10.12, plotting the amplitude against the parameter. Presently we have two parameters, λ and μ, so we have to keep one fixed.

In the Figs. 10.29-10.32 we keep λ fixed and plot the amplitude (vertical coordinate as before) against μ, ranging in the interval [-0.5,0.5]. We see

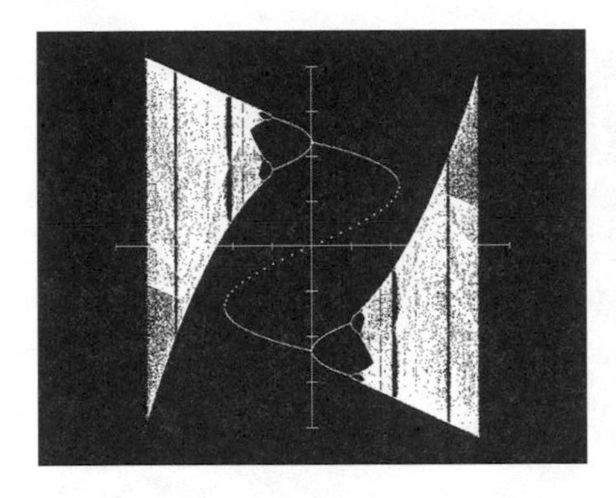

Fig. 10.31. Amplitude versus constant. $\lambda = 2.00$

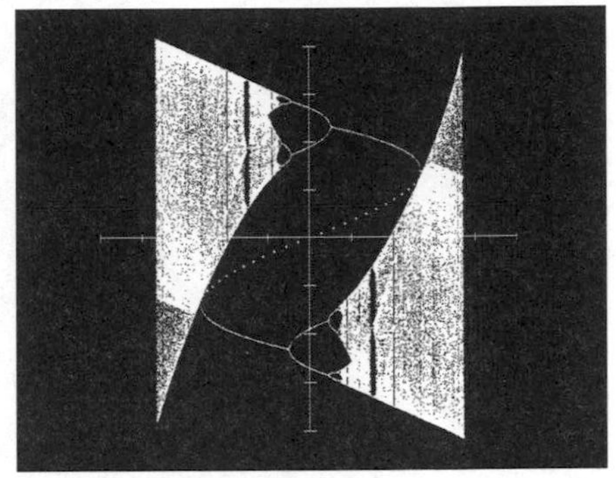

Fig. 10.32. Amplitude versus constant. $\lambda = 2.15$

that in each of the four pictures there are two copies of the bifurcation diagram from Fig. 10.12. For a too large (positive or negative) μ we get chaotic response, and there is an easily recognizable period doubling route to chaos in both directions. For small absolute μ there is also the option of a fixed point.

This option stops to exist at very definite points, i.e., the turning points of a cubic, inserted as a dashed curve. The equation of this cubic is (10.38).

The pictures illustrate different cases as regards the position of the turning points of the cubic relative to the chaotic region. In Figs. 10.31 and 10.32 they lie on vertical lines through the interiors of the chaotic regions. In Figs. 10.29 and 10.30 they do not extend as far. In Fig. 10.32 they actually enter the chaotic region.

We can now understand how the different relaxation cycles depicted in Figs. 10.24 through 10.26 arouse. When the system has to leave the characteristic at its turning points it can jump to another branch of itself as in Figs. 10.24 and 10.29. It can also jump to a quadruple period cycle as in Figs. 10.25 and 10.30, or to the chaotic region as in Figs. 10.23 and 10.31. Finally, it can enter the chaotic region without having to jump, as in Figs. 10.26 and 10.32. Obviously the outcome is completely determined by the position of the turning points of the cubic with reference to the period doubling and chaotic regions.

The turning points are obtained from (10.38) by solving for μ, differentiating with respect to Z, and equating the derivative to zero. In this way we find that

$$Z = \pm\sqrt{\frac{\lambda - 1}{3(\lambda + 1)}} \tag{10.41}$$

Substituting back in (10.38) we find:

$$\mu = \pm\frac{2}{3}\sqrt{\frac{(\lambda - 1)^3}{3(\lambda + 1)}} \tag{10.42}$$

On the other hand the general definition of the fixed points of (10.36) was given in (10.38). These fixed points lose stability when the derivative of the mapping has unitary absolute value, i.e., where

$$f'_{\lambda\mu}(Z) = \lambda - 3(\lambda + 1)Z^2 = -1 \tag{10.43}$$

holds (we know that the derivative is negative). This yields the solutions

$$Z = \pm\frac{1}{\sqrt{3}} \tag{10.44}$$

Substituting into (10.38) we get the corresponding values of the constant:

$$\mu = \pm\frac{2}{3}\frac{(\lambda - 2)}{\sqrt{3}} \tag{10.45}$$

This tells us that the first period doubling points lie on the vertical axis whenever $\lambda = 2$, which indeed is confirmed by looking at Fig. 10.31. The case of a turning point lying exactly below/above the first bifurcation point is obtained by equating μ according to (10.42) and (10.45). Squaring, all powers higher than the linear cancel, and we obtain:

$$\lambda = \frac{5}{3} \tag{10.46}$$

Simulation experiments show that indeed the simple relaxation cycles ceases to exist for a parameter $\lambda = (v - s) \approx 1.67$. From Figs. 10.24 and 10.25 we already know that it happens in the interval between 1.5 and 1.85.

Figs. 10.31 and 10.32 are different in the respect that in the latter the turning points actually enter the chaotic region. We could expect that if they project still further, to the ends of the chaotic regions (where the model no longer converges) then the model explodes. This happens at a parameter value $\lambda \approx 2.27$ as can also be confirmed experimentally.

10.17 Lyapunov Exponents and Fractal Dimensions

In equation (10.18) we defined a Lyapunov exponent as a measure of the average separation of close initial conditions over a process, and hence as an indication of chaos according to the prerequisite of sensitive dependence. The definition concerned a one-dimensional process. We can do the same for two dimensions.

It, however, becomes a bit more tricky, because in two dimensions the equivalent of the derivative in one dimension, i.e. the gradient of our system (10.27)-(10.28), changes direction all the time. For this reason we have to project any new gradient at a new iteration upon the direction of the previous one, and measure the relative length of the projection of the new gradient to the length of the previous one in order to get at the relative increase in separation at the particular stage. That being done, we can take logarithms, sum, and divide by the number of observations, just as in equation (10.18).

Formally, define the direction θ_t and the unit direction vectors:

$$\mathbf{v}_t = \begin{bmatrix} \cos\theta_t \\ \sin\theta_t \end{bmatrix} \tag{10.47}$$

The Jacobian matrix of partial derivatives to equations (10.27)-(10.28):

$$\mathbf{J}_t = \begin{bmatrix} 1 & 1 \\ -\sigma & \lambda - 3(\lambda + 1)Z_t^2 \end{bmatrix} \tag{10.48}$$

is projected on the direction $\mathbf{v}_t$:

$$\mathbf{J}_t \cdot \mathbf{v}_t = \begin{bmatrix} \cos\theta_t + \sin\theta_t \\ -\sigma\cos\theta_t + \left(\lambda - 3(\lambda+1)Z_t^2\right)\sin\theta_t \end{bmatrix} \tag{10.49}$$

The largest Lyapunov Exponent now becomes:

$$L = \lim_{n\to\infty} \frac{1}{n} \sum_{i=1}^{i=n} \ln\left|\mathbf{J}_t \cdot \mathbf{v}_t\right| \tag{10.50}$$

The condition for updating the direction vectors is:

$$\mathbf{v}_{t+1} = \frac{\mathbf{J}_t \cdot \mathbf{v}_t}{\left|\mathbf{J}_t \cdot \mathbf{v}_t\right|} \tag{10.51}$$

and the choice of initial direction θ_0 is arbitrary.

We should also recall that in two dimensions there are always two Lyapunov exponents. Whenever the process does not just explode, but approaches some equilibrium, cycle, or chaotic attractor, then the sum of the Lyapunov exponents is negative, though one of them can be positive. This means that iterating a collection of initial conditions one step, say those enclosed within a circular disk, is stretched in one direction but compressed in another, the circle becoming an oblong ellipse, though of an area smaller than the circle. The sum of the Lyapunov exponents in fact measures the exponential rate of area shrinking.

What we have been aiming at is the largest Lyapunov exponent, because it measures the sensitivity to initial conditions, a positive value being an indication of chaos. In the following pictures we display that exponent by the symbol L.

Another measure of interest is the fractal dimension. There are many procedures once we get into detail, but they all aim at the same thing. We want the measure to report dimension zero for a collection of disjoint points, dimension one for a curve, and dimension two for a surface. For some dust of points almost accumulating to something curve like we expect a dimension between zero and one, for a tangle of curves almost accumulating to a patchy space filling pattern we expect a dimension between one and two. Typically,

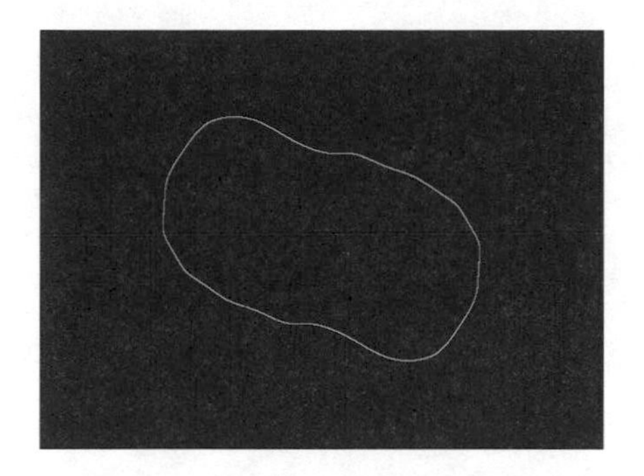

Fig. 10.33. Attractor F=1, *L*<0
$\lambda = 1.25 \quad \sigma = 0.225$

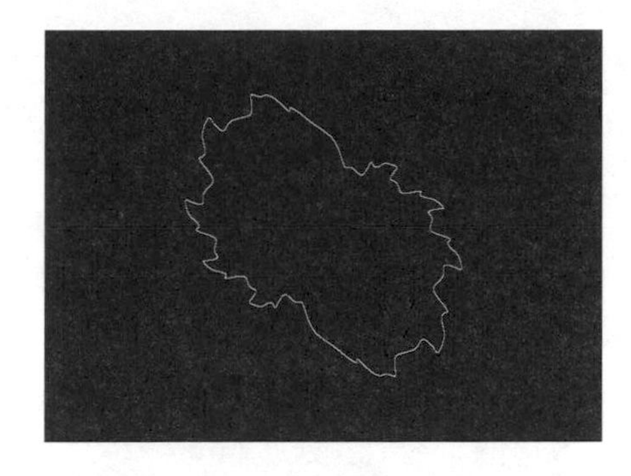

Fig. 10.34. Attractor F=1, *L*<0
$\lambda = 1.25 \quad \sigma = 0.375$

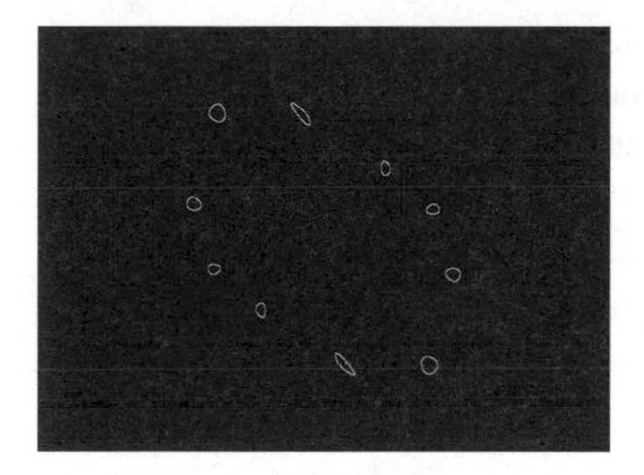

Fig. 10.35. Attractor F=1, *L*<0
$\lambda = 1.25 \quad \sigma = 0.435$

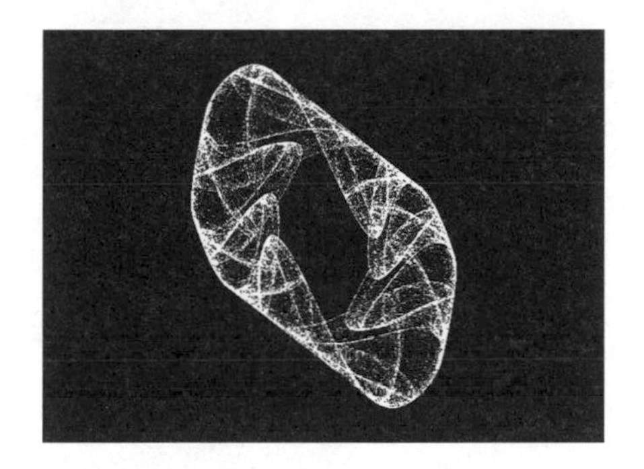

Fig. 10.36. Attractor F=1.3, *L*=0.12
$\lambda = 1.25 \quad \sigma = 0.525$

road networks approach fractal dimension two, and tracheas in the lungs even approach fractal dimension three, even though the topological dimension is one in both cases.

The measure we use counts the numbers of points in a circle centred on a point at the object and in another concentric circle of say double its radius. If the object is a curve then it is obvious that the number in the larger circle is double the number in the smaller. If we have a piece of the plane, then obviously there are four times as many points in the larger circle. Dividing the numbers and taking the logarithm we expect to get the fractal dimension. It is denoted F.

It should be noted that none of the definitions of either Lyapunov exponent or of fractal dimension are undisputed, and that the numerical estimates also

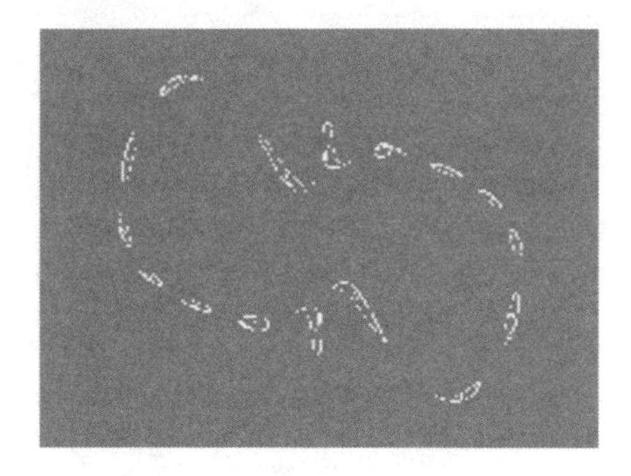

Fig. 10.37. One piece attractor
$\lambda = 1.5 \quad \sigma = 0.240$

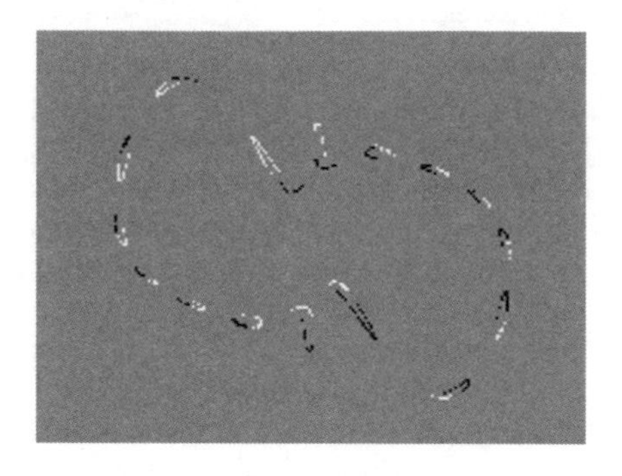

Fig. 10.38. Competing attractors
$\lambda = 1.5 \quad \sigma = 0.239$

have all the weaknesses of limited precision due to finite computation precision and time. Nevertheless they give interesting indications on the presence of chaos and on the fractal type of the attractor.

10.18 Non-Relaxation Cycles

Let us now take a look at the more general cases where the rate of saving is not vanishing and the cycles not of the relaxation type. In those cases we cannot get ahead with pure analysis, but we have to take recourse to computer simulations. Those reveal all sorts of bifurcations. The sequence of pictures in Figs. 10.33-10.36 display some cases when in $\lambda = 1.25$ in (10.28).

In Fig. 10.33, with $\sigma = 0.225$, we see that the fixed point (eternal equilibrium at the origin) has undergone a Hopf bifurcation to a cycle. In Fig. 10.34 at $\sigma = 0.375$ the cycle has got an irregular leaf like character with ten pointed corners. Those indicate an approaching global saddle-node bifurcation into a ten period regular cycle, which indeed occurs.

After the global saddle-node bifurcation the set of these ten nodes again at $\sigma = 0.435$ have undergone a global Hopf bifurcation, so that we arrive at the collection of ten cycles in Fig. 10.35. Finally, in Fig. 10.36 for $\sigma = 0.525$ we see a more normal type of chaotic attractor developed.

As indicated by the negative Lyapunov exponents in Figs. 10.33-10.35, the cases are not chaotic. On the other hand the positive $L = 0.12$ in Fig. 10.36

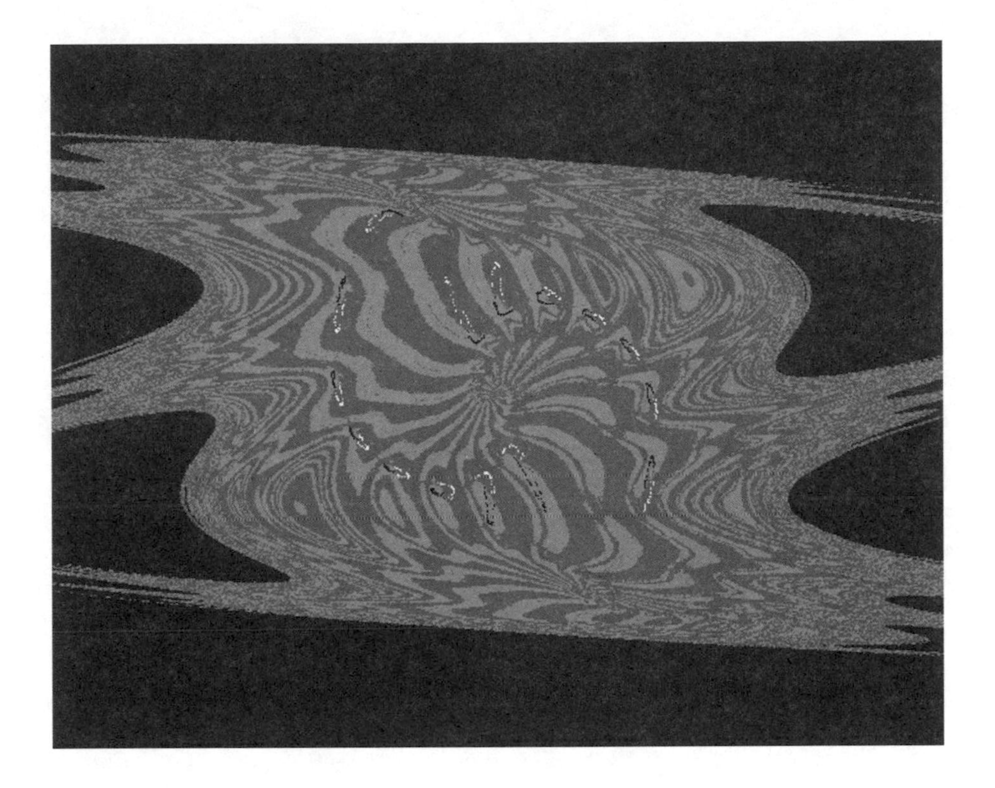

Fig. 10.39. Basins for competing attractors $\lambda = 1.5 \quad \sigma = 0.239$

indicates chaos. We also note that the fractal dimension in Figs. 10.33-10.35 is approximately unitary, as seems intuitively plausible due to the curve like character of the attractors. In Fig. 10.36 we have $F = 1.3$, so the attractor is definitely more than a curve, i.e. fractal. For the intermediate case between Figs. 10.34 and 10.35, which we do not display because of the triviality of a picture of just ten isolated points, we also in fact get F approximately zero, which too is reassuring.

There are, of course, many more attractor shapes to observe. The details of bifurcations are very intricate and occur at very tiny scales of parameter changes. As is always the case with fractals the development chain is broken by entirely different shapes, point sets, cycles, ribbons, and clouds succeeding each other in an infinity of sudden transitions.

In Figs. 10.37-10.39 we follow a different sequence of events with a slightly higher $\lambda = 1.5$. In Fig. 10.37 we see a case with $\sigma = 0.24$, when the attractor becomes a series of odd looking signatures. Just lowering the latter parameter a little bit, to $\sigma = 0.239$, we actually find two different coexistent attractors. The previous attractor has split in two, one shown in black and one in white. Depending on the initial conditions the process now goes to one or the other of the attractors.

From finding one of the attractors, which unlike the previous cases is not symmetric, we can infer the existence of its complementary mirror image. This is a general fact in the case of symmetric systems. From (10.27)-(10.28) we see that reversing the signs of all variables does not change the difference equations a bit, so they are symmetric. From general principles of symmetry breaking it now follows that whenever a nonsymmetric attractor exists then its complementary mirror image exists too.

Due to the coexistence of two attractors, there must also be two different basins of attraction, one for each of them. To be quite true there are three, because the point at infinity is an attractor too, for the cases when the model is not well behaved and explodes. The last case is indicated in black in Fig. 10.39, whereas the competing finite attractors produce the complex intertwined fractal areas. In the centre the attractors themselves are shown. The fractal character of the basins and of their boundary indicate that very tiny changes in initial conditions can make the system flip from one attractor to another. We also conclude that no nearness to an attractor can be taken as an indication that the process will finally end up there.

10.19 Critical Lines and Absorbing Areas

The method of critical curves makes it possible to analyse the general, nonrelaxation case in a somewhat more systematic manner. Consider the mapping of the system (10.27)-(10.28), which we restate here for convenience, deleting the indices:

$$U = Y + Z \tag{10.52}$$

$$V = \lambda Z - (\lambda + 1)Z^3 - \sigma Y \tag{10.53}$$

The map has a single fixed point at the origin. It is easy to calculate the Jacobian determinant at the origin as $J = \lambda + \sigma$, so if the sum of coefficients exceeds unity the fixed point turns unstable. As a rule it first bifurcates into a limit cycle as we saw in Fig. 10.33, which then undergoes further

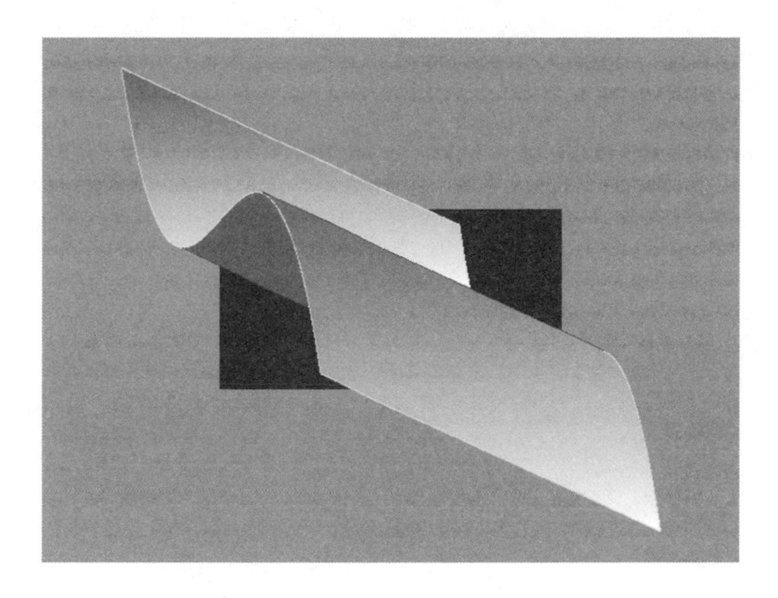

Fig. 10.40. Folding by the cubic map.

bifurcations. The series displayed in Figs. 10.33-10.36 displays snapshots in this series, and the critical curve method developed by Christian Mira around 1960 helps us to understand how the various attractors arise.

Consider how the system (10.52)-(10.53) acts upon a rectangle in the Y, Z - space. In Fig. 10.40 we display the image in U, V - space.

The spaces are shown one on top of the other. The black rectangle in the background is transformed into the twice folded area on the top. This folding is easy to understand in terms of the cubic involved. The significance of the folding is that though the map (10.52)-(10.53) has unique images for every point in Y, Z - space, the reverse does not hold. Every point in U, V - space does not have a unique preimage.

As a matter of fact there are three regions which are divided by the slanting fold lines of the top picture. Inbetween the folds all points have three preimages, above and below they only have one preimage.

We now note the important fact that the upper and lower edges of the folded surface, which are the images of the upper and lower edges of the black rectangle, are parallel to the folds. As the edges were chosen arbitrarily this means that all horizontal lines are mapped into lines parallel to the slanting folds.

For this reason we can expect that there are two particular horizontal lines in the original rectangle which are mapped into the folds. This indeed is the case, and we are next going to find them.

To this end solve (10.52) for Y and substitute into (10.53). In this way we obtain:

$$V = (\lambda + \sigma)Z - (\lambda + 1)Z^3 - \sigma U \tag{10.54}$$

Differentiate (10.54) with respect to Z and put the derivative equal to zero:

$$\frac{\partial V}{\partial Z} = (\lambda + \sigma) - 3(\lambda + 1)Z^2 = 0 \tag{10.55}$$

Solving we get:

$$Z = \pm \sqrt{\frac{\lambda + \sigma}{3(\lambda + 1)}} \tag{10.56}$$

These in fact give two horizontal lines in Y, Z - space whose images are the folds in U, V - space. Our next task then is to derive the equations of these lines themselves. These are readily obtained by substituting from (10.56) in (10.54):

$$V = \pm\left(\frac{2}{3}\lambda + \frac{2}{3}\sigma\right)\sqrt{\frac{\lambda + \sigma}{3(\lambda + 1)}} - \sigma U \tag{10.57}$$

The first term being a constant with positive or negative sign and the second being linear in U with a negative coefficient, we indeed see that we deal with a pair of downward slanting straight lines.

We have thus obtained the horizontal lines by (10.56), in the sequel denoted L_{-1}, L'_{-1}, the prime referring to the negative root, and the downward sloping lines L_0, L'_0 by (10.57). We recall that they are the folds and their preimages as shown in Fig 10.40.

We can now continue the process deriving various forward images of the lines in stead, and this we will do, though we cannot much further use the

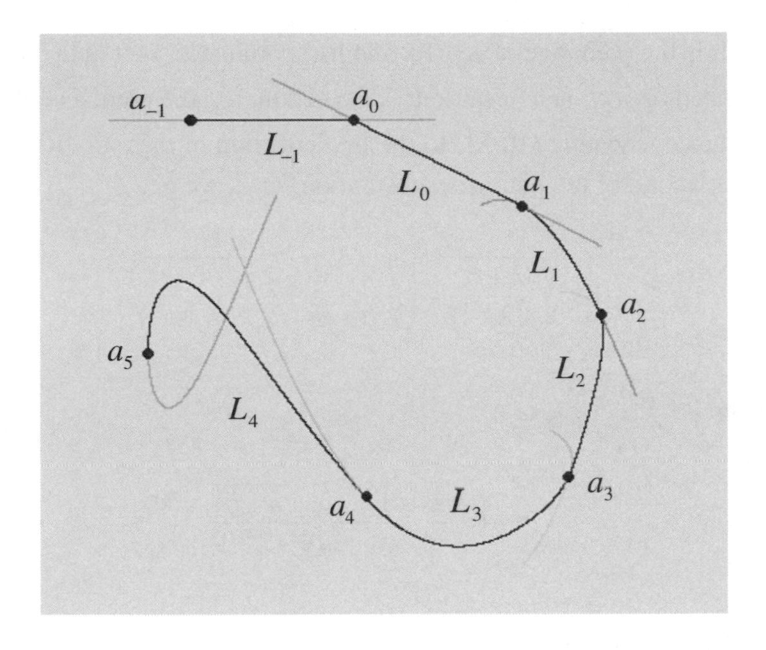

Fig. 10.41.Train of critical curve segments.

explicit forms such as (10.57), because the images of the lines L_1, L_1' already will be of the third order and very soon they become twisted in space. It is better to use parameterised forms, where the original Y coordinate provides an excellent parameter. We leave this for a moment.

Next consider the lines L_0, L_0' and their preimages L_{-1}, L_{-1}' in one and the same space, as in Fig. 10.40. We want to determine the intersections of the lines. For that purpose, identify Y with U and Z with V. Accordingly we have, equating (10.56) to (10.57), solving for U, and using (10.56) as it stands:

$$Y = U = \pm\left(\frac{2}{3}\frac{\lambda}{\sigma} - \frac{1}{\sigma} + \frac{2}{3}\right)\sqrt{\frac{\lambda+\sigma}{3(\lambda+1)}} \qquad Z = V = \pm\sqrt{\frac{\lambda+\sigma}{3(\lambda+1)}} \tag{10.58}$$

These intersection points are normally denoted a_0, a_0'. We can see the first of them along with the lines L_{-1} and L_0 in Fig. 10.41, where we for the sake of visibility only display half the diagram. There is also the point a_{-1} in the picture. It is the preimage of a_0. To find its coordinates, first note that it has to be located on L_{-1}, just because it is located on L_0, the preimage of which L_{-1} is. Hence, equating (10.52) to the first equation of the pair (10.58), and using the second of the pair as a substitution for Z, we get:

$$Y+\sqrt{\frac{\lambda+\sigma}{3(\lambda+1)}}=\left(\frac{2}{3}\frac{\lambda}{\sigma}-\frac{1}{\sigma}+\frac{2}{3}\right)\sqrt{\frac{\lambda+\sigma}{3(\lambda+1)}} \tag{10.59}$$

or solving for Y:

$$Y=\left(\frac{2}{3}\frac{\lambda}{\sigma}-\frac{1}{\sigma}-\frac{1}{3}\right)\sqrt{\frac{\lambda+\sigma}{3(\lambda+1)}} \tag{10.60}$$

Accordingly, the coordinates of a_{-1} are:

$$Y=\left(\frac{2}{3}\frac{\lambda}{\sigma}-\frac{1}{\sigma}-\frac{1}{3}\right)\sqrt{\frac{\lambda+\sigma}{3(\lambda+1)}} \qquad Z=\sqrt{\frac{\lambda+\sigma}{3(\lambda+1)}} \tag{10.61}$$

What we now need to do in order to proceed is to take the section of L_{-1} located between the points a_{-1} and a_0, drawn in black in Fig. 10.41, and iterate it again. Formally we use (10.52)-(10.53) with the substitution of the positive root from (10.56) for Z, and then plug it back into (10.52)-(10.53) to iterate any number of times, always keeping the original Y as a free parameter.

The first image is the section of the slanting line L_0 located between the points a_0 and a_1, again drawn in black. And so we go on. The twice iterated original segment results in a portion of the cubic L_1 between the points a_1 and a_2.

The successive segments some of which are shown in Fig. 10.41 delineate a portion of the phase plane called "absorbing area". Consider the following detail: The point a_1 is the location at which the straight fold line is replaced by the cubic L_1. Note that this is a point of tangency. There is a fundamental reason for this: a_1 is the image of a_0, which is an intersection of the lines L_{-1}

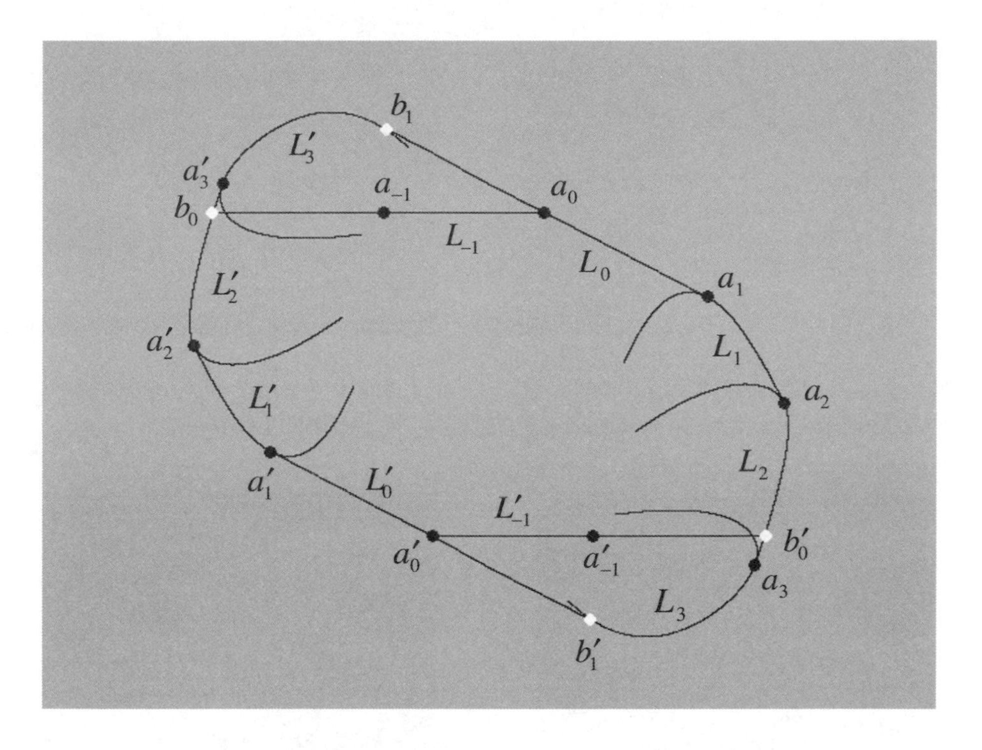

Fig. 10.42. Critical curves and absorbing area.

and L_0. As the plane is folded in L_{-1}, any intersecting curve, such as L_0 is mapped into a curve, such as L_1 which meets at a tangency. It is due to these tangencies that the curves bend back and so delimit an absorbing area.

In Fig. 10.41 we show 5 iterations of the segment (a_{-1}, a_0) of the line L_{-1}. Note that the train of segments does not close to an outer boundary of the absorbing area. As indicated in Chapter 4, we can change this situation by considering a larger segment than (a_{-1}, a_0) of L_{-1}, extended to the left in Fig. 10.41. The exact extension we need is to the point b_0 where L_{-1} meets some forward iterate of itself. But we must now remember that the pair of critical lines L_{-1}, L'_{-1} belong together, so we must find the intersection with the line which is encountered first. It happens that L'_{-1} is hit first by forward iterates of L_{-1} and vice versa. As we see from Figure 10.42, this occurs already at the third iterate of L_{-1}: L_2 intersects L'_{-1} in the point b'_0, and L'_2 intersects L_{-1} in b_0.

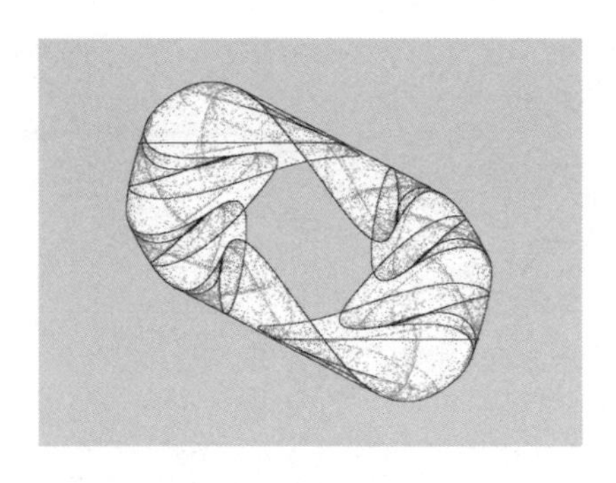

Fig. 10.43. Absorbing area with attractor $\lambda = 1.25$ $\sigma = 0.525$

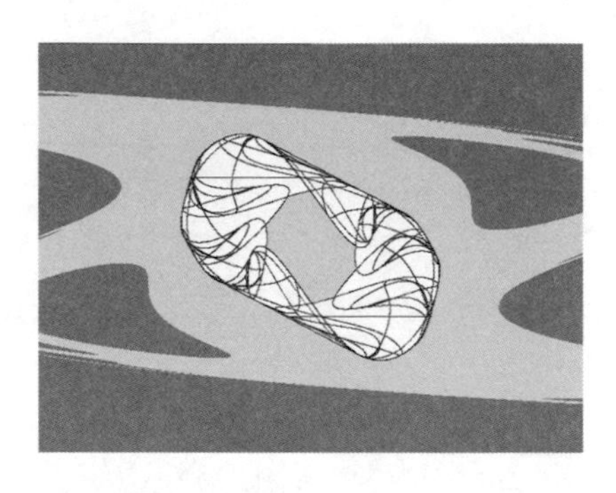

Fig. 10.44. Absorbing area and basin of attraction

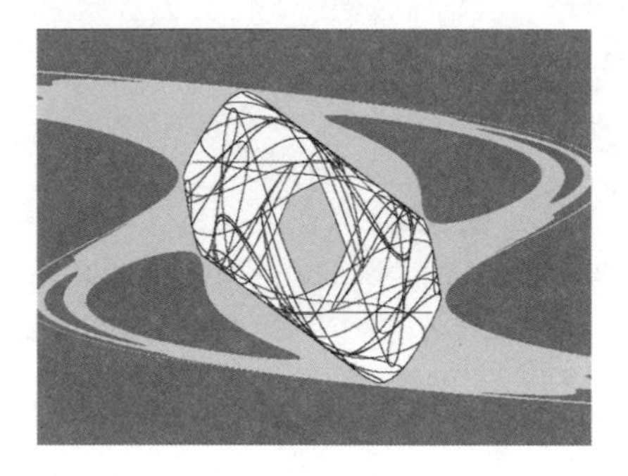

Fig. 10.45. Basin squeezing absorbing area $\lambda = 1.25$ $\sigma = 0.750$

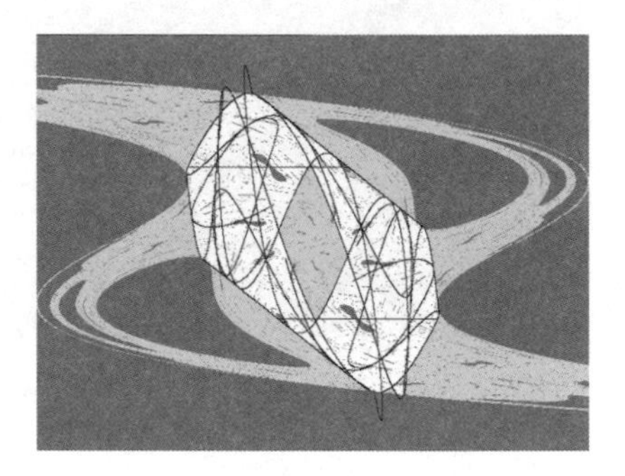

Fig. 10.46. Tongues and fractal basin $\lambda = 1.25$ $\sigma = 0.795$

We now have to recall that any transverse intersection with L_{-1} or L'_{-1} is mapped to a contact of tangency with L_0 or L'_0. Hence L'_3 becomes tangent to L_0 and L_3 tangent to L'_0, and so the outer boundary of the absorbing area becomes closed in only four iterations of the segments (b_0, a_0) and (b'_0, a'_0). All this is illustrated in Figure 10.42, where we mark the points b_i, b'_i in white.

In Figure 10.43 we use eight iterations, and, as we see, the inner boundary of the absorbing area is then nicely outlined as well. On top of the picture we put the attractor (the same as in Figure 10.36) to show how well it fits in the absorbing area.

Note that the absorbing area is not identical to the basin of attraction. While the former outlines the final attractor in as much detail as possible, the latter is the set of all initial points that eventually go to the attractor. In Fig. 10.44

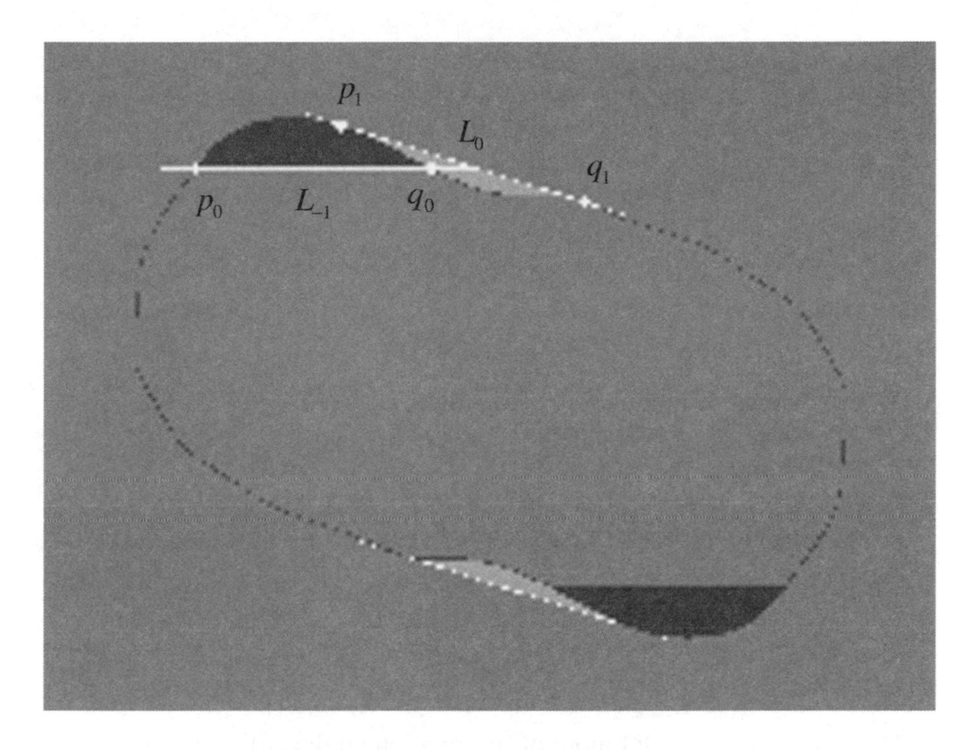

Fig. 10.47. Emerging waviness of limit cycle.

we repeat the picture from Fig. 10.43 but now we add the light shade gray basin of attraction which is surrounded by the dark shade gray area in which the model explodes to infinity.

We note that the absorbing area keeps well clear of the darker shade gray zone of Fig. 10.44. In Fig. 10.45 we see how with an increasing parameter the basin shrinks upon the absorbing area. Finally, in Fig. 10.46, the critical lines which delimit the absorbing area shoot so called "tongues" into the explosive region. As a consequence the basins become fractally intertwined, we see gray explosion spots developing everywhere in the attraction basin. As a matter of fact the picture was drawn for those initial conditions that explode in twenty iterations. More elaborate simulation would show that the now white area would be completely filled by the gray "islands", probably leaving only spots of zero area as initial conditions for convergence. In other words the chaotic attractor has turned unstable, as we are unlikely to hit a Cantor dust in choosing initial conditions.

We will give one more illustration to the usefulness of the critical line method. We saw how it could be used to outline the attractor in Figure 10.36 and to understand its further evolution (=explosion) as the parameter increases. But we can also use it to understand the shape of the closed loop attractors such as in Figures 10.33-10.34. Once the fixed point at the origin loses stability there is a Hopf bifurcation and a sort of limit cycle evolves. But the limit cycles we see are wavy in outline, whereas the first ones that arise are more or less smoothly elliptic. It is easy to explain the wavy nature of the cycles by the critical lines.

Consider Figure 10.47 where we reproduce the cycle from Figure 10.33. In addition we also see the lines L_{-1}, L'_{-1} and L_0, L'_0. Consider the intersection points between the cycle and L_{-1}, denoted p_0, q_0. As they are located on L_{-1} they are mapped to points on L_0 which are denoted p_1, q_1. But as p_0, q_0 arise from transverse intersections with L_{-1} they are mapped to tangencies with L_0.

But as the plane is folded in L_{-1} all points inside the dark area above L_{-1} are mapped below the line L_0 in the bright area, and the convex boundary segment is mapped to a concave one. Hence the wavy nature of the cycle. This waviness arises from a first bifurcation which occurs when the cycle becomes tangent to L_{-1} and then crosses it in two points.

The reader can find a lot more of information on the critical curve method in Abraham, Gardini and Mira (1997) and in Mira, Gardini, Barugola, and Cathala (1996).

10.20 Conclusion

We have seen how a very simple business cycle model of traditional outline with a cubic nonlinearity generates fixed points, regular cycles, or chaos in either one or two quadrants, depending on the parameters of the system. This motion is secluded to income differences, so that the income variable itself follows a random walk process. This seems to restore some kind of order, as repeated sampling processes from a stochastic (chaotic) population do. A general conjecture is that maybe chaotic/stochastic structures may alternate with ordered ones at different levels of reality, and may arise from each other. Chaotic processes generated by deterministic systems give one direction, synergetic self-organization of disordered systems give the reverse.

A deterministic process like the flow of air past an obstacle at low speeds creates a laminar flow, at high speeds turbulent (chaotic) eddies arise. Nevertheless these irregular eddies may organize themselves in time around the labium of a recorder so that 440 of them per second produce the musical note a' (i.e. an ordered structure). Anybody who has listened to a group of children treating recorders at school understands that chaos can again quite easily emerge at the next level.

An undesirable feature of the model proposed is that, whenever the process starts outside a specified interval, explosive spiralling motion will be created. The same happens whenever the parameter exceeds a certain critical value. This is to some extent a consequence of the use of difference equations in modelling. There is, however, a considerable advantage in working with discrete time processes as the tools for analysing chaotic motion are developed for those. For a system cast in terms of differential equations we first have to integrate over a "cycle" to obtain the return map on the Poincaré section, something that may often be impracticable.

An alternative continuous model was presented in Chapter 8, where such difficulties were avoided. As a rule, in order to produce chaos, the order has to be higher for a continuous system than for discrete one. This was obtained by coupling two regions, each with persistent cycles of their own, by interregional trade. This raises the order to the fourth, or, if a one-directional influence (the case of the small open economy) is treated, the third.

Except for the convenience of analysing discrete time processes we could defend the model on the grounds that it would be too much to demand from a scientific model that it works for all parameter values and all initial conditions. After all the situation is much better than with linear growth and business cycle modelling which does not work decently for any parameter values or initial conditions.

The cubic mapping provided an opportunity to generate pure growth as well as cyclic motion, something that has been impossible with the frequently applied quadratic mapping. At the same time the shift map of symbolic dynamics could be introduced by a trigonometric transformation, even more easily than it has been done in the study of quadratic mappings.

The simplified introductory model was based on a specific lag system for consumption expenditures, excluding a feed back mechanism from income to income differences that occurs under more general circumstances. As we saw the internal feed back present in the more general model causes bifurcations between ordered and chaotic behaviour in one and the same cycle.

Once we abandoned the specific lag structure, it was possible to investigate the relaxation case in some detail. For more general cases we had recourse to simulations, and detected some series interesting bifurcations: Hopf bifurcations, global saddle node bifurcations and further global Hopf bifurcations. We also saw the existence of competing attractors with fractal basins of high complexity.

The tools used to analyse the more general cases also included the computation of Lyapunov exponents and fractal dimension measures, as well as the Mira critical line method to find the absorbing areas of the attractors and their interactions with the attraction basins as causes of bifurcations.

11 Dynamics of Interregional Trade

11.1 Interregional Trade Models

Nonlinear dynamics does not by necessity lead to complex phenomena, such as chaos. As an example to the contrary we can take the elegant model of interregional trade equilibrium and dynamics as formulated by Martin Beckmann in 1952 and 1953. This gives us one more opportunity to study partial differential equation models in economics and to employ the concepts from vector analysis introduced in Chapter 3.

Beckmann considered the following problem: Given distributions, in continuous 2D-space, of local (price-dependent) excess demand/supply, and local (isotropic) freight rates, which are the equilibrium flow of trade and the corresponding distribution of prices? The answer was given in terms of two succinct PDE which completely described the problem.

This approach can be compared to the model Paul Samuelson by the same time suggested for the identical problem, formulated for a discrete set of locations, instead of for continuous two dimensional space. The solution was by necessity much more messy, and did not give the same intuitive understanding for the geometry of the phenomena in geographical space. Yet it is the Samuelson model, not the Beckmann model, which found its way to mainstream economics. Most likely this is a consequence of a general shortage of PDE and vector analysis tools in the analytical equipment of the average economist.

A quarter Century later, in 1976, Beckmann dynamised his essentially static model and showed that for given boundary conditions, the equilibrium was unique and further asymptotically stable. The only substantial assumption for the strong conclusions was that excess demand be a decreasing function of price. The conclusion held in spite of the fact that the whole model was nonlinear.

11.2 The Basic Model

The basics of the Beckmann model are the following: There is a closed region $R \subset \mathbf{R}^2$. Its boundary is denoted C, and the space coordinates are denoted x_1, x_2. On the region there are defined three scalar fields:

i) the local price distribution $\lambda(x_1, x_2)$ (11.1)

ii) the local excess demand distribution $z = f(\lambda, x_1, x_2)$ (11.2)

iii) the local freight rate distribution $k(x_1, x_2)$. (11.3)

Normally we take excess demand z as a function of price, with $f_\lambda < 0$ as usual, though we can also include the space coordinates separately to account for any uneven distribution of exogenous factors such as the availability of natural resources or of the consuming population. As for the freight rate, assuming dependence on the space coordinates, but not on the direction of transit, means an isotropic approximation. This is not an absolute necessity, though it helps to bring out the full elegance of the Beckmann model. In addition to the three scalar fields there is a vector field:

iv) the flow of traded commodities $\phi = \phi(\phi_1(x_1, x_2), \phi_2(x_1, x_2))$ (11.4)

In each location, the unit direction vector

$$\frac{\phi}{|\phi|} = \left(\frac{\phi_1}{\sqrt{\phi_1^2 + \phi_2^2}}, \frac{\phi_2}{\sqrt{\phi_1^2 + \phi_2^2}} \right) \tag{11.5}$$

gives the geographical direction of trade, whereas the norm

$$|\phi| = \sqrt{\phi_1^2 + \phi_2^2} \tag{11.6}$$

is the volume of flow.

In such a flow the divergence, i.e. the differential operator

$$\nabla \cdot \phi = \frac{\partial \phi_1}{\partial x_1} + \frac{\partial \phi_2}{\partial x_2} \tag{11.7}$$

as is well known from for instance hydrodynamics, denotes the change of flow volume due to local sources/sinks or excess supply/demand in economics terms. The condition, that local excess supply be entered into the flow, and local excess demand be withdrawn from it, states that the sum of flow divergence and excess demand be equal to zero. So

$$\text{Beckmann 1:} \quad \nabla \cdot \phi + z = 0 \tag{11.8}$$

is just a pure *equilibrium condition for interregional trade*. This is the first of Beckmann's equations.

The second equation states

$$\text{Beckmann 2:} \qquad k \frac{\phi}{|\phi|} = \nabla \lambda \tag{11.9}$$

Here

$$\nabla \lambda = \left(\frac{\partial \lambda}{\partial x_1}, \frac{\partial \lambda}{\partial x_2} \right) \tag{11.10}$$

is the standard notation for the gradient. Equation (11.9) says that *trade flows in the direction of the price gradient*, and, moreover, that *prices increase with accumulated transportation costs* in the flow direction. This is a vector equation. The economic sense is that traders everywhere ship goods in the direction where the price differences are largest, so as to make maximum profit, whereas competition among traders reduces the profit margins over transportation cost to zero.

Given suitable boundary conditions, (11.8) and (11.9) have a solution for the flow ϕ and the price distribution λ. The freight rate function k is a datum, and so is excess demand z, unless we take it as dependent on price $z = f(\lambda, x, y)$, as we will in the sequel. Then the boundary conditions with respect to exterior prices and demand/supply for exports/imports, along with the freight rate distribution (as representing properties of the transportation system on the region) are the only data. The configuration of trade flow and interior prices is the solution.

The model (11.8)-(11.9) is nonlinear, due to the occurrence of the norm $|\phi| = \sqrt{\phi_1^2 + \phi_2^2}$ in the denominator of (11.9). It is a static equilibrium model in the sense that time and temporal evolution are not included, though it is "dynamic" in the sense that modelling is in terms of differential equations involving space derivatives (second order as always).

To see the link to standard PDE more clearly, suppose $k = |\phi|$. The sense of this assumption could be that transportation costs vary due to congestion in some urban area, where, as a first linear approximation, transportation cost is taken as proportional to the volume of traffic flow. Note that any constant before the measure of flow could be dispensed with by rescaling the variables, so proportionality is the same as equality. Hence:

$$k = |\phi| \tag{11.11}$$

which inserted in (11.9) results in

$$\phi = \nabla \lambda \tag{11.12}$$

Accordingly, the vector field (11.4) becomes the gradient of the scalar field (11.1). Further, applying (11.8) to (11.12) we get:

$$\nabla \cdot \phi = \nabla \cdot \nabla \lambda = \nabla^2 \lambda = -z \tag{11.13}$$

Suppose excess demand only depends on local prices, i.e. $z = f(\lambda)$. Then we get

$$\nabla^2 \lambda + f(\lambda) = 0 \tag{11.14}$$

which is very similar to Laplace's Equation and to the eigenvalue problems for two dimensional space encountered in Chapter 2. Given some simple form for $z = f(\lambda)$, such as linear, (11.14) is easy to solve in closed form.

This, however, is very exceptional. As a rule we cannot boil down the two equations (11.8)-(11.9) to a single one such as (11.14).

Yet, there are things that can be said about the qualitative types of solutions to (11.8)-(11.9) that may turn up under considerations of structural stability. In order to apply such concepts we note that, in any case, (11.9) can be written:

$$k^2 = (\nabla\lambda)^2 \tag{11.15}$$

by taking squares of both sides. The square of the unit vector $\phi / |\phi|$ multiplies to scalar unity. Written out in derivatives (11.15) reads:

$$\left(\frac{\partial\lambda}{\partial x_1}\right)^2 + \left(\frac{\partial\lambda}{\partial x_2}\right)^2 = k^2 \tag{11.16}$$

It is important to note that (11.16) is a PDE in prices alone and in principle very easy to solve even by constructive methods. Imagine that we have a given line of constant prices in the geographical region of interest (a boundary condition). Then take a compass adjusted to the radius $1/k$ at any of the points of this line, and draw a circle. Draw as many circles as we wish, always adjusting the radius, sufficiently many to draw an envelope, which is a curve meeting this whole family of circles at tangency. This new tangency curve will then be the next line for constant prices in space, and the process can be repeated until the whole region of interest has been covered.

There is an intuitive reason for this procedure: If the radius of the circle is reciprocal to the freight rate, then the area of the circle represents all points of the region that can be reached by spending 1$ on transportation, and the envelope represents the most distant points that can be reached by spending 1$ when departing from all the different points of the original constant price line. Due to efficiency of competition the new line joins places where the local price in 1$ higher.

As a consequence of this construction, the lines of trade flow become orthogonal to the constant price lines, which can also be seen directly from equation (11.9): The flow direction coincides with that of the price gradient.

Given a price distribution $\lambda(x_1,x_2)$, we can next obtain the flow lines $x_1(t)$, $x_2(t)$ by solving a pair of ODE:

$$\frac{dx_1}{dt} = f(x_1, x_2) = \frac{\partial \lambda(x_1, x_2)}{\partial x_1} \tag{11.17}$$

$$\frac{dx_2}{dt} = g(x_1, x_2) = \frac{\partial \lambda(x_1, x_2)}{\partial x_2} \tag{11.18}$$

Note that the flow vector ϕ will be tangent to the solution curves $x_1(t), x_2(t)$ to (11.17)-(11.18). Also note that t has no interpretation of time, it is just a parameterisation of the flow lines.

The system (11.17)-(11.18) is quite like the general system of a pair of ODE as considered in Section 2.3, but it is more specific, as the right hand sides are the components of a gradient field. Hence, the conclusions on structural stability still apply, but become even more specific. For a gradient field no eigenvalues can be complex, i.e. there are no spirals, so the conclusions become even more specific than those stated in Section 2.3. (For the same reason no limit cycles can occur.)

To see this let us linearise (11.17)-(11.18):

$$\frac{dx_1}{dt} = f_1 x_1 + f_2 x_2 = \lambda_{11} x_1 + \lambda_{12} x_2 \tag{11.19}$$

$$\frac{dx_2}{dt} = g_1 x_1 + g_2 x_2 = \lambda_{21} x_1 + \lambda_{22} x_2 \tag{11.20}$$

where we for convenience denote the various partial derivatives by indices. The eigenvalues according to (2.16) now are:

$$\mu_1, \mu_2 = \frac{\mathrm{Tr}}{2} \pm \frac{1}{2}\sqrt{\mathrm{Tr}^2 - 4\mathrm{Det}} \tag{11.21}$$

with

$$\mathrm{Tr} = \lambda_{11} + \lambda_{22} \tag{11.22}$$

$$\mathrm{Det} = \lambda_{11}\lambda_{22} - \lambda_{12}^2 \tag{11.23}$$

Note that we have to use μ_1, μ_2 for eigenvalues because λ is now already occupied. Substituting from (11.22)-(11.23) in (11.21) we get:

$$\mu_1, \mu_2 = \frac{\lambda_{11} + \lambda_{22}}{2} \pm \frac{1}{2}\sqrt{(\lambda_{11} - \lambda_{22})^2 + \lambda_{12}^2} \tag{11.24}$$

As the expression under the root sign is a sum of squares it can never become negative, and hence the eigenvalues are always real.

There is an intuitive explanation for this. The flow lines are solutions of an optimization problem - minimisation of transportation cost for each transportation programme. A spiral singularity, being equivalent to an infinity of whirls around the final destination, or a limit cycle, never stopping at the destination at all, cannot be solutions. (Note that this is not to say that short arcs of spiral or closed loop routes cannot occur.)

Due to this reduction of possibilities, and *given structural stability*, we are left with three types of possible singularities: *saddle points, stable nodes, and unstable nodes*. The conclusion that there can be *no heteroclinic (or homoclinic) saddle connections* also remains.

The saddle points and their ingoing trajectories, as usual, provide basin boundaries, and now the basins can be interpreted as trade areas. The stable and unstable nodes, located in the trade areas can then be interpreted as productive or consumptive centres. Using Peixoto's characterisation theorem for structurally stable flows, we easily find the regular tessellations which possess this property. These tessellations are either square, or else triangular/hexagonal.

11.3 Structural Stability

Recall from Chapter 2 that, whereas the singularities of linear systems were simple and easily classified, we did not know anything about the singularities of the general nonlinear systems. The more remarkable is the fact that the mere assumption of structural stability reduces them to a finite number of the same type as in linear systems (and even puts global constraints on how they can be connected). The procedure we follow is the same as that suggested by Paul Samuelson in his formulation of the "correspondance principle". In his case the issue was the stability and comparative statics of a multimarket equilibrium price system. Just assuming stability of equilibrium made it possible to tell how this equilibrium would be affected through an exogenous disturbance, though the system itself was so general that nothing at all could be concluded without the stability assumption. Samuelson argued that the assumption was nonrestrictive, as without stability an equilibrium would not be worthwhile studying.

This brings the Arnol'd point from Chapter 2 in mind: Scientific modelling is subject to, hopefully small, mistakes and errors. We should therefore avoid making models that do not remain approximately true should there be some slight mistake in their specification, i.e. we should make them structurally stable. The present issue is stability of the model structure, rather than just stability of an equilibrium point as in Samuelson's case. But the philosophy is the same: The assumption is nonrestrictive, involving only scientific prudence, i.e. to avoid studying equilibria or models that might evaporate under the slightest disturbance. - And there is a reward in terms of amazingly detailed information which would otherwise need some substantial factual details to supplement the general model.

To be quite true it must be admitted that structural stability in its mathematical sense is not always an nonrestrictive assumption to make. Peixoto studied plane flows, but later work by Stephen Smale demonstrated that such structural stability was not a "generic" property in dimensions higher than two. To express matters in an intuitive way, structurally unstable systems in for instance three-dimensional space were not everywhere densely surrounded by structurally stable systems. The more fortunate then it is that in considering the geographical layout of a market equilibrium we no doubt deal with two dimensions. Hence structural stability indeed is a generic property, and the facts from Peixoto's Characterisation Theorem apply for all robust systems.

There is even more to it: Once we have characterised the structurally stable static patterns that may emerge in the Beckmann model, we can also characterize their transitions from one to another by use of catastrophe theory, more precisely the Elliptic Umblic Catastrophe, which was discussed in Chapter 5 above. The causes of such bifurcations can be any changes to the infrastructure or fuel prices which affect transportation costs $k(x_1,x_2)$ in (11.3), or any changes to consumer preferences and production techniques, or to the location of consumers and producers which influence the excess demand functions $z = f(\lambda, x_1, x_2)$ in (11.2).

11.4 The Square Flow Grid

As we know there are a finite number of singular points in a structurally stable flow. Now, it is the singular points of a flow that are really interesting. The trajectories outside the singular points are trivial as they are topologically equivalent to parallel straight lines, though possibly distorted. The singular points of a flow are the organizing elements, and once we know their configuration, we can fill in the trajectories by free hand. What we thus aim at is a kind of graph or grid made up of the singular points and the trajectories that connect them.

At this stage it is appropriate to recall that only two kinds of singular points were admitted for a structurally stable flow: nodes and saddles. As for the nodes, all trajectories in a neighbourhood are incident, either issuing or ending there. The infinity of such trajectories therefore is no good organizing element. As for saddles there are only two pairs of incident trajectories, one pair issuing at the saddle point and one pair ending there. All the other trajectories miss the saddle point, and go like hyperbolas in the four sectors into which those incident to the saddles split its neighbourhood. The trajectories incident to saddles are thus the organizing elements we need - they trace all the nodes anyhow, because, saddle connections being forbidden, they must issue or end up at nodes.

We thus use the singularities, along with the trajectories incident to saddle points, for our grid, shown in Fig. 11.1. To make things easy we use a special iconology: Stable nodes, or sinks, at which all surrounding trajectories end are black (think of black holes that even capture light). Unstable nodes, or sources, from which all neighbouring trajectories issue, are white (think of radiant suns). Saddle points are half black / half white (having one stable and one unstable direction).

Now, consider the saddle point in the centre of the bright square of Fig. 11.1. The outgoing trajectories East and West cannot end at saddle points, they must end at nodes, more precisely sinks. Likewise, the ingoing trajectories North and South must again issue from nodes, this time sources. Now, suppose we continue to organize the grid in terms of horizontal and vertical lines. Then, NE, SE, SW, and NW from the original saddle point, trajectories meet which are both ingoing and outgoing: there are hence new singular points, and the only type admitted having both stable and unstable directions is the saddle.

The diagonal corner singularities in the bright central square of Fig. 11.1 have thus been identified. We can therefore start the reasoning anew, starting

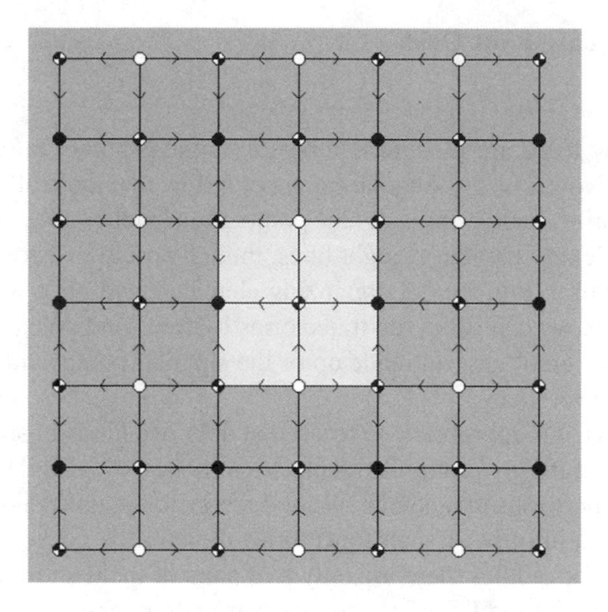

Fig. 11.1. The basic structurally stable grid.

Fig, 11.2. Flow fitted to the square grid.

Fig. 11.3. Square price landscape.

from each of these new saddles, and so orient the entire graph, identifying each singularity. We note that the graph is a square grid with as many saddle points as nodes. The nodes are divided in equal numbers of sources and sinks.

Several comments are in place now. First of all: our picture is topological. In topology continuous transformations do not count. The pattern we outline, with straight lines and right angles, can be drawn on a rubber sheet and stretched in any way we wish (without tearing). Alternatively, we can also put the picture on the bottom of a filled bath tub and stir the water moderately. Then all the pictures we see are topological equivalents. If one is stable, all of them are.

Once we have the oriented grid it is easy to fill in trajectories by free hand. They have to issue at sources, pass by saddle points, as if attracted but missing them, and end at sinks. Fig. 11.2 illustrates such a family of trajectories fitted to the square grid.

Observe that the picture, as compared to the previous one, has been tilted by 45 degrees. The sinks and sources have been marked, but the saddle points, whose locations are obvious, are not especially marked in Fig. 11.2.

Despite the tilting of the flow, there is still a chessboard pattern observable which is not tilted at all, so what is it? The answer is: constant price contours. We observe that they, as well as the more rounded inner boundaries, are everywhere orthogonal to the flow trajectories. Corresponding to a square grid of the flow is hence also a dual square pattern of price contours.

The square pattern of flow lines, where the tiles are bounded by incident saddle trajectories, can be said to represent market areas, because no trajectory ever crosses these lines, so all trade is confined within them. As the nodes in the centres of these tiles represent minimum and maximum prices, they can be regarded as centres of production and consumption. What about the saddle points? Except for the mathematical fact that in a landscape with maxima and minima there must be saddle points that separate them (see Fig. 11.3), there is an economic interpretation. In the saddles almost no trajectories are incident (four out of a continuous infinity), but all trajectories in some neighbourhood are attracted to them. They can thus be interpreted as locations without economic significance but with good transportation facilities.

11.5 Triangular / Hexagonal Grids

Let us now start out from Fig. 11.1, in part replicated in Fig. 11.4, and delete the singularities not shown in the new picture. We can never remove just one singularity, if we do, either the orientations no longer work, or we get forbidden saddle connections. What we are however allowed to do is to remove pairs of adjacent nodes and saddles.

As we see, three such node-saddle pairs have been removed in Fig 11.4. Comparing to Fig. 11.1, we find that three sinks, and three saddles have gone. Removing three such pairs can never leave equal numbers of sources and sinks. In the way we have done it, no sources, only sinks have gone. The whole construction may seem arbitrary to the reader, but it is not, as we will see shortly. The alternative would have been to remove only sources, along with the saddles.

Of course there must arise new trajectories in the diagonal directions between the saddles and the nodes left, but we do not complicate the picture by displaying them. For identifying the character of this new picture we have inserted the letters, *A* through *L*, for the source points located on the boundary of the bright area. It is now possible to bend the set of squares in Fig. 11.4 into the pattern of three hexagons in Fig. 11.5. All singularities in Fig. 11.5 have their companions in Fig. 11.4, and all trajectories too, except those new ones mentioned, which we did not care to set out.

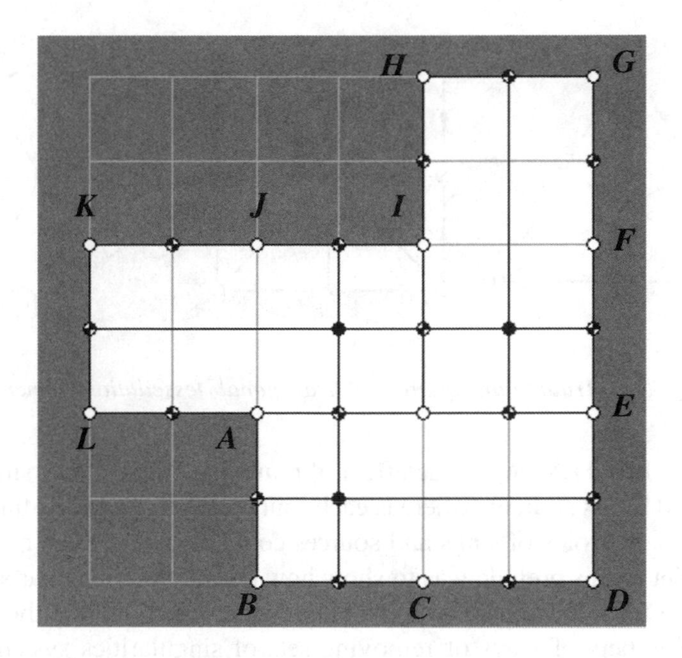

Fig. 11.4. Removing saddle-node pairs of singularities.

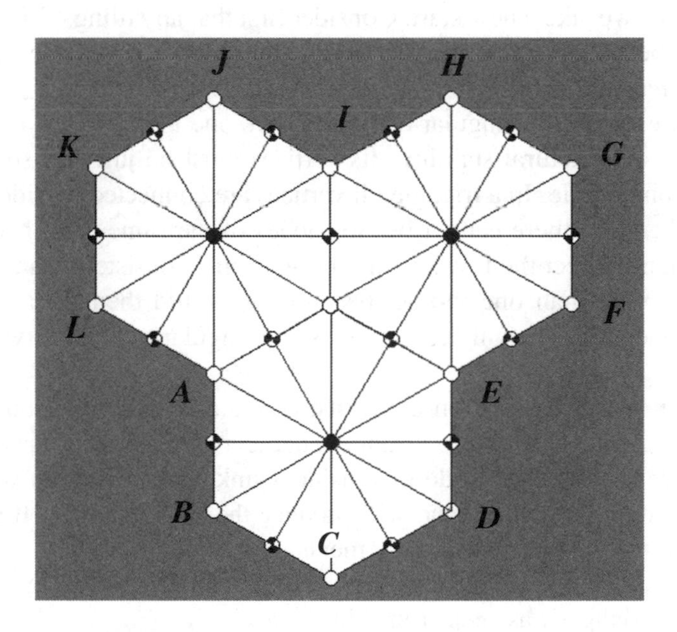

Fig. 11.5. Transforming to hexagonal / triangular grid.

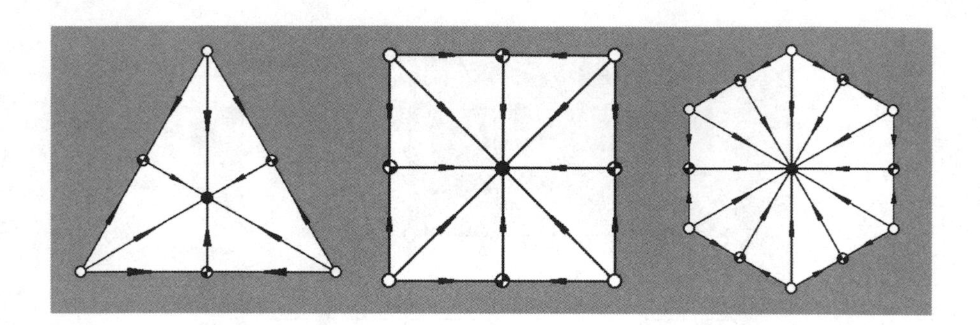

Fig. 11.6. Triangular, square and heaxagonal tessellation elements.

Looking at Fig. 11.5 we see the effect of removing sinks. Every sink is now surrounded by six sources, whereas each source is surrounded by three sinks. As we said, the roles of sinks and sources could also have been reversed.

The point in this prelude was to show how hexagonal/triangular structures could arise from the square pattern. The reader may think that there are indefinite numbers of ways of removing sets of singularities systematically, and that they might all produce new regular patterns. The first half of the conjecture is correct, but not the second. We have already found out all there is to find out.

To see this we take a new start. Consider first that any tiling of the plane by polygons, be it square or triangular or hexagonal or whatever, can be triangulated, i.e. reduced to an aggregate of triangular cells.

Second, each such triangular cell must have one look in view of the considerations of structural stability: Its vertices must be just one source, one sink, and one saddle. In a triangle all vertices are connected by sides (trajectories). Therefore there cannot be two nodes of the same type, because we could then not direct the flows along the sides in a consistent manner. There cannot be more than one saddle, because we would then get a forbidden saddle connection. The only case open is one singularity of each type, just as suggested.

To produce the tessellation elements, equilateral triangles, squares, and regular hexagons, we have to arrange a number of elementary triangles cyclically. We have chosen to do this, using a sink as the common centre, but any of the vertices will do, and will produce the same patterns. It is a good exercise to try the different arrangements.

For the triangle we need 6, for the square 8, and for the hexagon 12 such elementary triangles, as seen from Fig. 11.6.

Now, using the square of Fig. 11.6 we, of course, arrive at the square tessellation shown in Fig. 11.1. By using the hexagon of Fig. 11.6, we arrive at the tessellation illustrated in Fig. 11.5, which also was the same as the intermediate stage displayed in Fig. 11.4 when we deleted pairs of singularities from Fig. 11.1. So far there is nothing new.

What about the triangles of Fig. 11.6? If we tile the plane with them we find out that we arrive at the same tessellation as that displayed in Fig. 11.5, only with the roles of sinks and sources reversed. The black dots are interchanged with the white dots, but nothing else happens.

We thus must have either a square tessellation, with equal numbers of sinks and sources, or else a hexagonal/triangular one with the number of sinks either half or double the number of sources.

11.6 Changes of Structure

We have seen that two, or to be more precise, three different layouts for the space economy were compatible with structurally stable patterns of flow. They were either square, with equal numbers of sinks and sources, or triangular/hexagonal, with twice as many sinks as sources, or twice as many sources as sinks. Everything else (among regular patterns) is unstable. This is unexpectedly much to know for a general flow model without introducing any substantial assumptions at all.

The natural question now is: Can we do better than that, can we say anything about the transitions from one stable pattern to another? The answer is yes. Like we could use the generic theory of differential equations for characterizing the stable flows, we can study the transitions by means of catastrophe theory.

Locally, at each singular point, transitions between the stable spatial patterns, the square and triangular/hexagonal, occur through an emergence of the monkey saddle flow. They can hence be modelled by the Elliptic Umblic catastrophe. Later we shall see how simultaneous catastrophe can occur at all the singular points at once by using spatially periodic (trigonometric) functions, whose Taylor series at the singularities start out with the standard monkey saddle potential. The standard formula for a monkey saddle is:

$$\lambda = x_1^3 - 3x_1x_2^2 \tag{11.25}$$

Fig. 11.7. The elliptic umblic catastrophe.

Note that we take the space coordinates as phase variables, and the price distribution itself as the potential. To universally unfold this degenerate potential, as we saw in Chapter 5, we add terms involving three different parameters, which will represent all the factors involving external change to the model: redistribution in space of producers/consumers, changing technology/tastes, and changes in infrastructure/fuel costs. These parameters will be denoted a, b, and c in the sequel.

As we know the most unstable situation that can occur in our model is the monkey saddle itself. In Chapter 2, Figs. 2.11-2.12 we saw how it, through changes of the unfolding parameters, was first split in two ordinary saddles, and then had the heteroclinic saddle connection broken. If we compute the eigenvalues of a dynamical system that is gradient to a monkey saddle function, we find that both eigenvalues are zero, as shown in Chapter 5. The universal unfolding reads:

$$\lambda = x_1^3 - 3x_1x_1^2 + ax_1 + bx_2 + c(x_1^2 + x_2^2) \tag{11.26}$$

which becomes the monkey saddle if all parameters are zero, $a = b = c = 0$.

To get a picture of this function we would need a five dimensional space, so we have to find some easier strategy.

The strategy therefore is to portray the flows in the small medallions in the lower part of Fig. 11.7. The central piece shows the flow when $a = b = c = 0$, which is the monkey saddle case. To the right and left (with only $c \neq 0$) we see cases where the monkey saddle splits in one node and three saddles. The cases left and right are different as there is a central sink on one side and a central source on the other. The locations of the saddles seem to be rotated too when the coefficient c passes from negative to positive values over the degenerate monkey saddle case.

If we keep $c = 0$, but make a and b nonzero, then the monkey saddle splits in two disjoint saddles, a scenario we know from Figs. 2.11-2.12. If only one of those parameters is nonzero, then the monkey saddle splits, but there remains a heteroclinic saddle connection.

So much about the x_1, x_2-space, displayed in the medallions. On top of Fig. 11.7 we see the bifurcation manifold. It is a surface in the three dimensional a, b, c-space. Each time the parameter combination traverses one of the three sides of this cuspoid figure the pattern bifurcates. Outside the bifurcation manifold there are two disconnected saddles, inside are the cases with one node and three saddles, a sink on one side and a source on the other. The inner sides are separated by the narrow waist of the bifurcation manifold where it contracts to a single point - the origin of coordinate space, corresponding to the monkey saddle flow.

We have now been able to classify not only the stable flow patterns, but the typical ways of transitions between them as well.

The present characterization has been concerned with the local organisation around a single singular point of the flow, whereas we above portrayed global structures of the space economy.

It is, however, not difficult to extend the discussion to global regular tilings. To this end we use trigonometric functions that are periodic over space.

The triangular structure inherent in these phase transitions is characterized by three directions in space, which we can put equal to: $u = x_1 + \sqrt{3}x_2$, $v = x_1 - \sqrt{3}x_2$, and $w = 2x_1$. We can portray a global spatially periodic monkey saddle pattern by $\lambda = \sin(u)\sin(v)\sin(w)$, or

Fig. 11.8. The degenerate global hexagonal pattern.

$$\lambda = \sin(x_1 + \sqrt{3}x_2)\sin(x_1 - \sqrt{3}x_2)\sin(2x_1) \tag{11.27}$$

which is displayed in Fig. 11.8. If we compute the Taylor series at any of the singular points we obtain (11.25). These singular points are characterised by the fact that two of the component functions are zero, say for instance that $x_1 + \sqrt{3}x_2 = i\pi$ *and* $x_1 - \sqrt{3}x_2 = j\pi$, where i and j are integers. Then it is easy to see that the third component function too must zero. By adding the two previous expressions we get $2x_1 = (i + j)\pi$. The functions were simply put up this way.

Note that this pattern is nothing like the stable triangular/hexagonal tessellations we have been talking of. There are as many sinks as sources, each source is surrounded by six sinks, and each sink by six sources. There are no ordinary saddles, just degenerate monkey saddles.

To split this pattern of monkey saddles we have to add functions, periodic over space in the same way as (11.27), but working like the quadratic and linear terms in equation (11.26). For the term corresponding to the quadratic

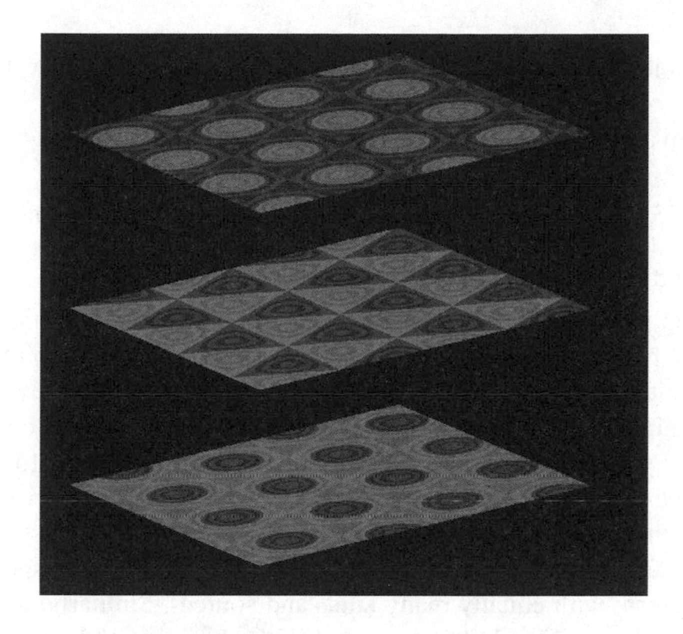

Fig. 11.9. First global bifurcation.

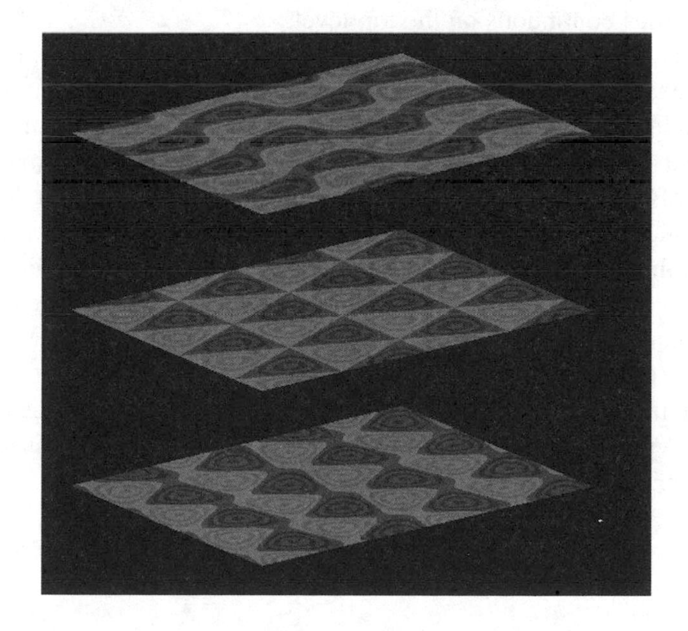

Fig. 11.10. Second global bifurcation.

we note that $\cos\left(x_1+\sqrt{3}x_2\right)\cos\left(x_1-\sqrt{3}x_2\right)\cos(2x_1)\approx 1-3\left(x_1^2+x_2^2\right)$. For the linear terms we have $\sin\left(x_1+\sqrt{3}x_2\right)\cos\left(x_1-\sqrt{3}x_2\right)\cos(2x_1)\approx x_1+\sqrt{3}x_2$ and $\cos\left(x_1+\sqrt{3}x_2\right)\sin\left(x_1-\sqrt{3}x_2\right)\cos(2x_1)\approx x_1-\sqrt{3}x_2$, so suitable linear combinations of the latter expressions produce Taylor Series with only x_1 or x_2 as the leading term. These Taylor Series are computed at the origin, but due to the spatial periodicity they also work at all the other singular points.

The result is displayed in Figs. 11.9 and 11.10. On the middle levels we see the degenerate monkey saddle flow of Fig. 11.8. On the top and bottom levels we see the changes to the global patterns, corresponding to the local phenomena that we experience in Fig. 11.7 when we move horizontally or vertically through the central medallion.

The top and bottom pictures of Fig. 11.10 topologically correspond to the square pattern, with equally many sinks and sources. Similarly, the top and bottom pictures in Fig. 11.9 correspond to the hexagonal/triangular pattern with one type of nodes double the other.

We see that, as the structure passes through the singular case in Fig. 11.9, the number of one type of nodes doubles whereas the other halves. We also see that whereas the bright area is continuous on the bottom level the dark one becomes continuous on the top level.

Of course it is also possible to make a direct transit from the square pattern to the hexagonal/triangular. After all, the bifurcations or catastrophes illustrated in Figs. 11.9 and 11.10 mean passing through the waist of the surface displayed on top of Fig. 11.8. Those cases are theoretically interesting as they involve transition through the most degenerate pattern. But, looking at a slow motion of the point in parameter space, which describes a curve, it is unlikely that it hits the waist of the bifurcation manifold. Most likely there is a passage through any of the three side surfaces, and that means a direct bifurcation from triangular/hexagonal to square, or the reverse.

We are, however, not going to follow this track any longer. It is not really dynamic in a temporary sense, even if structural stability and the emergence of spatial patterns are at issue. We rather jump to the dynamisation of his model which Beckmann presented in 1976.

11.7 Dynamisation of Beckmann's Model

Consider another flow field $\psi = (\psi_1(x_1, x_2), \psi_2(x_1, x_2))$ along with ϕ, another price distribution $\mu(x_1, x_2)$ along with λ, and another excess demand distribution w along with z. These fields, unlike the previous ones, are supposed to be *disequilibrium* fields. The freight rate distribution k, of course, remains the same. Also, provided $z = f(\lambda, x_1, x_2)$ and $w = f(\mu, x_1, x_2)$, given $f_\lambda < 0$ as assumed, we have from the mean value theorem

$$(\mu - \lambda)(w - z) < 0 \tag{11.28}$$

whenever $\mu \neq \lambda$.

For the disequilibrium pair of fields μ, ψ we can now easily formulate a first order dynamics. For local prices we suppose they are driven up by (i) the local excess demand w and (ii) the drain from interregional trade $\nabla \cdot \psi$ alike, so we write:

$$\dot{\mu} = \nabla \cdot \psi + w \tag{11.29}$$

We might have expected an adjustment speed multiplier in the right hand side, but we have so many degrees of freedom, the unit of time, the unit of commodities, and the currency for accounting the prices, so we can reduce this adjustment speed, and the one to appear in the next equation, to unity.

In exactly the same way we may dynamise the second equation:

$$\dot{\psi} = \nabla \mu - k \frac{\psi}{|\psi|} \tag{11.30}$$

To interpret, recall that ψ is a two component vector, so we just stated two differential equations in vector notation.

We can again consider things in terms of flow direction and flow volume. If the direction is correct, i.e., that of the price gradient, then flow direction does not change, whereas flow volume is increased or decreased to the extent the price gradient exceeds or falls short of the freight rate. If the direction is not correct, then the vector equation also tells how it is reoriented towards the price gradient direction.

11.8 Stability

It is obvious that the two differential equations (11.29)-(11.30) provide a quite complex system, because the latter, despite its first order nature, is nonlinear (due to the occurrence of the vector norm). A full investigation is out of question, but we can say some nice things about stability of equilibrium.

Consider the nonnegative quantity:

$$E = (\mu - \lambda)^2 + (\psi - \phi)^2 \tag{11.31}$$

The squared expressions are the deviations of the disequilibrium price and flow patterns from the equilibrium ones. To get things right, note that the second square is a dot product of the difference vector with itself

$$(\psi - \phi)^2 = (\psi_1 - \phi_1)^2 + (\psi_2 - \phi_2)^2 \tag{11.32}$$

Observe that the sum of squares (11.31) is zero only when the system is in equilibrium, and positive otherwise. Let us now take the time derivative of (11.31):

$$\frac{1}{2}\frac{d}{dt}E = (\mu - \lambda)\dot{\mu} + (\psi - \phi)\cdot\dot{\psi} \tag{11.33}$$

We might have expected the time derivatives $\dot{\lambda}, \dot{\phi}$ as well, but they are zero as they represent an equilibrium pattern. It would now seem appropriate to substitute from (11.29)-(11.30) into (11.33), but as a preliminary step we first form the differences (11.29) minus (11.8) and (11.30) minus (11.9), slightly rearranged. Thus:

$$\dot{\mu} = \nabla\cdot(\psi - \phi) + (w - z) \tag{11.34}$$

$$\dot{\psi} = \nabla(\mu - \lambda) - k\left(\frac{\psi}{|\psi|} - \frac{\phi}{|\phi|}\right) \tag{11.35}$$

The procedure is fully legitimate because we can always subtract zero expressions whenever we wish. We also profit from the linearity of the divergence and gradient operators to collect the terms in pairs.

It is now time to substitute from (11.34)-(11.35) in (11.33). Doing this we get:

$$\frac{1}{2}\frac{d}{dt}E = (\mu - \lambda)\nabla \cdot (\psi - \phi) + (\mu - \lambda)(w - z) \tag{11.36}$$

$$+ (\psi - \phi) \cdot \nabla(\mu - \lambda) - k(\psi - \phi) \cdot \left(\frac{\psi}{|\psi|} - \frac{\phi}{|\phi|} \right)$$

Expression (11.36) may seem rather complex, but we will interpret the terms one by one. Let us start with the last one. Multiply the two parentheses together. Note that $\phi^2 = |\phi|^2, \quad \psi^2 = |\psi|^2$. Hence:

$$(\psi - \phi) \cdot \left(\frac{\psi}{|\psi|} - \frac{\phi}{|\phi|} \right) = \left(|\psi| + |\phi| \right) \left(1 - \frac{\psi}{|\psi|} \cdot \frac{\phi}{|\phi|} \right) > 0 \tag{11.37}$$

This strict inequality holds unless $\phi / |\phi| = \psi / |\psi|$. On the right hand side, the first parenthesis is the sum of two positive flow volumes, and the second parenthesis is the difference between unity and the product of two unit vectors. Such a product is at most unity, in the case the flow directions coincide, otherwise it is less. The expression can hence never become negative, and it is positive whenever the flow fields are different.

Let us next deal with the second term in (11.36):

$$(\mu - \lambda)(w - z) = (\mu - \lambda)(f(\mu) - f(\lambda)) < 0 \tag{11.38}$$

where the first parenthesis is the difference of two prices, and the second is the difference of the corresponding excess demands. It is negative as already discussed unless prices are equal, i.e. $\mu = \lambda$. Due to the general fact that excess demand functions are decreasing, the sign is as stated.

Using (11.37)-(11.38) in (11.36) we get:

$$\frac{1}{2}\frac{d}{dt}E < (\mu - \lambda)\nabla \cdot (\psi - \phi) + (\psi - \phi) \cdot \nabla(\mu - \lambda) \tag{11.39}$$

The strong inequality holds unless disequilibrium prices *and* flows agree with the equilibrium ones.

From the basic "chain rule" of vector analysis, we can collect terms in the right hand side and write:

$$\frac{1}{2}\frac{d}{dt}E < \nabla \cdot ((\mu - \lambda)(\psi - \phi)) \tag{11.40}$$

Condition (11.40) applies in each single location of R. Next, integrate (11.40) with respect to the space coordinates over the entire region R, and use Gauss's Integral Theorem:

$$\begin{aligned} \frac{1}{2}\frac{d}{dt}\iint_R E dx_1 dx_2 &< \iint_R \nabla \cdot ((\mu - \lambda)(\psi - \phi)) dx_1 dx_2 \\ &= \oint_C (\mu - \lambda)(\psi - \phi) \cdot \mathbf{n} ds \leq 0 \end{aligned} \tag{11.41}$$

The divergence theorem - one of the most important tools in vector analysis - states that the double integral of the divergence of any vector field on a closed and bounded area equals the curve integral of the outward normal projection of the field taken along the boundary. The outward normal is denoted **n**, and any projection on it, such as $\psi \cdot \mathbf{n}$ or $\phi \cdot \mathbf{n}$, is hence exports, or imports if negative, at the boundary point. The curve integral along the boundary C in (11.41), integrates the product of the differences of prices in the disequilibrium and the equilibrium patterns, and the corresponding boundary normal components of the flows. As the latter denote exports from the region, or imports if negative, we could directly apply the decreasing property of excess demand functions to exports/imports and prices in points on the boundary. In that case the integrand in every boundary point and hence the whole curve integral in (11.41) becomes nonpositive, or even negative if prices do not agree along the whole boundary. But we do not need the strong inequality as there is one already.

Also note that, as we are now dealing with an integral, the strong inequality would apply unless prices and flows agree with equilibrium prices *in every point of the entire region*. We thus get:

$$\frac{1}{2}\frac{d}{dt}\iint_R E dx_1 dx_2 < 0 \tag{11.42}$$

provided the disequilibrium data are at least somewhere different from the equilibrium data. We conclude that as time passes the integral over space of the nonnegative quantity E, as defined in (11.31), decreases monotonically toward zero. As E did measure the deviation from equilibrium, this means that the equilibrium is globally, asymptotically stable.

11.9 Uniqueness

Along the same lines we can prove uniqueness of equilibrium. So, suppose that the alternative pair of fields μ, ψ, like λ, ϕ, represents another equilibrium state. We hence restate equation (11.29) as

$$\nabla \cdot \psi + w = 0 \tag{11.43}$$

and (11.30) as

$$k\frac{\psi}{|\psi|} = \nabla\mu \tag{11.44}$$

in exact parallel to (11.8) and (11.9). Now, subtract (11.43) from (11.8). As the divergence operator is linear, we can write the result:

$$\nabla \cdot (\phi - \psi) = -(z - w) \tag{11.45}$$

and, multiplying through by $(\lambda - \mu)$ we get:

$$(\lambda - \mu)\nabla \cdot (\phi - \psi) = -(\lambda - \mu)(z - w) \geq 0 \tag{11.46}$$

The sign is as in (11.38), though we prefer to state it as a weak inequality so that no special qualifications are needed.

In the same way we subtract (11.44) from (11.9), obtaining:

$$\nabla(\lambda-\mu)=k\left(\frac{\phi}{|\phi|}-\frac{\psi}{|\psi|}\right) \tag{11.47}$$

and multiply through by $(\phi-\psi)$. In this way we obtain:

$$(\phi-\psi)\cdot\left(\frac{\phi}{|\phi|}-\frac{\psi}{|\psi|}\right)=\left(|\phi|+|\psi|\right)\left(1-\frac{\phi}{|\phi|}\cdot\frac{\psi}{|\psi|}\right)\geq 0 \tag{11.48}$$

as in (11.37), though we prefer the weak inequality even here.

As the freight rate $k>0$, we get from (11.47)-(11.48):

$$(\phi-\psi)\cdot\nabla(\lambda-\mu)\geq 0 \tag{11.49}$$

whereas from (11.46)

$$(\lambda-\mu)\nabla\cdot(\phi-\psi)\geq 0 \tag{11.50}$$

Now, consider the identity already referred to:

$$\begin{aligned}\nabla\cdot\left((\lambda-\mu)(\phi-\psi)\right)&=\nabla(\lambda-\mu)\cdot(\phi-\psi)\\&\quad+(\lambda-\mu)\nabla\cdot(\phi-\psi)\end{aligned} \tag{11.51}$$

So, from (11.49)-(11.50), and (11.51):

$$\nabla\cdot\left((\lambda-\mu)(\phi-\psi)\right)\geq 0 \tag{11.52}$$

Next apply Gauss's Integral Theorem to (11.52):

$$\iint_R \nabla \cdot ((\lambda - \mu)(\phi - \psi)) dx_1 dx_2 = \oint_C (\lambda - \mu)(\phi - \psi) \cdot \mathbf{n} dt \leq 0 \quad (11.53)$$

Again the most reasonable boundary condition prescribes that exports are lower and imports higher when boundary prices are higher, so the integrand and the whole curve integral are again nonpositive.

In (11.52) we have a nonpositive double integral of a nonnegative integrand. The integrand must hence be zero everywhere:

$$\nabla \cdot ((\lambda - \mu)(\phi - \psi)) \equiv 0 \quad (11.54)$$

But, according to (11.51), (11.54) is the sum of two expressions, which each, by (11.49) and (11.50), are nonnegative. Hence, both terms have to be zero separately. Thus:

$$(\lambda - \mu)(z - w) \equiv 0$$

and, according to (11.47)-(11.48), (11.55)

$$(|\phi| + |\psi|)\left(1 - \frac{\phi}{|\phi|} \cdot \frac{\psi}{|\psi|}\right) \equiv 0 \quad (11.56)$$

By (11.55) the value of excess demand must be zero everywhere, and this is possible only if the prices are equal. By (11.56), the flows must be co-directional everywhere (unless they are zero). So, indeed, the assumed solutions must be identical, and this is to say that the equilibrium is unique.

This model illustrates that, as is so often the case in economics without explicit consideration of space, the price/trade tâtonnement processes are well behaved and without surprising dynamic phenomena. This, however, is so as long as we model the processes in continuous time.

In discrete time, as Chiarella and Hommes have demonstrated, it can be sufficient that a supply function has a slight S-shape (without back-bending) for an adjustment process to go chaotic.

12 Development: Increasing Complexity

The most obvious aspect of economic development is that of increasing mass and volume: of people, households, commodities, services, firms, communications.

There is also another salient feature, a qualitative aspect of increasing complexity, diversity, and specialization of function. Typically, the various components of a more complex whole seem to become internally more complex, more complex the more specialized their function is.

This development towards increasing specialization and internal complexity almost seems to be a natural law, like the degradation of energy in thermodynamics, though it has a direction reverse to energy degradation, towards higher levels of order and organization.

In this respect economic development is similar to the development of the species in zoology: from a few primitive organisms at a low level of specialization and internal complexity, to highly organized ecosystems of specialized and complex organisms.

It can be maintained that this difference of direction is a salient difference between matter living and dead. Human society in this respect then is just one aspect of life.

The economics of development focuses increasing quantity rather than increasing diversity. This is natural because quantitative increase in well defined sets of objects is easier to analyse than increasing diversity of the set structure itself.

Yet the latter feature is the more interesting one in the development process. It therefore seems necessary to take a new start for modelling economic development focusing qualitative issues, even if it may have to be a simplistic one.

In the history of economics many literary descriptions of development through qualitative change have been advanced, but none has been translated into a successful formal model. There are two excellent such examples:

Böhm-Bawerk's theory of roundabout production, and Adam Smith's theory of labour division.

Böhm-Bawerk (1900), discussed how step by step the production process becomes more and more roundabout, and pays off in terms of an increased productivity which more than compensates for the time spent on capital formation:

"*A farmer needs and wants water for drinking. The spring bubbles at some distance from his house. In order to satisfy his need for water he can choose different ways. Either, he goes each time to the spring and drinks from his hollowed hand ...*

Or ... - the farmer makes a jug from a block of wood and carries in it his daily use at once from the spring to his living place ... but in order to achieve this it would have been necessary to make a not insignificant detour: the man might have to carve on this jug a whole day; he should, in order to be able to carve it ... fell a tree ... even before that produce an axe, and so on ...

Finally ... in stead of one tree he could fell many, hollow them all in the centre, build a pipe from them and bring an abundant stream of water to his house. Obviously roundaboutness is now even more considerable. As a compensation it leads to a much enhanced achievement."

Adam Smith (1776) spoke almost lyrically of the productivity increases resulting from specialization of labour. His famous example was that of specialized pin-making, where both organisation and training, combined with specialization in accordance with natural talent, result in improved productivity:

"*One man draws out the wire, another straights it, a third cuts it, a fourth points it, a fifth grinds it at the top for receiving the head; to make the head requires two or three distinct operations; to put it on, is a peculiar business, to whiten the pins is another; it is even a trade by itself to put them into paper; and the important business of making a pin is, in this manner, divided into about eighteen distinct operations ...*

Each person, therefore, ... might be considered as making four thousand eight hundred pins a day. But if they had all wrought separately and independently, and without any of them having been educated to this peculiar business, they certainly could not each of them have made twenty, perhaps not one pin in a day."

It is interesting to consider how economists would nowadays model the benefits of roundaboutness or of specialization. Roundaboutness would be introduced as an increase of capital K in the production function $Q = F(K,L)$, as a proxy for the heterogeneous set of jugs and pipes, whereas division of labour would be represented by an increasing productivity constant, multiplied with the labour input L or just with output Q.

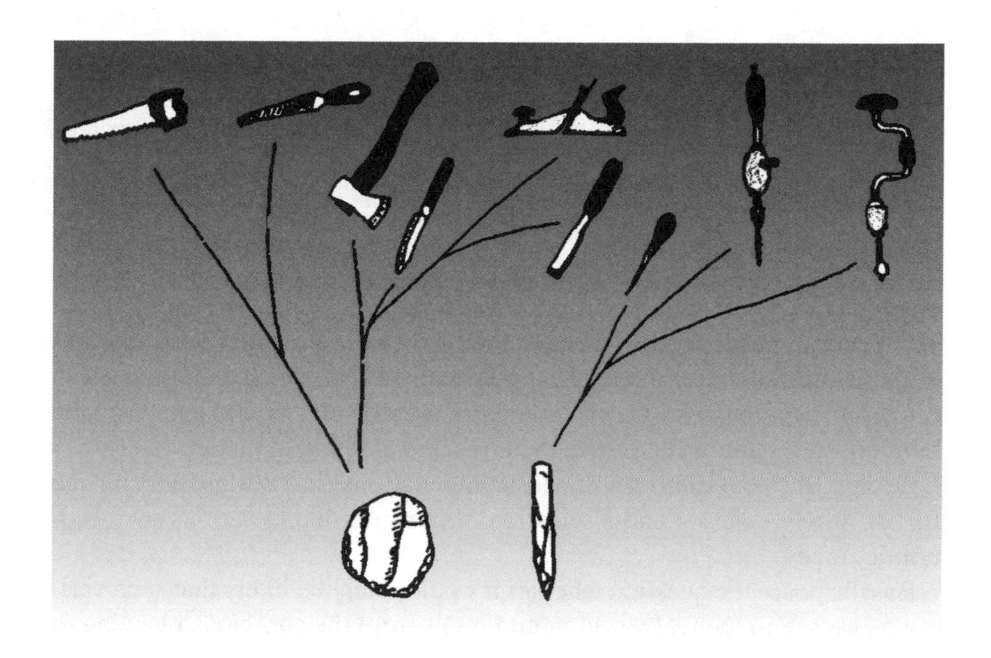

Fig. 12.1 Darwinian development tree for human implements.

This practice must be regarded as very crude, doing no justice at all to the original ideas. It is obvious that not much of the phenomena discussed by Böhm-Bawerk and Smith are caught in this way.

12.1 The Development Tree

As we said, a new start would be needed for focusing increasing diversity. Such a starting point could be an analogy to the Darwinian development tree. As can be seen from archeological findings, the development of the tools of primitive man started with multipurpose stone tools, for instance, not differentiating between the operations of cutting, scraping, and sawing, just like zoological development started with primitive organisms of low differentiation, such as amoebas. See Figure 12.1.

In the course of time the saw developed as distinct from the knife and from the scraper, i.e., the lines branched exactly as they do in the biological development tree. Some reflection convinces us that this process of branching and specialization is universal in the development of whatever human implements we study. The question is what the reason for this increasing specialization is.

The idea of regarding the development of human implements as analogous to the development of biological organisms is far from new. A few contemporaries of Charles Darwin actually saw things this way. Most prominent among those was Karl Marx in *Das Kapital* 1867, who, however found no use for parallels to the ideas of mutations and survival of the fittest. Accordingly he did not detect the greatest potential for Darwinian evolutionism in economic development. If development is a result of wilful planning by the humans, for better or for worse, there remains nothing to model.

George Basalla (1988) suggested both mutations (innovations just for the fun of "*homo ludens*") and a selection mechanism due to various socioeconomic forces.

Basalla poses the question whether it would really be likely that such variety as we see in the real world would result from the conscious planning to satisfy human needs concerning shelter, nourishment, and the like.

He asks whether for instance the automobile really was necessary and claims that the automobile essentially was a toy in its first decade 1895-1905, in no way triggered by any kind of shortage of horsepower.

But no matter what the reasons, the interesting thing is that these things were invented without any guarantee for being useful at all, and hence became more like random mutations in Biology. They might linger on, if only as toys, and their use in practice might then explode when a niche is opened up.

Basalla pursues the analogy between the evolution of living organisms and that of human artifacts in great detail. First of all he compares the number of different species, estimated to 1,500,000, with the number of patents, as representing different artifacts, of which only those taken in the US from 1790 to 1980, amounted to 4,700,000. So, he concludes that the world of artifacts is hardly less complex than the world of living organisms.

Earlier writings on these topics classified the mechanical "life" in genera, species, and varieties, and attempted to sketch a development tree in analogy to Darwin's. There also was an absurd preoccupation with the lack of reproductive capacity among the machines, and attempts to make the analogy complete by considering machines making machines on their own. Coupled to this there was a fear that machines would somehow "take over" the role of humanity. At this stage the analogy ceases to be fruitful. We do not need self-reproduction because humans by reinvestment see to it that viable things are reproduced.

Basalla also claims that there is missing a clear idea of the selection mechanism in these early writings, and that they also flip between the idea of technological evolution as a continuous, collectively social process, and as a discrete process of inventions with great names labelled to each one of them. He stresses that there is nothing exclusive in the outlooks, and that in reality there is discontinuity and continuity as well.

Being a pure historian, he, however, has no idea of mathematical bifurcations, competing attractors, basins of attraction, niches for the survival of human implements, or the general way the totality of technology feeds back to provide these niches.

12.2 Continuous Evolution

Basalla gives very detailed evidence for the inadequacy of the common idea that technological development is a sequence of discrete revolutionary inventions, and he claims that there is much more of continuous evolution than generally admitted.

He takes a starting point in James Watt's invention of the steam engine in 1775, popularly represented as springing out of a dream of a boy watching a boiling tea kettle. He points at the preexistence of Thomas Newcomen's steam engine for pumping water out from mines, invented in 1712. Though the principles are quite different, Newcomen's working with single action on vacuum resulting from condensating steam, Watt's with double action on expanding steam, it is a fact that Watt's invention arouse in direct connection with repair and improvement of one of the Newcomen engines at the University of Glasgow where he was instrument maker.

Basalla also points at the fact that neither Newcomen's invention was out of the blue sky, as all the mechanical elements, such as piston pumps, linkages, and steam displacement devices were well known and used in Europe in the preceding Century.

He also points at links to the later internal combustion engine, as it in 1759 already was proposed to replace steam by hot air in the engines, and as in 1791 an internal combustion "steam engine", working on turpentine vapour was actually patented. Though there still was two way action remaining from the steam engine, nevertheless the giant leap to Nikolaus Otto's combustion engine is thus bridged over.

More surprising is that even early electric motors were modelled directly after the steam engine, with pistons moving back and forth in cylindrical coils. The rotary motion we take as a matter of course in the electric motor hence was not the only alternative, and, by the way, it also had its obvious mechanical analog in the steam turbine, which itself, of course, goes back to water wheels known since ancient times.

All this demonstrates the usefulness of the biological analogy. First, there no doubt exists a continuous evolution process in the realm of human artifacts quite as in that of living organisms. Second, there is such an overwhelming richness of "species" that they must have arisen through a process of branching and diversification, rather than by some act of premeditated "creation", which is the obvious alternative.

A natural question would be whether bifurcations (i. e. successive splitting in two) could really account for the large number of artifacts which we mentioned. The process of "splitting in two" however is an exponential one, and it easily leads to large numbers. We mentioned the number of US patents, 4,700,000 in the period 1790 to 1980. However $2^{22} = 4{,}194{,}304$, so supposing there was only one implement in 1790, we would on average need no more than around one bifurcation per decade in order to arrive at the required order of magnitude.

12.3 Diversification

So let us try to pin down the facts pertaining to the development of human artifacts, keeping in mind the bifurcation idea. We may observe something like a curious natural law: At each state of technology, an implement can be made more efficient the more specific its operation is. The original all-purpose tool can never become as efficient as a saw for sawing, or a knife for cutting. On the other hand an all-purpose tool is better for sawing than a knife.

As a consequence, in the process of harvesting increased efficiency by specialization of function something in terms of the versatility in secondary functions is sacrificed. Only when the different specialized tools are again assembled in a joiner’s workshop, or in a modern combination machine, is there an overall improvement of efficiency.

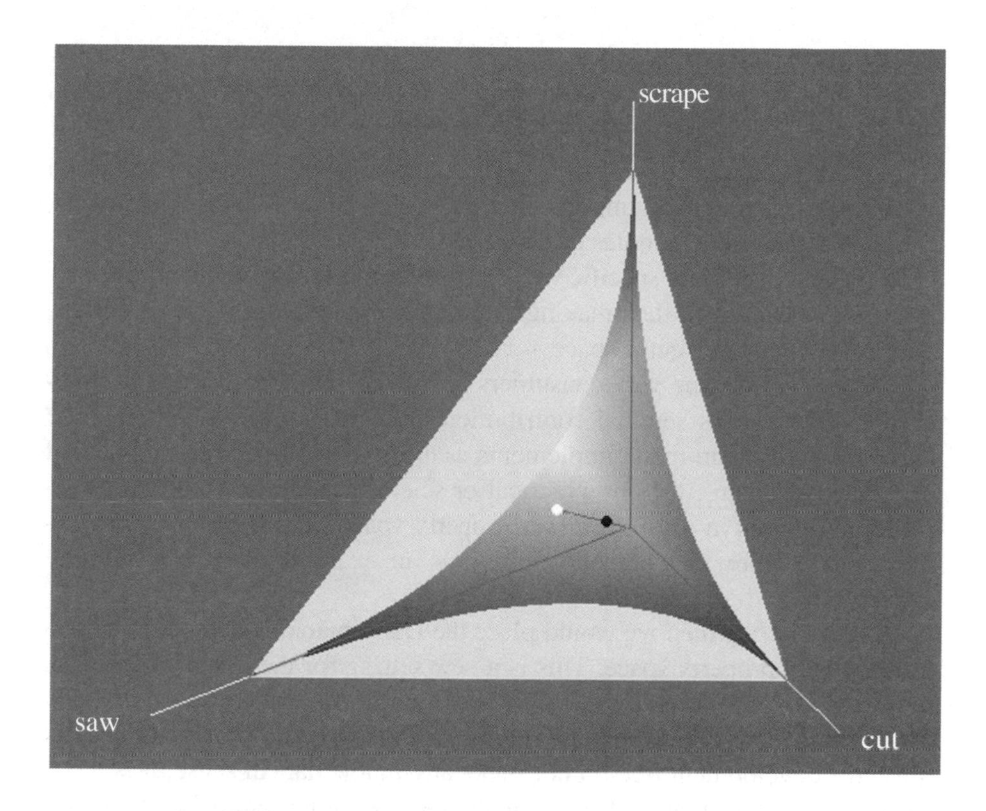

Fig. 12.2. Advantages of diversification and property space.

It is very unlikely that such a development would in general ensue according to some premeditated blueprint. Rather we would expect that a more knifelike tool is produced just when efficient cutting is highly needed, and the sawlike tool when sawing is particularly essential (in terms of economics: when the prices for these services are high).

The complete workshop or combination machine is likely to be organized first when most of the specialized tools are already at hand. This will then be the result of recognizing the complementarity property of various specialized tools when they are used in combination. See Figure 12.2 for an illustration to the technological advantages of diversification, the dark shade technological possibility frontier being convex to the origin and extended along the axes. The advantages of combining complementary specialized tools is illustrated in the bright shade plane of convex combinations, and actual dominance by the white spot as compared to the black located on the possibility frontier.

12.4 Lancaster's Property Space

In mentioning the properties of a tool, exemplified by the ability to cut, to saw, and to scrape, it is implicit that tools, and other human implements, could be regarded as bundles of properties.

To make things more specific we can treat them as measurable efficiencies in various operations, thus making the implements representable as coordinate points in some vector space.

This is precisely the way consumers' goods were once regarded by Lancaster (1971) in his seminal contribution. It seems fairly obvious that the performance of man-made implements as discussed above, exactly like that of living organisms, for any practical or scientific purpose could be represented by points in a Lancasterian property space of sufficiently many dimensions. Figure 12.2 actually represents an example of such a property space in three dimensions.

As the next step then we would place the Darwinian development tree in a Lancasterian property space. This is a new setting for the development tree, the branching tree structure remains, but is placed in a vector space of many dimensions. The dimensions of the space in which the biologists usually draw the development tree do not make any mathematical sense, it is just a two-dimensional picture of a set of three-dimensional animals. In this respect our use of the analogy is different.

The essential characteristic of the development tree is its branching structure. By the branchings or bifurcations we can best understand the nature of the emergence of new implements with more specialized function. Later we will focus on mathematical descriptions of such branching structure.

12.5 Branching Points

We understand that a new direction in the design of implements is most often caused by changes of needs and tastes. Design ideas occur quite frequently and even erratically over time. Some of them become successful, some are skipped from the beginning.

There are two issues involved. First, the innovation must be transformed from a blueprint idea to a material implement, and second, it must remain at the stage as useful over its lifetime and even be reproduced when it is worn out.

Again, taking the development tree analogy as support, we would say that they arise like mutations. Exactly like the mutations they most often do not remain permanent; only when there is a macro environment with a suitable niche for them, do they appear and remain. The two stages are the same too. First, the genetic change in a chromosome must result in an individual organism that can at all live. Second, it must survive in terms of being propagated to further generations.

Such a permanence for implements results, as we know, from the recognition of the superiority of specialized tools, and of their complementarity, providing for an overall productivity increase when using them in combination.

The emergence of new implements is also a matter of transgressing certain critical threshold values, and the nature of the branching points, as sudden dramatic events contrasted to smooth continuous evolution becomes understandable.

For instance, it is well known that the aeroplane was invented many times, but that it became a reality only when the aerofoil wing could be combined with an engine of a sufficient power and a trunk of sufficient strength in respect to weight.

12.6 Bifurcations

Earlier models for dynamical systems were focused on smooth movement. In our days sudden transitions have been focused more and more.

René Thom opened up what used to be called Catastrophe Theory. His intention, as we saw in Chapter 5 was to formulate a method by which structural change such as the development of a bud into a leave or into a flower could be mathematically modelled.

Bifurcation, as we saw, is more general than catastrophe. The latter applies to gradient systems, i.e., such that optimise something, whereas the former applies to dynamical systems in general, whether they optimise anything or not.

If we assume that economic development is the result of some sort of optimisation, subject to changing macro environment constraints in terms of changing prices and general know-how, then we can restrict our interest to catastrophe theory.

In our case the state space would be the Lancasterian property space, and the parameters would be represented by such things as the imputed prices that reflect the current tastes and needs.

Under certain parameter changes a previously stable optimum might lose stability and the slowly evolving optimum jump to any of the new optima into which the previous one bifurcates. Assuming that several agents operate in the same system they may together recover several or all of the new stable optima. This explains bifurcations of the type we have been talking about.

Optimising might be represented by an explicit gradient dynamics, and if we wish we can assume adjustment to be momentary. The evolution of the parameters, on the other hand, is slow, and as a result we may see how an equilibrium changes slowly for considerable time, and then bifurcates at an instant. Two words of warning are in place if we want to make an explicit model.

First, we assumed that the Lancasterian property space would need to have a considerable number of state variables, whereas all catastrophe theory attained was a classification of systems with one or two state variables and no more than four parameters.

Second, even though the mathematical proofs, that all possible phenomena can be modelled in terms of the "canonical forms" of the catastrophes and their "universal unfoldings", are safe, it is extremely hard to actually transfer substantial models into these canonical forms by the proper nonlinear coordinate changes. This is why there were so many trivial catastrophe models. As for the first argument, about the multitude of variables, the Synergetics school of Haken (1976, 1983) provides a good argument why those low dimensional problems are not so bad after all.

In reality, despite the multitude of variables involved, the probability that more than very few of them are involved in actual phase transitions is very small. The stability of a certain variable is measured by its eigenvalue, obtained at linearising the system in the neighbourhood of an equilibrium.

The eigenvalues can be zero, negative, or positive. Positive eigenvalues signify instability, and they can hence not exist for more than very short periods, otherwise the system will blow up. Distinctively negative eigenvalues mean stability, and in the perspective of what eigenvalues close to zero can cause, their effect can be regarded as momentary damping to equilibrium.

The interesting things, catastrophes and bifurcations, in a system are triggered by those few eigenvalues that are very close to zero and that for a while can become positive.

By probability considerations it is very unlikely that more than a few ones among the negative eigenvalues are close to zero. The philosophical argument is that this movement of the very few "slow" variables is the reason why we find relatively much order and relatively few phase transitions in reality despite its complexity. As a result, low dimensional models may still be useful for high-dimensional systems.

12.7 Consumers

According to Lancaster one should not enter the commodities themselves in consumers' preferences, but rather their properties. In this way it would be easier to deal with different brands of a commodity, such as automobiles. The consumer would judge each vehicle according to top speed, reliability, durability, safety, and so forth, and any substitution in the preferences would be between these properties.

Suppose we consider n properties according to a list referring to the various qualities of composition and performance, measured as scores x_i, for $i = 1,\ldots n$. A single object would then represent a fixed bundle of such performance scores, and an item j would be represented as a vector of values a_{ij} for $i = 1,\ldots n$.

A consumer could then, by acquiring a collection of quantities q_j of the objects, obtain a combined representation of the performance dimensions. Suppose there are in all m different objects. Thus the overall performance of the collection would be:

$$x_i = \sum_{j=1}^{j=m} a_{ij} q_j \quad i = 1,\ldots n \tag{12.1}$$

For an individual consumer many quantities could, and should, of course, be entered as zeros.

If we now had prices for the x_i variables we could get ahead with analysis, doing things such as evaluating different collections and choosing the best among them.

The x_i variables would be included in the utility functions of various consumers, and so establish demand for the various objects. Provided there is any truth in this type of representation, i.e. that consumers roughly agree about the properties of the objects, though perhaps not about their valuation, this will establish a set of imputed prices for the properties themselves.

There is one problem: We have no guarantee that the number of objects equals the number of variables.

If the objects are fewer than the variables, we cannot determine all the imputed prices. This is to be expected in the process of evolution when the number of objects is increasing and at any moment is less than the dimension of the huge embedding property space. As a result only operative subspaces can be reached.

The imputed prices might also be over-determined, so that we get different sets of imputed prices, one for each selection of basis, which also can happen in the operative subspace.

However, if the consumers agree about how to characterize the objects there will be linear dependence. The rank of the matrix $\left[a_{ij}\right]$ in the operative subspace will be full. This may seem to be a bold assumption but it is implied by the very approach itself. We are going to comment on the reasonable objections against this below.

For the moment let us put all this up formally. Suppose we want to maximize:

$$U(x_1,\dots x_n) \tag{12.2}$$

subject to (12.1) and to:

$$\sum_{j=1}^{j=n} p_j q_j = C \tag{12.3}$$

where p_j denote the commodity prices as actually determined on the market, and where C denotes a given cost budget.

We then associate the Lagrangian multipliers λ with (12.3) and λ_i with each of the equations (12.1), and formulate:

$$U(x_1,\dots x_n) + \sum_{i=1}^{i=n} \lambda_i \left(\sum_{j=1}^{j=n} a_{ij} q_j - x_i \right) + \lambda \left(C - \sum_{j=1}^{j=n} p_j q_j \right) \tag{12.4}$$

Differentiating with respect to x_i and q_j we obtain:

$$\frac{\partial U}{\partial x_i} = \lambda_i \tag{12.5}$$

and

$$\sum_{i=1}^{i=n} a_{ij}\lambda_i = \lambda p_j \tag{12.6}$$

Equations (12.5) are of no particular interest for the moment, but (12.6) are. First, note that multiplying (12.6) by x_i, summing over index i, and substituting from (12.1) and (12.3), we obtain:

$$\sum_{i=1}^{i=n} \lambda_i x_i = \lambda C \tag{12.7}$$

Thus λ is only a proportionality numéraire for the vector of the Lagrange multipliers, which we can set equal to unity. To do this is reasonable, because in this way (12.7) tells us just that the total cost budget is spent on a mix of properties evaluated at their imputed prices. Then (12.7) becomes:

$$\sum_{i=1}^{i=n} \lambda_i x_i = C \tag{12.8}$$

Moreover (12.6) becomes:

$$\sum_{i=1}^{i=n} a_{ij}\lambda_i = p_j \tag{12.9}$$

which says the value of the j:th implement in terms of its properties and the imputed prices for these properties equals it market price. The interpretation of the λ_i as imputed prices is also obvious from their association with the scarcity constraints.

It is easy to invert the matrix $\left[a_{ij}\right]$ and obtain the solution:

$$\lambda_i = \sum_{j=1}^{j=n} \frac{D_{ij}}{D} p_j \tag{12.10}$$

Here $D = \left|a_{ij}\right|$ denotes the determinant of the coefficient matrix, and D_{ij} denote its co-factors. The quotients in (12.10) have the interpretation of the numbers of units of each commodity in the set forming the basis, that go into one unit of some given property. Thus (12.10) explains how we, using commodity prices, arrive at the imputed prices for the properties.

Equation (12.9) can be interpreted as a definition of the imputed prices. It says that all the commodities included in the basis represent property combinations, such that their values, according to the imputed prices for these properties, add up to exactly the market prices for these commodities.

From what we said it should then also be the case that the remaining commodities share this property of fulfilling equation (12.9). If their values fall short of market price, they would not be worth their price, if they overshoot, they would have an excess value. The situation is, however, not too bad for the following reasons: Whenever we are dealing with durables which may be numerous and not objects of daily trade, the latest transaction prices for them may in many cases be out of date. In any collection of capital goods there will be obsolete ones, which have expired economically but not physically, and which are still retained in service for various reasons.

Let us now consider how change occurs in this framework. Considering a development of the product, its characteristic coefficients evolve over time, so accordingly we put $a_{ij}(t)$ as functions of time t.

12.8 Producers

It is now useful to draw attention to an idea proposed in Johansen (1972). Johansen distinguishes between the "ex ante" production function, which admits continuous substitution, such as represented by for instance the Cobb-Douglas or the CES-function, and the "ex post" production function, which is of the fixed proportions Leontief type, the choice being made and incorporated in one given setup of the production process.

The idea is that, though the classical production function may be a good representation of the long run technological options, they do not all stay open indefinitely. A choice has to be made, and once it is made there remains no substitution at all.

Given new conditions for production in terms of the relative prices, a different choice might become appropriate, but before being materialized it has to wait until the embodiment of the old one has become technologically or economically obsolete and will have to be replaced.

Before continuing on this track let us specify how different evaluations arise. In the format we introduced, the subjective preferences are represented by the utility function (12.2).

To make things very simple, suppose we assume (12.2) to be a social utility function, which we make time dependent: $U(x_1,\ldots x_n,t)$. Then the imputed prices, interpreted as marginal utilities according to (12.5) will change over time and thereby provide the changing environment for the producers.

How would we now fit this in the framework proposed by Johansen? He was concerned with one product isoquants for inputs, but there is no reason why we could not shift focus to multiple product production possibility frontiers instead, though keeping Johansen's long run and short run distinction in mind.

Moreover, suppose we formulate the production in terms of those very properties in Lancaster's sense which were ultimately included in consumers' preferences. Such a possibility frontier could be written as an implicit function:

$$P(x_1,\ldots x_n)=0 \tag{12.11}$$

and all the m different products, represented by their coefficients a_{ij}, would then have to satisfy (12.11). Thus:

$$P(a_{1j},\ldots a_{nj})=0 \quad j=1,\ldots m \tag{12.12}$$

The natural question is: Given the assumptions, why not just maximize (12.2) subject to (12.11), and devise the single perfect product which best satisfies the preferences?

There are several reasons why not, apart from the weakness of the social utility concept. The possibility frontier is largely unknown. The only certain points on it are the goods actually produced. It can be assumed to be successively exploited, presumably in the neighbourhoods of the existing products. It may not even be in one piece, if it were we would see cars that are computers and dishwashers at the same time.

We might or might not assume that the possibility frontier shifts over time. If we decide to define it in terms of physical constraints, the laws of mechanics, electromagnetism, and thermodynamics it stays constant, but we can also take a more psychological approach. No matter which way we choose, development on the possibility frontier can be expected to manifest itself in terms of curves, such as the parameterised curves:

$$\left[a_{1j}(t),\ldots a_{nj}(t)\right] \quad j=1,\ldots m \tag{12.13}$$

all embedded in the possibility surface according to (12.12).

Let us now turn to the development tree put into the framework of the Lancasterian hyperspace.

In the tree there are two features, continuous development along the lines and bifurcations at the branching points. A development with branching points cannot be represented by development curves such as those stated in (12.13), because they split. We therefore need a different paradigm to represent this phenomenon.

12.9 Catastrophe

Modern dynamics, in terms of bifurcation and catastrophe theories, fortunately, provides an ample supply of models for such phenomena.

An evolving point such as $\left[a_{1j}(t),\ldots a_{nj}(t)\right]$ traces a curve in n-dimensional phase space. If at any moment this represents the optimal choice of a design out of a multitude of possibilities, the evolving point will in a mathematical sense represent an attractor in phase space.

The movement towards such an attracting point can then be represented by a dynamical system:

$$\dot{x}_i = F_i(x_1,\ldots x_n,t) \quad i=1,\ldots n \tag{12.14}$$

where all the $\left[a_{1j}(t),\ldots a_{nj}(t)\right]$ are attracting equilibrium points, i.e.

$$F_i(a_{1j}(t),\ldots a_{nj}(t),t) \equiv 0 \quad \forall i,j \tag{12.15}$$

If we prefer to think that (12.14) represents the result of optimization, we can think of the process as being a more or less instantaneous transit to equilibrium. Then the $F_i = -\partial V / \partial x_i$ are the partial derivatives of some potential function $V(x_1,\ldots x_n,t)$. Even in the case of optimization it does no harm to think of a dynamical process attracted by the optimal solution.

Hence a fast and a slow dynamics is implied, a fast approach to the optimum design, as determined by prices and by other characteristics of the macro environment, and a slow dynamics, changing the optimal design as those characteristics are themselves changed due to external causes.

It may now happen in this slow evolution that at a certain stage the previous solution, though still existing, loses stability. As a rule it is then replaced by several new attracting equilibria.

The old equilibrium, that turned into a repellor from having been an attractor, becomes a kind of watershed, separating the basins of attraction of the new stable equilibria. Depending on the initial conditions, the fast process will now go to one or the other of these. In our case there is no reason to think of just one dynamic process. It is perfectly possible that different agents proceed simultaneously from different initial points, or that even one single agent does this, thus tracing the entire bifurcated line.

12.10 Simple Branching in 1 D

The simplest catastrophe type of any interest here is the "cusp". It lives in a one dimensional phase space, so we have an object with just one characteristic feature. To be specific:

$$\dot{x} = F(x) = -x^3 + kx \tag{12.16}$$

where there is a parameter k, which can vary. The equilibria of this equation are:

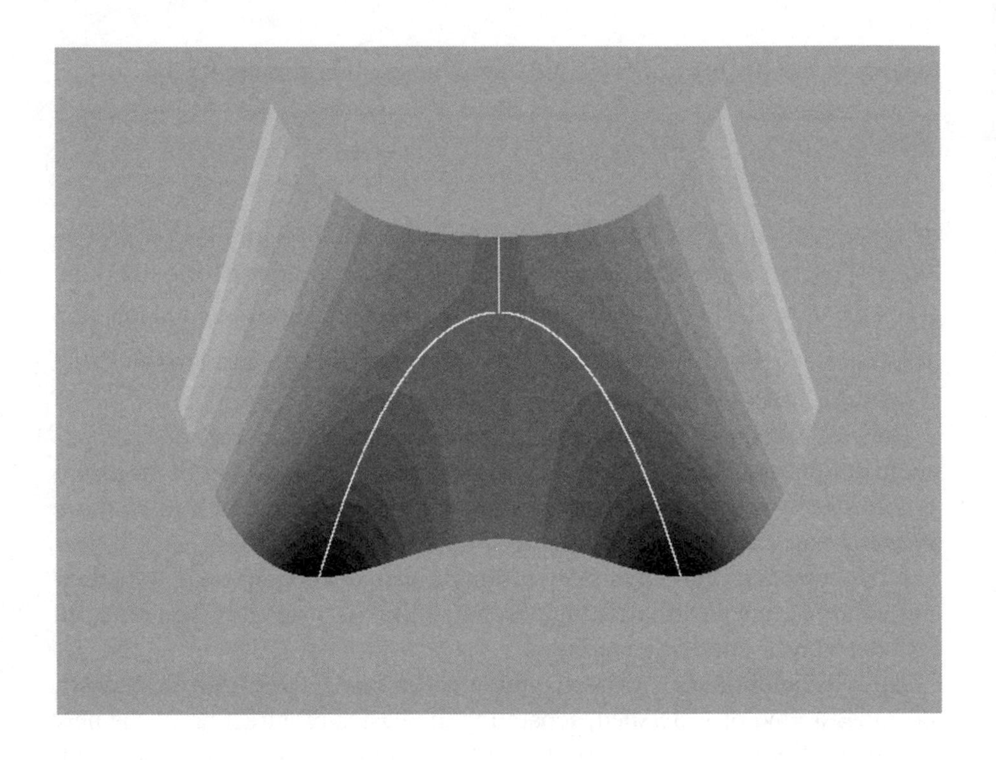

Fig. 12.3. Simplest bifurcation: the cusp.

$$x = 0, \quad \pm\sqrt{k} \tag{12.17}$$

The latter pair, of course, only exist if $k > 0$. As for stability, the normal investigation procedure is to linearise (12.16), i.e. to calculate its derivative:

$$F'(x) = -3x^2 + k \tag{12.18}$$

In the equilibrium point $x = 0$ we have:

$$F'(0) = k \tag{12.19}$$

whereas for $x = \pm\sqrt{k}$ we have:

$$F'\left(\pm\sqrt{k}\right) = -2k \tag{12.20}$$

Stability requires a negative derivative, so the solution at the origin of phase space is stable for $k < 0$ and unstable for $k > 0$.

The nonzero solutions according to (12.17) do not exist for $k < 0$, but due to (12.20) they acquire stability the very moment they arise, and at that very moment the solution at the origin also loses stability.

So, if we let k pass through zero from negative to positive values, the old equilibrium at the origin loses stability and is replaced by two newborn stable equilibria. The only thing we now need is to let the parameter k vary slowly over time, i.e. vary at a time scale different from the one represented in the dynamical process (12.16).

It is interesting to note that (12.16) represents a so called gradient dynamics for minimization of the potential:

$$V = \frac{x^4}{4} - k\frac{x^2}{2} \tag{12.21}$$

or, if we prefer, maximisation of its negative. In one dimension, with an integrable right hand side, it is always possible to find such a potential.

In higher dimensions this is no longer true. Thus gradient dynamics becomes a more restrictive type in higher dimension, and catastrophe theory becomes a restricted variant of bifurcation theory. However, whenever we deal with optimization, we can always derive the dynamics from some potential, no matter what the dimension is.

Figure 12.3 shows the potential curves for a continuum of parameter values, with k increasing from the back to the foreground. We can see how the single minimum of the potential well splits in two. The white curve traces the path of the local minimum and we see how it too splits in two.

12.11 Branching and Emergence of New Implements in 1 D

To see something more let us raise the power of the potential function (12.21) and suppose:

$$V = \frac{x^6}{6} - (2k-1)\frac{x^4}{4} + k(k-1)\frac{x^2}{2} \tag{12.22}$$

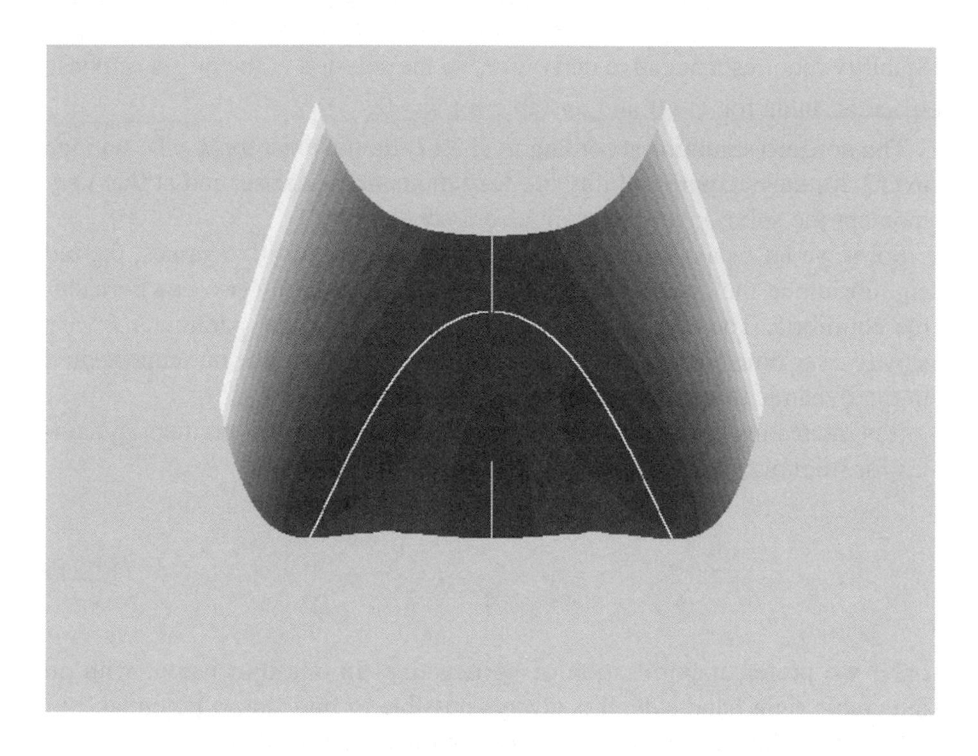

Fig. 12.4. Combined cusp and blue sky catastrophe.

While the previous catastrophe was called "cusp", the present one has the poetic name "butterfly". It should be said that, as we saw in Chapter 5, the general form of this catastrophe would involve two independent coefficients for the quartic and quadratic terms, and that usually a cubic and linear term will be included as well.

In all, four different terms involving four different parameters would have to be added to the sixth order term in order to represent the butterfly in full generality. This is however, not what we intend to do. Rather we have been using the catastrophes as mere illustrations of what kinds of bifurcations can happen along a one parameter path.

The corresponding potential surface is shown in Figure 12.4. Observe that it is a bit more complex than the previous surface. Looking at the front ridge we see that there seem to be three minima separated by two maxima, whereas in the back there is just one minimum.

The (fast) dynamics corresponding to this potential is easily obtained by differentiation:

$$\dot{x} = F(x) = -x^5 + (2k-1)x^3 - k(k-1)x \tag{12.23}$$

We have got rid of the denominators, but there is more to the way of introducing the coefficients. The expression easily factorizes into:

$$\dot{x} = -x(x^2 - k)(x^2 - k + 1) \tag{12.24}$$

so that we can immediately find the zeros:

$$x = 0, \quad \pm\sqrt{k}, \quad \pm\sqrt{k-1} \tag{12.25}$$

They are in all at most five, as can be expected with a fifth degree polynomial. As before, stability is determined by the first derivative of (12.23):

$$F'(x) = -5x^4 + 3(2k-1)x^2 - k(k-1) \tag{12.26}$$

Substituting the various roots we get the following expressions:

$$F'(0) = -k(k-1) \tag{12.27}$$

$$F'\left(\pm\sqrt{k}\right) = -2k \tag{12.28}$$

$$F'\left(\pm\sqrt{k-1}\right) = 2(k-1) \tag{12.29}$$

Stability requires a negative derivative. We therefore note that $x = 0$ is stable whenever $k < 0$ or $k > 1$. Next, we see that $x = \pm\sqrt{k}$ is stable whenever $k > 0$, i.e. whenever the square root exists. Finally, $x = \pm\sqrt{k-1}$ is stable when $k < 1$, i.e. never, because stability implies that the root is complex.

Thus, we can sort out the last pair of roots $x = \pm\sqrt{k-1}$. They simply represent the local watershed repellors once they appear on the stage. We also note that the second pair $x = \pm\sqrt{k}$ always represent stability once they become real. As to the origin it has two intervals of stability, when the parameter is negative, and when it exceeds unity.

Therefore, bifurcations occur precisely at the points $k = 0$ and $k = 1$. At the first, the point at the origin loses stability, and is replaced by the two new emergent real roots. At the second, it again gains stability, while the two already existent ones continue to coexist.

The first bifurcation is of the same type as in the case of the cusp, it means splitting of an earlier path. The second is different. The new attractor just arises out of the blue sky. It is also an example of a hard bifurcation, as compared to the previous soft bifurcation, in the sense that the newly born attractor emerges at a finite distance from the previous path. In the same manner that the attractor emerges out of the blue sky, it can just disappear.

If the reader would prefer further splitting of the branches in development, this can be arranged, but not with the present catastrophe. As we have a sixth order potential, i.e. a fifth order polynomial in the dynamical system, it at most has five real zeros. Having three stable equilibria we must reserve two as watersheds separating their basins. A further splitting of an already bifurcated path would mean four attractors, so this is not possible!

To get four stable equilibria, we would also need three separating unstable ones, i.e. seven zeros in all, so the dynamical system would have to be a seventh order polynomial, and the potential of order eight. That far the classification of the elementary catastrophes never went, but if we are just after constructing examples without reference to a general theory, they can easily be produced.

12.12 Catastrophe Cascade in 1 D

Suppose we have the potential:

$$V = \frac{x^8}{8} - (5k-2)\frac{x^6}{6} + (4k^2 - 2k + 1)\frac{x^4}{4} - k\frac{x^2}{2} \tag{12.30}$$

The dynamics corresponding to this potential is given by:

$$\dot{x} = F(x) = -x^7 + (5k-2)x^5 - (4k^2 - 2k + 1)x^3 + kx \tag{12.31}$$

with the zeros:

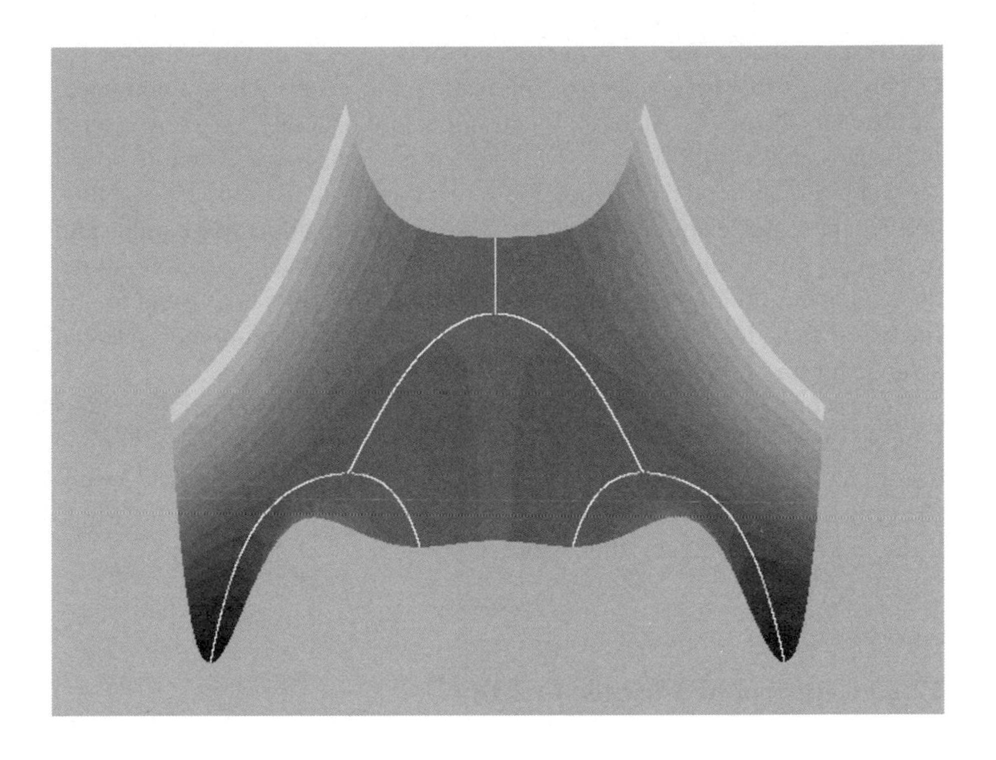

Fig. 12.5. Bifurcating cascade.

$$x = 0, \quad x = \pm\sqrt{k}, \quad x = \pm\sqrt{k} \pm \sqrt{k-1} \tag{12.32}$$

The derivative of the right hand side of the differential equation is:

$$F'(x) = -7x^6 + (25k - 10)x^4 - (12k^2 - 6k + 3)x^2 + k \tag{12.33}$$

which for the various roots takes on the values:

$$F'(0) = k \tag{12.34}$$

$$F'\left(\pm\sqrt{k}\right) = 2k(k-1)(3k+1) \tag{12.35}$$

$$F'\left(\pm\sqrt{k} \pm \sqrt{k-1}\right) = -8\sqrt{k}(k-1)\left(2\sqrt{k}(3k-2) \pm \sqrt{k-1}(6k-1)\right) \tag{12.36}$$

The first is negative for $k < 0$, the second for $0 < k < 1$, and the third for $1 < k$. This means that before the first bifurcation at $k = 0$ there is just one stable equilibrium: the origin. At that point the origin loses stability, and there emerge two new stable equilibria just as in the two previous cases. There is now a second bifurcation at $k = 1$, when these two equilibria as well lose stability, and four new stable equilibria emerge. The previously bifurcated equilibrium points thus split again. Figure 12.5 shows how, from the back to the foreground the simple minimum well of the potential curve first splits in two and finally further in four. By raising the power of the potential appropriately we can arrange for any number of successive bifurcations in a cascade.

There is one problem with all these cases, provided we interpret them literally: As there is just one phase variable its value would either be good or bad according to consumers' preferences, and so several equilibria would be unlikely to coexist. This problem is, however removed if we put the whole thing in a space of higher dimension.

12.13 Catastrophe Cascade in 2 D

As an example suppose we have the potential.

$$V = \frac{x_1^4 + x_2^4}{4} - (k-a)\frac{x_1^2}{2} - (k-b)\frac{x_2^2}{2} \tag{12.37}$$

We now have two state variables, but still only one parameter, as we want to consider a single development path through parameter space. The coefficients are, however, not equal. The reason is that we want the bifurcations to occur in succession, not simultaneously.

Comparing (12.21) to (12.37) we see that the latter is just the sum of two independent expressions each of the type of the former. We can therefore expect one bifurcation to occur at $k = a$. Likewise we can conjecture the other to occur at $k = b$. It is easy to check that this indeed is so. The gradient dynamics corresponding to (12.37) is:

$$\dot{x}_1 = -x_1^3 + (k-a)x_1 \tag{12.38}$$

$$\dot{x}_2 = -x_2^3 + (k-b)x_2 \tag{12.39}$$

Calculating the fixed points we have:

$$x_1 = 0, \pm\sqrt{k-a} \tag{12.40}$$

$$x_2 = 0, \pm\sqrt{k-b} \tag{12.41}$$

Note that there may be nine fixed points in all. Supposing $a < b$, we find that for $k < a$ only one is real, the origin. Further, for $a < k < b$ there are three, and finally for $b < k$ there are nine.

To classify the critical points we need the matrix of the derivatives of the system (12.38)-(12.39):

$$\begin{bmatrix} -3x_1^2 + k - a & 0 \\ 0 & -3x_2^2 + k - b \end{bmatrix} \tag{12.42}$$

As the matrix happens to be diagonalised already the elements of the main diagonal are the eigenvalues:

$$\mu_1 = -3x_1^2 + k - a \tag{12.43}$$

$$\mu_2 = -3x_2^2 + k - b \tag{12.44}$$

If both eigenvalues are negative we have an attracting node, if both are positive we have a repelling node, if one is positive and one is negative we have a saddle point. So let us check the signs. For the origin we have:

$$\mu_1(0) = k - a, \quad \mu_2(0) = k - b \tag{12.45}$$

For the pair of critical points on the x_1-axis:

$$\mu_1\left(\pm\sqrt{k-a}\right) = -2(k-a), \quad \mu_2(0) = k - b \tag{12.46}$$

for the pair on the x_2-axis:

$$\mu_1(0) = k - a, \quad \mu_2\left(\pm\sqrt{k-b}\right) = -2(k-b) \tag{12.47}$$

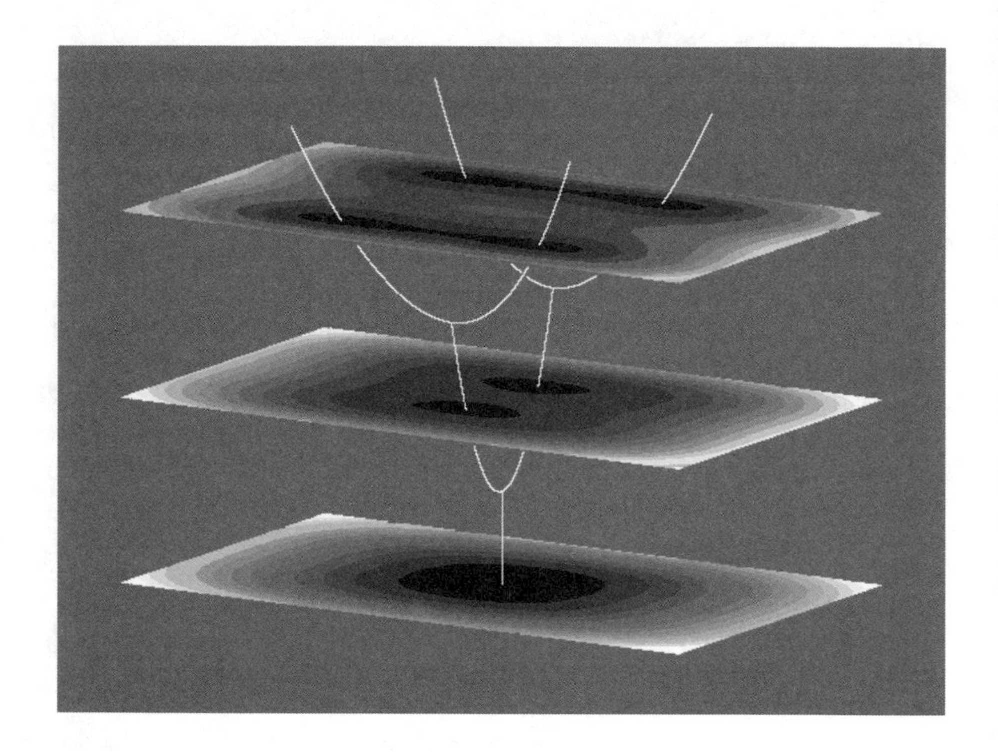

Fig. 12.6. Potential surfaces for different parameter values, and bifurcating tree.

and for the four off the axes:

$$\mu_1\left(\pm\sqrt{k-a}\right)=-2(k-a),\quad \mu_2\left(\pm\sqrt{k-b}\right)=-2(k-b) \tag{12.48}$$

Suppose $k < a$, for which case, as we saw, only the origin exists as a critical point. From (12.45) both eigenvalues are negative, so the point is an attractor.

For the interval $a < k < b$, the pair of points on the x_1-axis become real, and according to (12.46) they immediately get both eigenvalues negative, so that they are born attractors. The origin on the other hand, according to (12.45) gets one positive and one negative eigenvalue and hence is converted into a saddle.

Finally, for $b < k$, all the points off the axes become real and, according to (12.48), become attractors, whereas the points on the x_1-axis are converted into saddles, as we see from (12.47). Moreover, the new pair of critical points on the x_2-axis become real, but from (12.47) they are born saddles.

So there are two bifurcations. Before the first occurs at $k = a$, only the origin is an attractor. After the first bifurcation the origin loses stability and becomes a saddle point. Two new attractors are born and their basins of attraction are separated by that saddle point. We could also say that the previously unique attractor has split in two.

At the second bifurcation, when $k = b$, the attractors, emergent at the first bifurcation, lose stability and turn into saddles. There arise four new attractors and two new saddles. The point at origin is converted to a repellor, and the four attractors are separated by four saddles. Another way of telling what happens is that the formerly split up attractors undergo a further splitting. This is the beginning of a cascade of bifurcations which may result in a development tree. By raising the dimension of the phase space we can produce any number of further splitting. The same result can be attained by increasing the order of the polynomial potential.

In Fig. 12.6 we display what happens graphically. Having two phase variables we now need planes to show the potential surfaces. There are three such specimens displayed for various values of the parameter. The height variation has been extremely small for reasons of improved visibility, and the height is indicated by means of shading. In the picture in full three dimension we also display the bifurcating line which traces the locations of potential minima as the parameter varies.

Simple results, such as illustrated, however, depend on keeping the number of parameters low. This might seem problematic, as in a high dimensional property space there are also a large number of parameters, for instance the prices.

12.14 Fast and Slow Processes

Synergetics, as developed by Hermann Haken (1976, 1983), however, gives an argument for why bifurcations in high dimensional systems are low dimensional: The local stability of a system in the neighbourhood of equilibrium is described by the eigenvalues. Strictly positive eigenvalues would blow up the system and do not exist. Strictly negative eigenvalues mean immediate damping of the system. According to Haken it is the eigenvalues close to zero, which may become positive for a short while, which also control structural change, and, by mere considerations of probability, they are very few as compared to all the damped dimensions. In reality the negative eigenvalue closest to zero controls the whole development, and establishes a slow dimension as contrasted to the enormous number of fast dimensions, which can be considered as being in constant equilibrium.

To see things more specifically by means of our last example, consider the system (12.38)-(12.39) at the bifurcation point $k = b$, more specifically in the neighbourhood of the fixed point $x_1 = \sqrt{k-a}, x_2 = 0$. From (12.46) we see that $\mu_1 = -2(b-a) < 0, \mu_2 = 0$. Accordingly x_2 is the slow dimension and controls everything that is happening. Let us look at the system (12.38)-(12.39) rewritten for the following deviation variables from the equilibrium point: $\xi_1 = x_1 - \sqrt{b-a}$, $\xi_2 = x_2 - 0$. Note that we used the fact that $k = b$. This also makes (12.24) lose its second term. Hence:

$$\dot{\xi}_1 = -\xi_1\left(\xi_1 + \sqrt{b-a}\right)\left(\xi_1 + 2\sqrt{b-a}\right) \tag{12.49}$$

and

$$\dot{\xi}_2 = -\xi_2^3 \tag{12.50}$$

Observe that this does not involve any linearisation or any other kind of approximation. These equations are exact in the new coordinates. To simplify, let us assume $a = 0$, $b = 1$. Then we get $\sqrt{b-a} = 1$, and (12.49) can be solved by the method of partial fractions:

$$\xi_1 = \frac{e^t}{\sqrt{e^{2t} - c_1}} - 1 \tag{12.51}$$

We can also solve (12.50) by direct integration:

$$\xi_2 = \frac{1}{\sqrt{c_2 + 2t}} \tag{12.52}$$

After the rescaling which puts the fixed point at the origin, zero is the equilibrium for both (12.51) and (12.52).

To compare the speeds of approach suppose we put both $\xi_1 = 1$ and $\xi_2 = 1$ initially. From (12.51) and (12.52) we can compute how long it takes for the variables to get down to 0.01 and 0.001 respectively. For (12.51) this happens at $t = 1.8$ and $t = 2.9$ respectively, whereas it for (12.52) takes $t = 5{,}000$ and $t = 500{,}000$ respectively. This is what the difference in time scales actually means, and it gives the full reason why the fast dynamics can be ignored.

12.15 Alternative Futures

There is one final issue we can deal with using the group of models introduced up to now: The question of alternative potential futures as advocated by Brian Arthur (1990), according to whom the generally rationalistic lookout of the economics profession would seem to imply that every development follows the unique path of some sort of global optimality. An evolutionary Panglossian philosophy if we wish. His most convincing counterexample is the QWERTY layout of today's computer keyboards. According to Arthur it was devised in 1873 to be sufficiently awkward for slowing the typists down so as to not make the old mechanical typewriters jam too often. Habit has kept the layout, though the rationale for it has evaporated long ago.

Another interesting example is given by Basalla: All oriental hand tools work on the opposite action as compared to the occidental, on drawing the tool towards you instead of on pushing the tool away from you. The interesting thing is that both systems work provided the craftsmen are accustomed to their action. The present discussion with multiple local equilibria in a generally nonlinear world seem to be well suited to accommodate these facts.

Consider Fig. 12.7. It represents the same type of bifurcation cascade as Fig. 12.5, though we added one further step.

The polynomial for the potential now has degree 16, and that for the differential equation, its derivative, accordingly degree 15, so that there may be at most 8 local minima and 7 local maxima as watersheds, dividing the attraction basins for the minima. We note that each time we want to double the number of minima we also have to double the degree of the polynomial defining the potential. The full expression becomes awkward, so we do not reproduce the formula. We just note that the maxima and minima of the

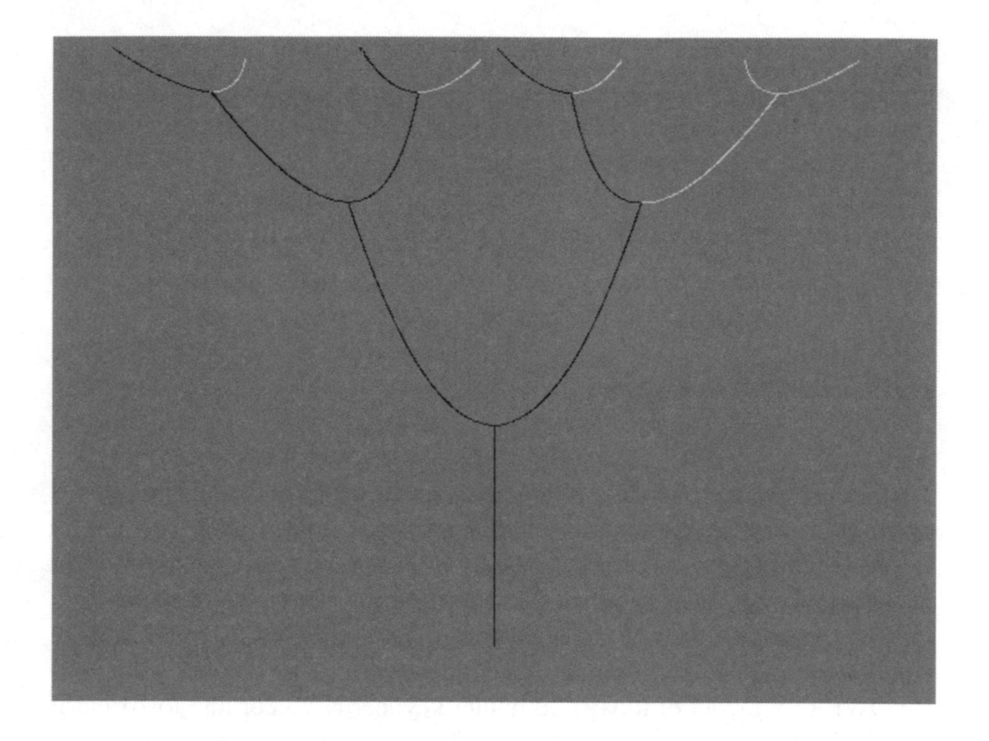

Fig. 12.7. Bifurcating tree and alternative potential futures.

system represented in Fig. 12.7 are at the 1+2+4=7 locations given in equation (12.32) and in addition in the following 8 locations:

$$x = \pm\sqrt{k} \pm \sqrt{k-1} \pm \sqrt{k-3/2} \tag{12.53}$$

This time we did not display the potential curves, just the bifurcating tree of its minimum points.

Also note that we reverted to the one-dimensional case. Once we saw how to deal with two-dimensional state spaces, we note that this can easily be generalized to even higher dimensions. However, once we want to visualize things, we always have to project what happen into two dimensions. This was actually done in drawing Fig. 12.6. Essentially this means that, whatever the dimension of the problem, we can use a projection to see the bifurcating tree.

As one dimension has to be reserved for the parameter, only one is left for some representative state variable, or a projective combination of several state variables. This means that a one dimensional state space is no limitation at all for the things we want to visualize, and so we can safely revert to the simpler case from the outset. Curves are easy to see in two dimensions, even if projected from higher dimension. The only problem is that in a projection a set of non-intersecting curves may seem to intersect, just like the trajectory in twospace of a forced (three-dimensional) oscillator. But this problem is not serious.

There are different features on display in Fig. 12.7. The full bifurcating tree is displayed in both dark and bright shade. The picture was first drawn just letting the computer trace all the minimum points and plot them in the bright shade.

The part of it drawn in dark shade is the outcome of a different procedure. The actual dynamic process, starting from various initial conditions, chosen at random, was simulated in 1000 steps, and the final points to which the system converged were then plotted in the dark shade. We see that not all the branches were traced.

Once a minimum is converted into a maximum and thus loses stability the process can go either way to the new equilibria left and right. Above we assumed that there would be agents tracing all the branches, but this is not necessarily so. It is fully possible that only part of the bifurcating tree is traced, just as in the picture.

The left side dominance shown is a result of introducing a bias in the random perturbation of the initial conditions. In this way we fully exploit the analogy to the idea of innovations as random "mutations". The outcome then is a partial tree, which is only one possible development among many, and another, very slightly different, sequence of random perturbations would result in a different development. The real surprise lies in the large number of partial trees, and in the accordingly small probability for each of them.

There are in all no less than 32,768 different possibilities even in the present case with only three bifurcation points! This may seem amazing, but there are 2 possibilities of including the basic stem of the tree or not, 4 possibilities of including one, both, or none of the first branches, 16 possibilities for the next level, and 256 for the last. The large number of combinations noted arises as the product of these four numbers. Add one more bifurcation and we end up at billions.

The exact formula for the number of partial trees reads:

$$\prod_{k=0}^{n}\sum_{i=0}^{2^k}\frac{\left(2^k\right)!}{i!\left(2^k-i\right)!}=2^{2^{n+1}-1} \tag{12.54}$$

where n is the number of levels in the tree, 0 being the number of the stem. The function is an exponential of an exponential, so the growth rate of numbers is extremely fast.

13 Development: Multiple Attractors

Most existent dynamic macroeconomic models share two features which may be strongly objected on grounds of realism: They produce *exponential growth, beyond any bound*, and they possess *unique paths of development*. Moreover, these paths are often *unstable*, so called knife edge balanced growth routes, a limitation clearly admitted by the originator Sir Roy Harrod in 1948, but forgotten by later growth theorists. This is just a modelling imperfection - a misspecification of the process due to choosing a too low order for it - which we will not further elaborate on in the present context.

We should always see unlimited growth as an anomaly in any model of reality, physical or social. This is at least true if the models are regarded as *global*, focusing asymptotic behaviour, which very often *is* the case. Locally in time a process can of course involve almost exponential growth in a phase of transition from a state that turned unstable, but this must sooner or later give way to damping as a stable attractor (point, cycle, or even strange) is approached. This trivial fact was pointed out in the introduction already, but deserves renewed emphasis. All real processes of change involve divergence from an unstable attractor, followed by convergence to some new stable attractor. Such processes can, however, not be produced by linear models.

There have been few nonlinear growth models produced in economic theory. A unique exception was Harold Hotelling's ingenious model of population growth from 1921. Hotelling took notice of the Malthusian argument that population tends to grow in "geometrical progression", i.e exponentially, whereas the means of subsistence increase in "arithmetical progression", i.e. linearly. Population growth was modelled as a logistic process, where the rate of growth was explosive at very small populations, but became damped once the sustainable level was approached. In case the actual population did overshoot the sustainable, for instance because the means of subsistence dropped due to a crop failure, the "positive" and "moral" checks would be set at work to decrease the population.

13.1 Population Dynamics

The type of model Hotelling used was

$$\dot{p} = \lambda(s - p)p \tag{13.1}$$

where p denotes population, and s denotes its given, sustainable value. As we have two degrees of freedom, the measurement units for time and population size, we can normalize both the adjustment speed and the sustainable population to unity, $\lambda = 1, s = 1$. Accordingly:

$$\dot{p} = (1 - p)p \tag{13.2}$$

In Chapter 2 we solved such logistic equations by separation of variables and partial fractions, obtaining the closed form solution

$$p(t) = \frac{1}{\left(1 - ce^{-t}\right)} \tag{13.3}$$

We also saw in Chapter 4 all the weird things that could happen if the process was put in discrete time.

We illustrate population growth in continuous time according to the solution (13.3) from various initial conditions in Figure 13.1. We see how all evolution paths are attracted to the sustainable value 1 as time goes to infinity.

The most ingenious feature in the Hotelling model was that spatial dispersion, away from densely to sparsely populated areas, due to diminishing returns, was added to growth, so that the combination became a model of nonlinear growth and dispersion of population. Sadly enough the model made no success at all, until it was re-invented in application to non-human populations 30 years later, and became one of the most basic tools in mathematical ecology.

We take the logistic growth process as a starting point rather than the classical linear, but raise the degree of nonlinearity, so obtaining multiple stable and unstable steady states.

All we need to this end is a production process with initially increasing and ultimately decreasing returns to scale. In terms of polynomial represen-

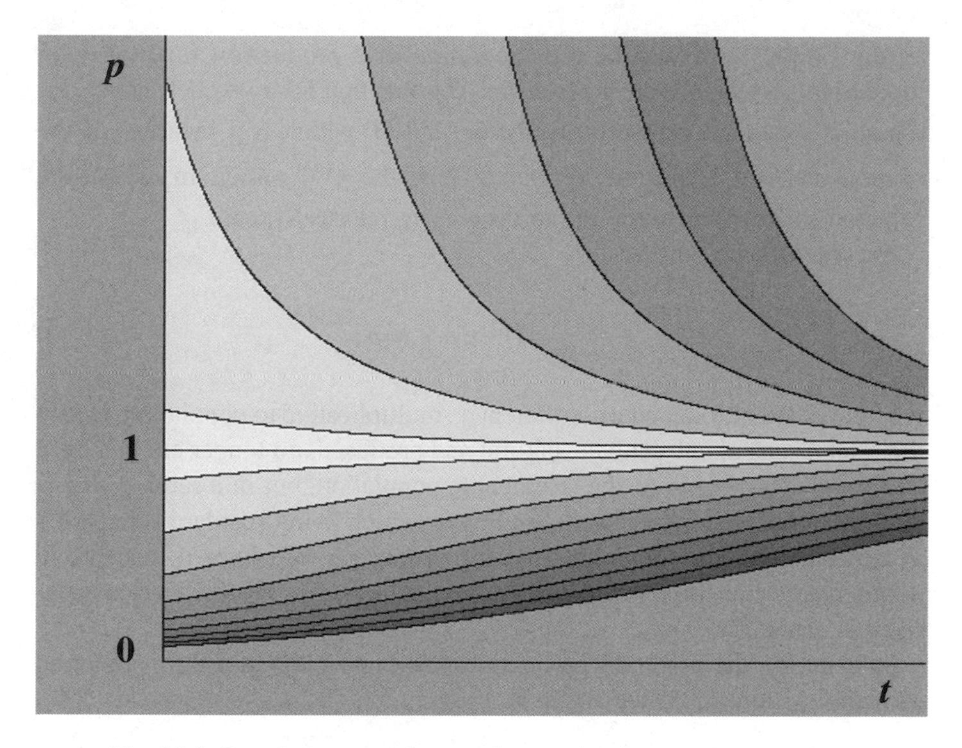

Fig. 13.1. Population growth according to the orginal Hotelling model.

tation, the production function has to be third order, and this is exactly what is needed to produce a multiplicity of coexistent attractors and interesting bifurcations.

In this way we can also take account of Brian Arthur's objection to economic development modelling. Development no longer has to take just one single possible course, and alternative histories and futures, due to minor divergences in the actual development process, become a possibility.

Suppose first that we just complement the given means of subsistence (fruit, game, and fish, captured by humankind at the early stages of development), by produced means q. In the society without capital, labour is the only input, and, as it is not our intention to enter the more sophisticated facts of demography, it is reasonable to just equate population to labour force. So, q depends on p, and we take

$$q = \alpha\left(\beta p^2 - p^3\right) \tag{13.4}$$

as the simplest polynomial representation of a production function with increasing/decreasing returns to scale. The function has two coefficients, α, which is a general productivity factor, and β, which is a measure of the optimal scale of operation, as $p = \beta / 2$ at the *AVC*-minimum, where the function shifts from increasing to decreasing returns to scale.

Accordingly we now write

$$\dot{p} = (1 + q - \gamma\, p)p \tag{13.5}$$

where we also introduced a coefficient γ, multiplicative to population. Hence the total means of subsistence, natural and produced, no longer are in a once and for all given ratio to the sustainable population, but in a ratio γ that is decided by humankind itself, depending on which living standard is regarded as acceptable in the long run. It is for instance a well known fact that in demographic transitions in developed economies the γ coefficient usually rises dramatically.

Substituting the assumed production function (13.4) into the population growth function (13.5) we arrive at

$$\dot{p} = p - \gamma\, p^2 + \alpha\beta\, p^3 - \alpha\, p^4 \tag{13.6}$$

which is fourth order and has three parameters, representing production efficiency, optimal production scale, and accepted living standard. This growth function can have at most four real equilibria, one of which obviously is the origin. It is however unstable as a simple linearisation shows (just delete all terms but the first). In addition there can be three nonzero equilibria, two attractors, necessarily separated by another repellor.

We may hence have two coexistent attractors, but one of them can merge with the repellor and be annihilated, which represents a bifurcation or catastrophe. As the order of the differential equation is four, it corresponds to a fifth order potential. Fifth order potential and three parameters means the swallowtail catastrophe, as we saw in Chapter 5. Note that in order to arrive at the universal swallowtail form we would have to dispense with the quartic term of the potential, i.e. the cubic of the dynamic equation (13.6), but this would not remove its coefficient. The parameter would turn up independently in another coefficient of the polynomial. However, there is no particular need to transform the potential to the universal form, which would result in a much more messy analysis, so we just integrate (13.6), sign reversed, and write:

$$V(p) = \alpha \frac{p^5}{5} - \alpha\beta \frac{p^4}{4} + \gamma \frac{p^3}{3} - \frac{p^2}{2} \quad (13.7)$$

The equilibrium set is now defined by:

$$V'(p) = \alpha p^4 - \alpha\beta p^3 + \gamma p^2 - p = 0 \quad (13.8)$$

and the singularity set by:

$$V''(p) = 4\alpha\, p^3 - 3\alpha\beta\, p^2 + \gamma\, p - 1 = 0 \quad (13.9)$$

As $p = 0$ can be factored out from (13.8), both equations (13.8)-(13.9) become third order polynomials, and it is not too hard, though necessarily a bit messy, to eliminate p, and so obtain an implicit equation for the bifurcation set:

$$4\alpha^2\beta^3 - \alpha\beta^2\gamma^2 - 18\alpha\beta\gamma + 4\gamma^3 + 27\alpha = 0 \quad (13.10)$$

Observe that we did not even attempt such elimination while presenting the swallowtail catastrophe in Chapter 5. The equilibrium set (13.8) contains three nonzero real roots if the left hand side of (13.10) is negative, but only one if it is positive.

Whenever we deal with three distinct roots then (13.6) factorises into:

$$\dot{p} = \alpha p(\bar{p}_1 - p)(\bar{p}_2 - p)(\bar{p}_3 - p) \quad (13.11)$$

where $\alpha = 1/(\bar{p}_1\bar{p}_2\bar{p}_3)$, $\beta = (\bar{p}_1 + \bar{p}_2 + \bar{p}_3)$, and $\gamma = (1/\bar{p}_1 + 1/\bar{p}_2 + 1/\bar{p}_3)$. Before proceeding with this it is more instructive to introduce numerical coefficients. In the illustrations we used $\alpha = 1$ and $\beta = \gamma = 3.5$. Then we get $\bar{p}_1 = 0.5$, $\bar{p}_2 = 1$, and $\bar{p}_3 = 2$. We can now separate variables as in Chapter 2 and write:

$$\frac{dp}{p(0.5 - p)(1 - p)(2 - p)} = dt \quad (13.12)$$

Expanding in partial fractions we have:

$$\frac{dp}{p} + 2\frac{dp}{p-1} - \frac{8}{3}\frac{dp}{p-0.5} - \frac{1}{3}\frac{1}{p-2} = dt \tag{13.13}$$

This equation is readily integrated, and we obtain:

$$\ln(p) + 2\ln(p-1) - \tfrac{8}{3}\ln(p-0.5) - \tfrac{1}{3}\ln(p-2) = \tfrac{1}{3}\ln(K) + t \tag{13.14}$$

where $\frac{1}{3}\ln K$ is the integration constant, introduced in this way for convenience. Multiplying (13.14) through by 3 and exponentiating, we finally get:

$$\frac{p^3(p-1)^6}{(p-0.5)^8(p-2)} = K\mathrm{e}^{3t} \tag{13.15}$$

This time there is no chance to solve for population p as a closed form expression of time t, because (13.15) results in a polynomial equation for p of the ninth degree.

We can, however, discuss the stable and unstable equilibria in general terms. When $t \to \infty$, then the right hand side of (13.15) goes to infinity. This must be true for the left hand side as well, so the denominator must go to zero, i.e. we must have either $p \to 0.5$ or $p \to 2$. As this is what happens asymptotically, $p = 0.5$ and $p = 2$ must be the stable equilibria of (13.11). If we, on the other hand, consider the case $t \to -\infty$, i.e. the situation infinitely far back in history, then the right hand side of (13.15) goes to zero, and this again must also hold for the left hand side. To this end the numerator must go to zero, i.e. either $p \to 0$ or $p \to 1$ must hold. As this is the state infinitely far back in history, the system diverges from these states as time proceeds in the usual positive direction, and so we conclude that $p = 0$ and $p = 1$ are unstable steady states.

We could have shown this for the general case (13.11). Whenever it is true that $0 < \overline{p}_1 < \overline{p}_2 < \overline{p}_3$, then zero and the intermediate positive root are unstable, while the remaining two are stable, but the formulas are very clumsy, and the numerical example shows sufficiently well what is at issue.

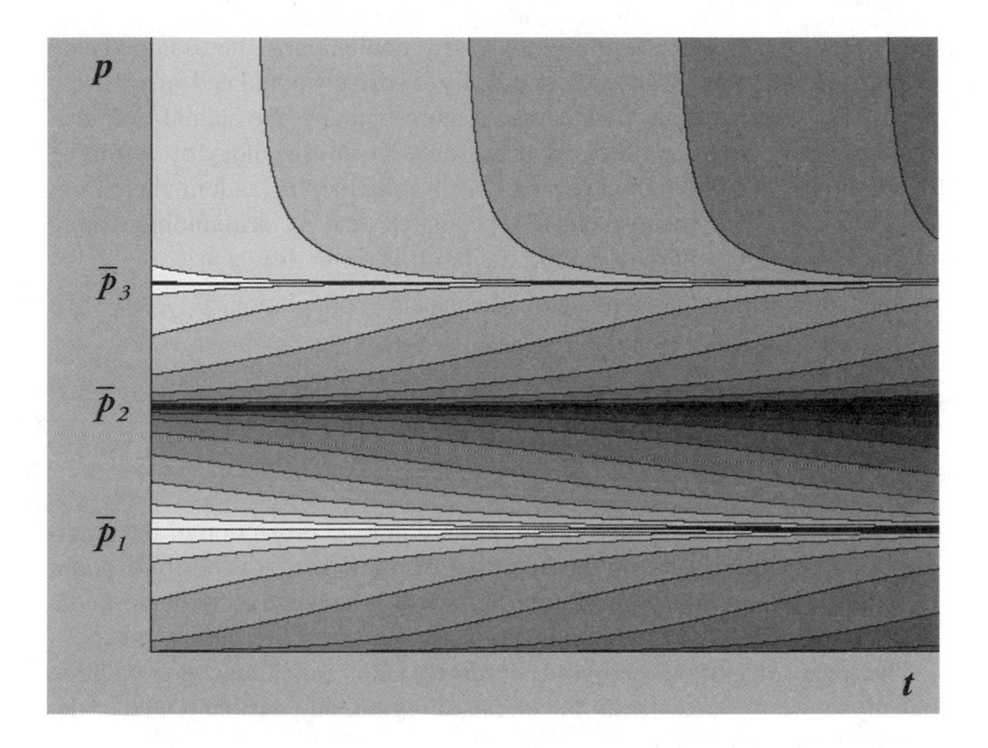

Fig. 13.2. Population growth according to the Hotelling model with production.

It is also easy to use (13.15) for drawing pictures on the computer. Just reverse the roles of p and t, regarding time as the dependent variable. Then, for each fixed constant K, we get a trajectory by assigning various values to population and solving for the time at which this occurs. Varying K then results in a family of curves as displayed in Figure 13.2

13.2 Diffusion

There has been no mention of space yet. Hotelling incorporated space in terms of an analogy to heat diffusion. Assuming decreasing returns to scale in production, he noted that the per capita standard of living would be in a negative relation to population density. Like heat in any medium diffuses away from local concentrations to nearby cooler parts, he assumed that population would move away from densely to sparsely populated areas (given decreasing returns) from poor areas to more opulent. We should note that Skellam thirty years later arrived at the same result in ecology by assuming random movement of animal populations: If migration is random, then, from any location which is more densely populated than its surroundings, more animals will move out than in, and the result is again diffusion.

To put things formally, consider the Laplacian of population: $\nabla^2 p = \nabla \cdot \nabla p$ According to Gauss's Integral Theorem we have:

$$\iint_R \nabla^2 p \, dx_1 \, dx_2 = \oint_C \nabla p \cdot \mathbf{n} \, ds \tag{13.16}$$

The right hand side integrand is the outward normal projection of the gradient of population density on the boundary curve. Where it is positive, population increases as we leave the region, where it is negative, population decreases, the absolute value in either case giving a measure of how much.

The curve integral on the right accordingly is the net change of population density as we move out from the enclosed region, all possible points of departure being considered.

If we now shrink the region to a point, we get the interpretation that the Laplacian is a measure of net increase of population density when we leave a point, all possible directions being considered. Thus it is natural that it is the Laplacian that is the cornerstone of diffusion processes, be it of heat or population, human or non-human.

Hotelling's complete model can now be stated: Population growth at any point is the sum of local growth there and net immigration, which is assumed proportionate to the Laplacian. From densely populated points there is emigration, to sparsely populated points there is immigration. Formally:

$$\dot{p} = (1 - p)p + \delta \nabla^2 p \tag{13.17}$$

where δ denotes the diffusivity constant, which can be taken as a (positive) spatial invariant.

The dynamics of this equation is largely unknown, so the discussions have been limited to the stationary solutions, i.e. to the solutions of the partial differential equation:

$$(1-p)p+\delta\nabla^2 p=0 \tag{13.18}$$

Hotelling himself only considered the linearised cases where p is close to zero, or close to the saturation population. The linearisation of the growth term being $(1-2p)$, local growth rate is approximated by 1 for $p\approx 0$ and by -1 for $p\approx 1$. Such linear cases are thoroughly studied in the classical literature on heat diffusion. About the nonlinear model, thousands of publications have been written in ecology.

It is easiest to study (13.18) if we rewrite it for a one dimensional region:

$$p-p^2+\delta\frac{\partial^2 p}{\partial x^2}=0 \tag{13.19}$$

Multiply (13.19) through by the first derivative $\partial p/\partial x$ and we get:

$$p\frac{\partial p}{\partial x}-p^2\frac{\partial p}{\partial x}+\delta\frac{\partial^2 p}{\partial x^2}\frac{\partial p}{\partial x}=0 \tag{13.20}$$

This, however, equals:

$$\frac{\partial}{\partial x}\left(\frac{p^2}{2}-\frac{p^3}{3}+\delta\frac{1}{2}\left(\frac{\partial p}{\partial x}\right)^2\right)=0 \tag{13.21}$$

Integrating we get:

$$\frac{p^2}{2}-\frac{p^3}{3}+\delta\frac{1}{2}\left(\frac{\partial p}{\partial x}\right)^2=c \tag{13.22}$$

Such first integral solutions can be portrayed in a phase diagram, such as Figure 13.3.

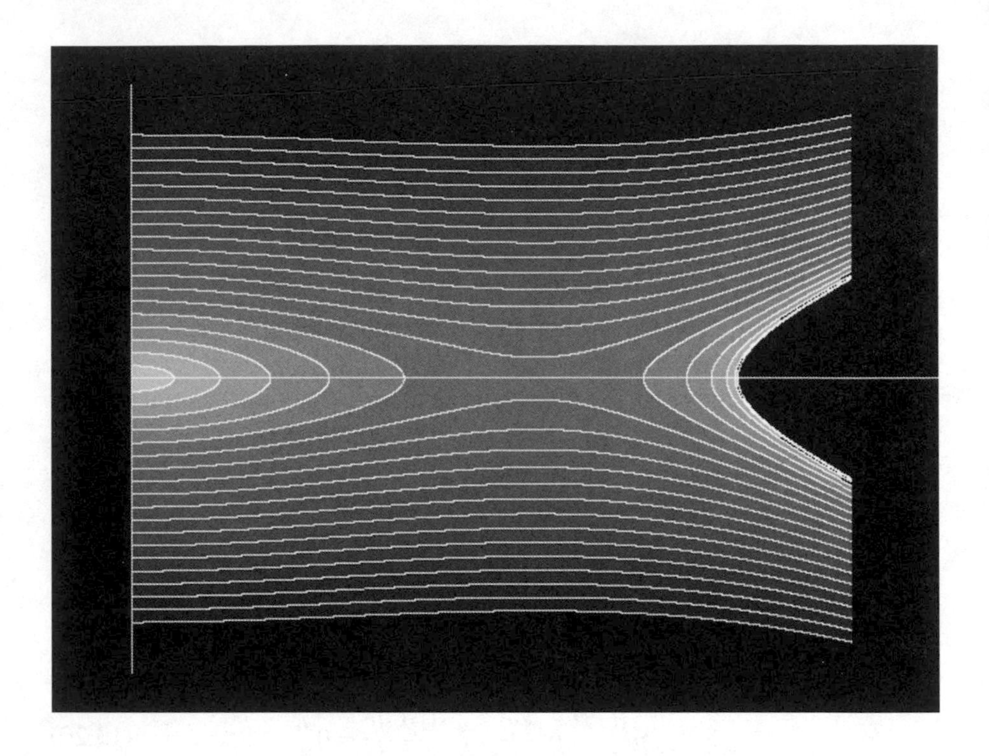

Fig. 13.3. Spatial phase diagram $p,\ \partial p / \partial x$ *for Hotelling model without production.*

Phase diagrams usually portray a phase variable and its first *time* derivative. Presently, we are, however, studying stationary solutions, so we deal with a *space* derivative. Each first integral curve represents combinations of population density and its spatial derivative.

Exactly as in the case of temporal systems, a closed orbit means a periodic solution. So, there are solutions for equilibrium settlements, such that densely and sparsely populated regions alternate. Some regions exhibit permanent population growth combined with emigration, whereas others exhibit population decrease combined with immigration. Yet population is stationary everywhere. It is remarkable that this comes out as a stationary solution under assumptions of complete spatial homogeneity.

However, as we see, all closed orbits are partly located half to the left of the vertical axis, i.e. in the range of negative populations. We conclude that there are no meaningful wavy settlement patterns in the original Hotelling model over the whole extension of any given region. Ecologists therefore assumed a patchy structure of "habitats" with positive populations.

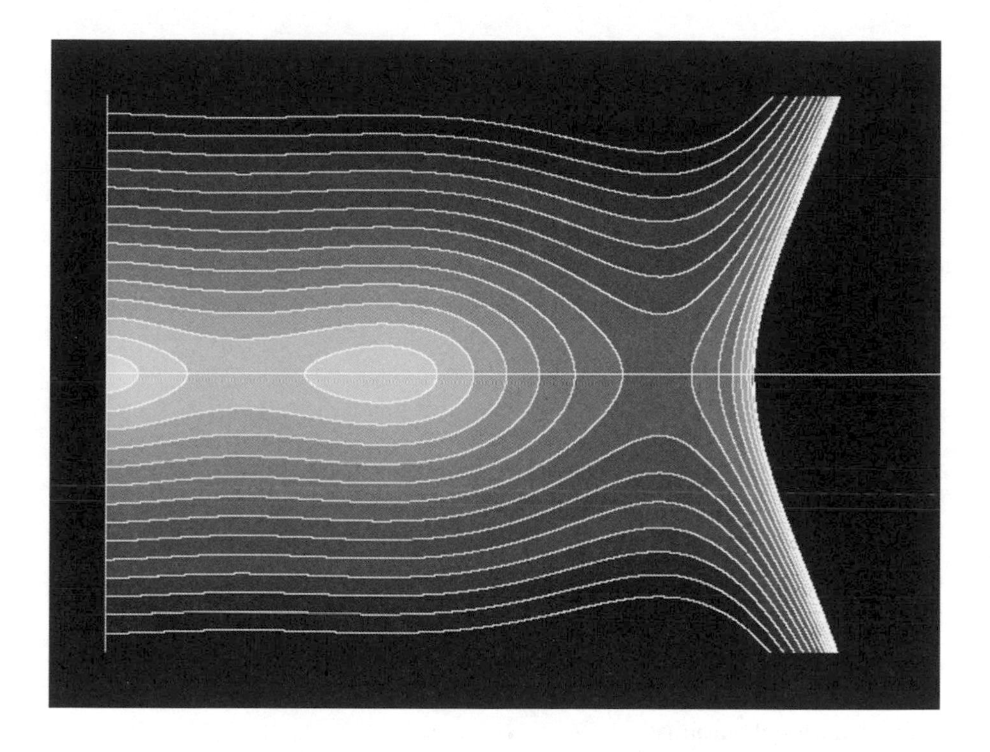

Fig. 13.4. Spatial phase diagram p, $\partial p / \partial x$ for Hotelling model with production.

Things become different when we consider production. Adding diffusion to the right hand side of (13.6) and putting the time derivative equal to zero we have:

$$p - \gamma p^2 + \alpha\beta p^3 - \alpha p^4 + \delta \frac{\partial^2 p}{\partial x^2} = 0 \tag{13.23}$$

Multiply (13.23) through by the first (space) derivative $\partial p / \partial x$ and integrate, exactly as before:

$$\frac{p^2}{2} - \gamma \frac{p^3}{3} + \alpha\beta \frac{p^4}{4} - \alpha \frac{p^5}{5} + \delta \frac{1}{2}\left(\frac{\partial p}{\partial x}\right)^2 = c \tag{13.24}$$

Under the assumption that we have two stable population equilibria in the growth equation, we find that there now exist spatially periodic stationary solutions with alternating dense and sparse populations, located entirely in the range of positive populations, as shown in Fig. 13.4.

The half orbits still are there, so, corresponding to the lower sustainable population, we can still have a patterns of isolated "habitats", but we can also have a strictly positive undulating population density pattern around the higher sustainable level. Note that all this difference is made by including production.

13.3 Stability

The equilibrium patterns as defined by (13.22) and (13.24) are without interest unless they are stable. Therefore we have to check stability. As the models obviously have numerous equilibria, as we see from Figs. 13.3 and 13.4, several of which may be stable, there is no point in attempting to try to prove more than local stability.

To this end let us consider (13.17) for the original Hotelling model together with an equilibrium pattern that fulfils (13.18). To distinguish the disequilibrium from the equilibrium, we denote the disequilibrium population by p and the equilibrium population by $\overline{p}$. Then, subtracting (13.18) from (13.17) we get:

$$\frac{d}{dt}(p-\overline{p})=(p-\overline{p}-p^2+\overline{p}^2)+\delta\nabla^2(p-\overline{p}) \tag{13.25}$$

Consider the identity

$$-p^2+\overline{p}^2=-(p-\overline{p})^2-2\overline{p}(p-\overline{p}) \tag{13.26}$$

Dealing with local stability, we, however, only consider small deviations from equilibrium, and so we can ignore the squared term in the right hand side of (13.26) which is bound to be very small if the deviation itself is small. Thus:

$$-p^2 + \bar{p}^2 = -2\bar{p}(p - \bar{p}) \tag{13.27}$$

This is the technique of linearisation in the neighbourhood of an equilibrium. Substituting from (13.27) into (13.25) we get:

$$\frac{d}{dt}(p - \bar{p}) = (1 - 2\bar{p})(p - \bar{p}) + \delta\nabla^2(p - \bar{p}) \tag{13.28}$$

For convenience let us introduce the abbreviation $z = p - \bar{p}$. Then (13.28) can be written:

$$\frac{dz}{dt} = (1 - 2\bar{p})z + \delta\nabla^2 z \tag{13.29}$$

Next, consider the nonnegative quantity z^2, and take its time derivative:

$$\frac{1}{2}\frac{d}{dt}z^2 = z\frac{dz}{dt} \tag{13.30}$$

Therefore, multiply (13.29) through by z and substitute into (13.30). This yields:

$$\frac{1}{2}\frac{d}{dt}z^2 = (1 - 2\bar{p})z^2 + \delta z\nabla^2 z \tag{13.31}$$

We are now going to integrate (13.31) over the region in space that we consider, but first note that (from the chain rule for vector analysis):

$$\nabla \cdot (z\nabla z) = z\nabla^2 z + (\nabla z)^2 \tag{13.32}$$

Apply Gauss's Integral Theorem to (13.32):

$$\iint_R \nabla \cdot (z\nabla z) dx_1 dx_2 = \oint_C z\nabla z \cdot \mathbf{n} ds = 0 \tag{13.33}$$

This integral is zero, as stated, because in the right hand side of (13.33) we have $p = \overline{p}$, i.e. $z = 0$ on the boundary. This is just to state that the equilibrium and disequilibrium solutions have to obey the same boundary conditions. Therefore, integrating (13.32) over space, and making use of (13.33) we get:

$$\iint_R z\nabla^2 z dx_1 dx_2 = -\iint_R (\nabla z)^2 dx_1 dx_2 \leq 0 \tag{13.34}$$

After this prelude we are ready to deal with (13.31). Integrating over space, taking note of (13.34), and of the fact that the diffusivity constant is positive, i.e. $\delta > 0$, we finally get:

$$\frac{1}{2}\frac{d}{dt}\iint_R z^2 dx_1 dx_2 \leq \iint_R (1-2\overline{p}) z^2 dx_1 dx_2 \tag{13.35}$$

The left hand side contains an integral over space of a squared and hence nonnegative quantity. It is zero only when there is no deviation from the equilibrium pattern anywhere. The right hand side contains the same nonnegative quantity preceded by the factor $1-2\overline{p}$. Suppose now that the equilibrium distribution of population we consider is such that $\overline{p} > 1/2$ holds everywhere on the region. Then the right hand side of (13.35) is negative whenever the system is out of equilibrium, and the equilibrium is hence locally asymptotically stable.

We already noted that the solutions to Hotelling's original model included undulating patterns which dropped to negative populations, so that we had to assume a patchy system of habitats in order to avoid negativity. We now see that if we want to be sure that the patterns are dynamically stable, then we have to assume that population is not only positive, but that it exceeds half of the unitary sustainable population according to the growth term.

Note that we dealt with stability for the general model, instead of for the one-dimensional case, because this was in no way tougher.

It is not difficult to deal with the case with production as stated in (13.23). We never wrote down the complete dynamic equation, but never mind, we only have to equate the left hand side of (13.23) to the time derivative of population instead of to zero in order to retrieve it. The procedure is exactly as before. Note that the diffusion term does not need any renewed treatment. Equation (13.34) still applies.

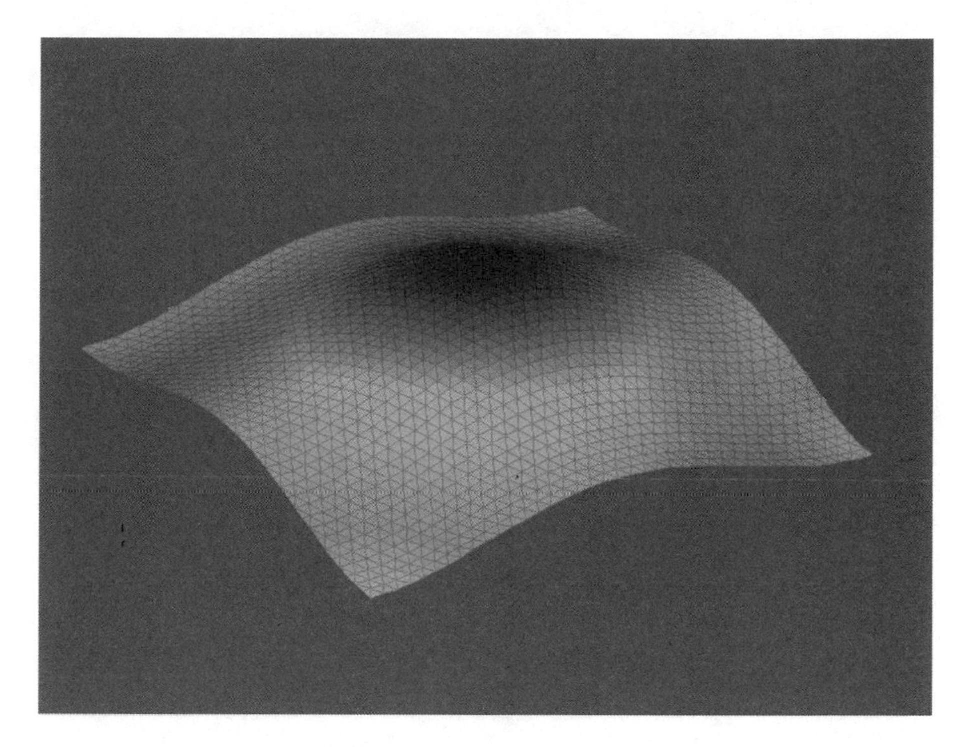

Fig. 13.5. Evolution of initial sinoid population density with high diffusion.

Only the growth term, which is now more complex, needs a new treatment. Linearising it we get the equivalent to (13.35):

$$\frac{1}{2}\frac{d}{dt}\iint_R z^2 dx_1 dx_2 \leq \iint_R \left(1 - 2\gamma\overline{p} + 3\alpha\beta\overline{p}^2 - 4\alpha\overline{p}^3\right) z^2 dx_1 dx_2 \quad (13.36)$$

We see that without production, i.e. with $\alpha = 0$, the condition for stability is as before, which indeed it should. With production, the parenthesis in (13.36) is negative in two intervals, i.e. the stability conditions are fulfilled in two intervals. Referring to Figure 13.4, the nasty thing is that the stability intervals are centred around the two saddle points which we see to the left and right in the picture. The saddle points, by the way, represent the stable equilibria of the pure non-spatial growth model, whereas the centres represent the unstable equilibria of that model. Hence none of the closed orbits around the centres, which represent the undulating patterns (over unlimited space), are completely contained in the intervals of stability. Hence, it would seem that we again have to consider limited patchy habitats if we want a stable non-uniform distribution of population as an outcome.

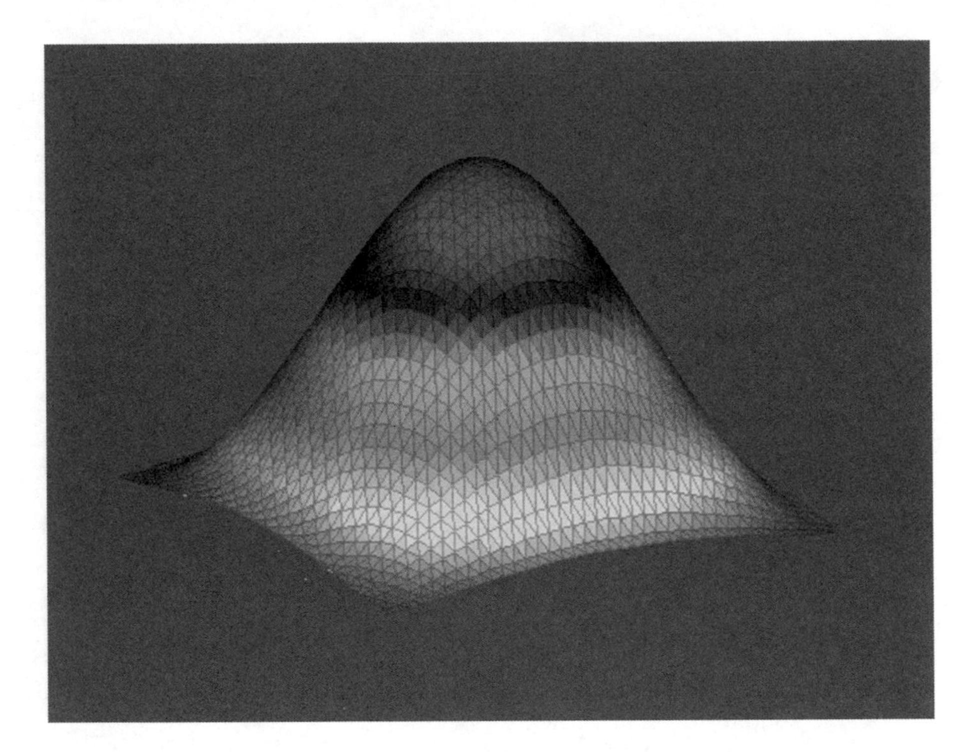

Fig. 13.6. Evolution of initial sinoid population density with low diffusion.

On the other hand we cannot be quite sure because (13.36), like (13.35) only gives us *sufficient* conditions for stability. The full dynamics of the Hotelling model, either in its original form or with production, is bound to be very complex.

With the recent Macsyma PDE software, referred to above in Chapter 3, it is possible to explore the system numerically in full 2D. Such simulation experiments invariably also indicate that the wavy patterns around the unstable positive equilibrium are unstable - any pattern of this character tends to be torn in patchy areas with uniform population at the higher and lower stable equilibrium levels. At a certain stage the emergent discontinuities make the software stop running.

It is not obvious how to deal theoretically with these emergent boundaries of the patchy spatial pattern. Maybe the most likely assumption is that the emergent boundaries become closed barriers to immigration from areas of lower to those of higher living standard. If so, we have a theory of the emergence of closed regions with varying living standard and migration barriers. We should however be careful with such conclusions, because we retained the linear diffusion from the original Hotelling model, which is a very simplistic representation of the facts of migration.

A striking feature in experiment is that diffusion temporarily stabilises even unstable equilibria. If we run the software with *strong diffusion* and a bumpy initial distribution to one side of an unstable equilibrium, then we first see population become equal in density at that unstable equilibrium. But this is deceptive! Once population has become sufficiently uniform, it starts to grow everywhere towards one of the stable equilibria.

This to be expected, because the Laplacian for a uniform distribution is zero, and then only the growth part of the model exerts any influence. We see an intermediate stage of the evolution of population on the unit square in this scenario in Fig. 13.5.

If we start the same process with *weak diffusion*, then the bumpiness increases towards the closest stable equilibrium, and only eventually are the fringes dragged towards that value. Fig. 13.6 displays the intermediate evolution in this scenario.

These are the facts of an initial distribution deviating *to one side* of an unstable equilibrium. We already said that if it initially deviates *to both sides*, then the distribution eventually tends to be torn in pieces.

An issue to settle in simulation experiments is that the outcomes differ very much due to which kind of boundary conditions we assume. The most reasonable conditions seem to be free boundary conditions, because even if we deal with an isolated island there is no reason to prescribe some given population density around the seashore. It is more natural to let boundary population evolve without constraint quite as that of the interior.

13.4 The Dynamics of Capital and Labour

It is now time to state the complete model involving capital as well. To feel more familiar we replace population p by labour force, denoted l, and identify them for reasons already given. It enters along with capital, denoted k, in the production function, which we write:

$$q = \alpha\left(\beta(k+l)^2 - k^3 - l^3\right) \tag{13.37}$$

It has about the same form as (13.4). By taking the square of the sum of capital and labour, but subtracting the cubes of them separately, we provide for synergy among the inputs and avoid making the isoquants straight lines.

The meaning of the coefficients is as before, α represents overall productivity, whereas β is associated with the optimal scale. This is shown by the fact that the locus of unitary scale elasticity points is:

$$k^3 + l^3 = \frac{\beta}{2}(k+l)^2 \tag{13.38}$$

which only depends on the parameter β and now becomes a closed curve.

Next question to consider is how the total product is shared between capital and labour. We could let one of the factors get its share according to the marginal condition and the other take the residual, or we could prescribe a certain ratio of saving (available for investment) to consumption.

We will do neither, but assume that in the development process capital and labour both estimate their shares according to the marginal conditions. This is not altogether satisfactory, because in a production function of variable returns to scale the factor shares do not add up exactly to the output. There are however also the non-produced means of subsistence in the case of a deficit, whereas in the opposite case a net profit to industrialists does not necessarily trigger either population growth or new capital formation. Accordingly, we take the following factor incomes as potentials for growth:

$$rk = \frac{\partial q}{\partial k}k = 2\alpha\beta(k+l)k - 3k^3 \tag{13.39}$$

$$wl = \frac{\partial q}{\partial l}l = 2\alpha\beta(k+l)l - 3l^3 \tag{13.40}$$

where r and w denote the capital rent and wage rate in real terms. The partial derivatives are readily obtained from the production function (13.37).

For the growth of capital stock we now propose the following dynamic equation:

$$\dot{k} = rk - \kappa k \tag{13.41}$$

where κ denotes the rate of capital depreciation. The equation (13.41) is quite straightforward: estimated real capital incomes are reinvested in capital

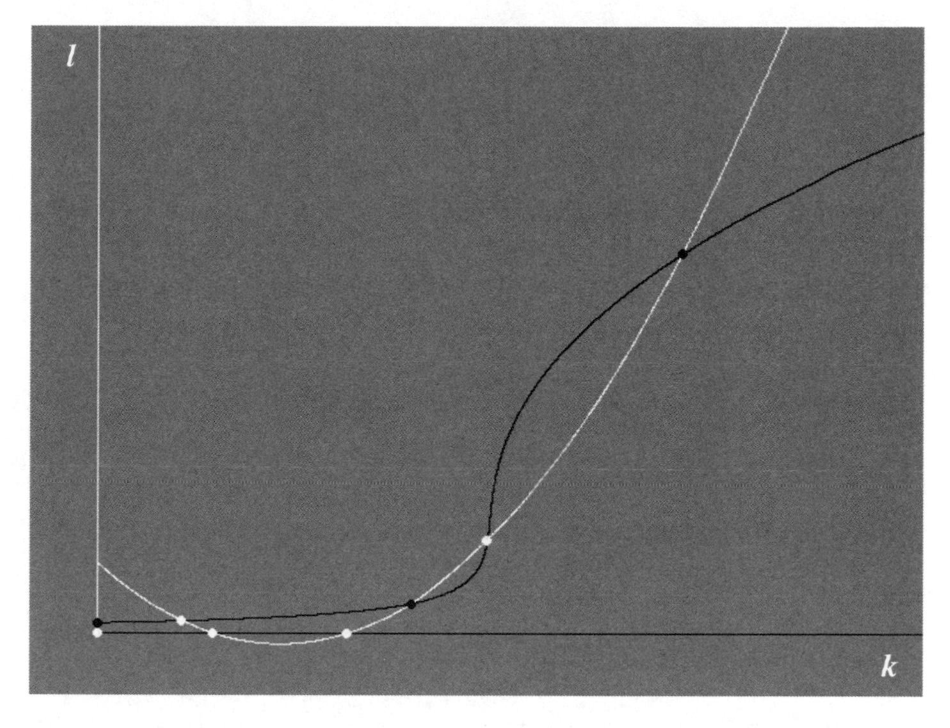

Fig. 13.7. Zero growth lines for capital (white) and labour (black).

formation but there is a deduction due to depreciation. For the stock of labour we similarly have:

$$\dot{l} = l(1 + wl - \gamma l) \tag{13.42}$$

It is different from the growth equation for capital, apart from the multiplicative factor l, in two more respects: i) There are the additional non-produced means of subsistence, and ii) there is the parameter γ multiplicative to labour stock, quite as in the model without capital.

The higher order for the growth of labour process as compared to that for capital, due to the extra factor l, among other things makes zero population an unstable steady state, as we have seen, and makes population start to multiply exponentially from any tiny initial number, as originally suggested by Malthus.

If we now substitute from (13.39)-(13.40) in (13.41)-(13.42) we obtain:

$$\dot{k} = 2\alpha\beta\,(k+l)k - 3\alpha\,k^3 - \kappa\,k \tag{13.43}$$

$$\dot{l} = l + 2\alpha\beta\,(k+l)l^2 - 3\alpha\,l^4 - \gamma\,l^2 \tag{13.44}$$

Considering the zeros of the system (13.43)-(13.44) we easily see that the rate of growth of capital is zero if either $k = 0$, or else the quadratic:

$$l = \frac{\kappa}{2\alpha\beta} - k + \frac{3}{2\beta}k^2 \tag{13.45}$$

is fulfilled. Likewise the rate of growth of labour is zero if either $l = 0$, or else

$$k = \frac{\gamma}{2\alpha\beta} - l + \frac{3}{2\beta}l^2 - \frac{1}{2\alpha\beta}\frac{1}{l} \tag{13.46}$$

The facts are illustrated in Figure 13.7 where we have drawn the zero growth lines for capital and labour, white for the first, black for the second. Note that the black S-shaped curve for zero labour growth is joined by the horizontal axis (and a branch in the negative quadrant which is factually irrelevant and therefore not shown). Likewise the parabolic zero growth curve for capital is joined by the vertical axis as representing another zero.

Possible steady states are the intersection points between black and white curves, and, as we see, in the position of the curves in Figure 13.7, there can be up to eight such states. If all are present, three will be attractors, four will be saddle points, and one a pure repellor.

In Figure 13.8 we see these facts. The zero lines from Figure 13.7 are still there, now drawn in modest gray shade. The eight intersection points have been marked, attractors a_1 through a_3 in black, the saddles s_1 through s_4 and the repellor r_1 in white.

The black lines are the incident trajectories, i.e. the stable and unstable manifolds of the saddle points. In particular those incident to s_2 and s_4 are interesting as their stable manifolds form basin boundaries for the three different attractors, the basins themselves drawn in different shades, whereas the unstable manifolds of these saddle points go right into the adjacent attractors on either side.

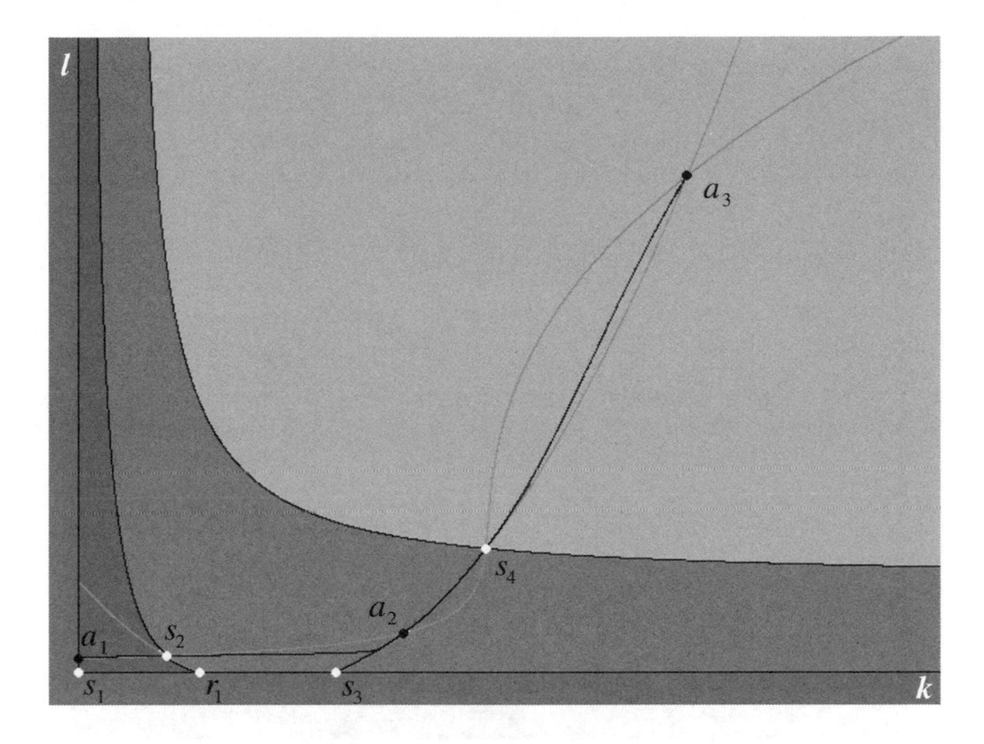

Fig.13.8. Attractors, basins, and saddle trajectories.

It is possible for any dynamic process starting close to the basin boundaries to go to one attractor or another dependent on a minor perturbation. To bring the system from the neighbourhood of one stable attractor to another would, however need a substantial finite initial displacement sufficient to take the system to the other side of a basin boundary. Such changes are not very likely. More interesting are bifurcations that are results of smooth developments of the parameters $\alpha, \beta, \gamma, \kappa$ such that an attractor suddenly loses stability. Such changes are frequent in the history of economic evolution. Technological efficiency α has been increasing over time, as has the scale parameter β. The demographic γ parameter as reflecting social attitudes has usually taken a sudden upward leap in developed societies, whereas the capital depreciation parameter κ has varied up and down, depending on the durability of capital goods.

In the series of illustrative pictures we actually varied this last parameter, though it is maybe the least interesting among the four. As we can see from (13.45) its effects are, however, easy to analyse, as it just causes the parabolic zero growth curve for capital to shift up or down. A similar (horizontal) shift is the effect of changing the demographic parameter as we see from

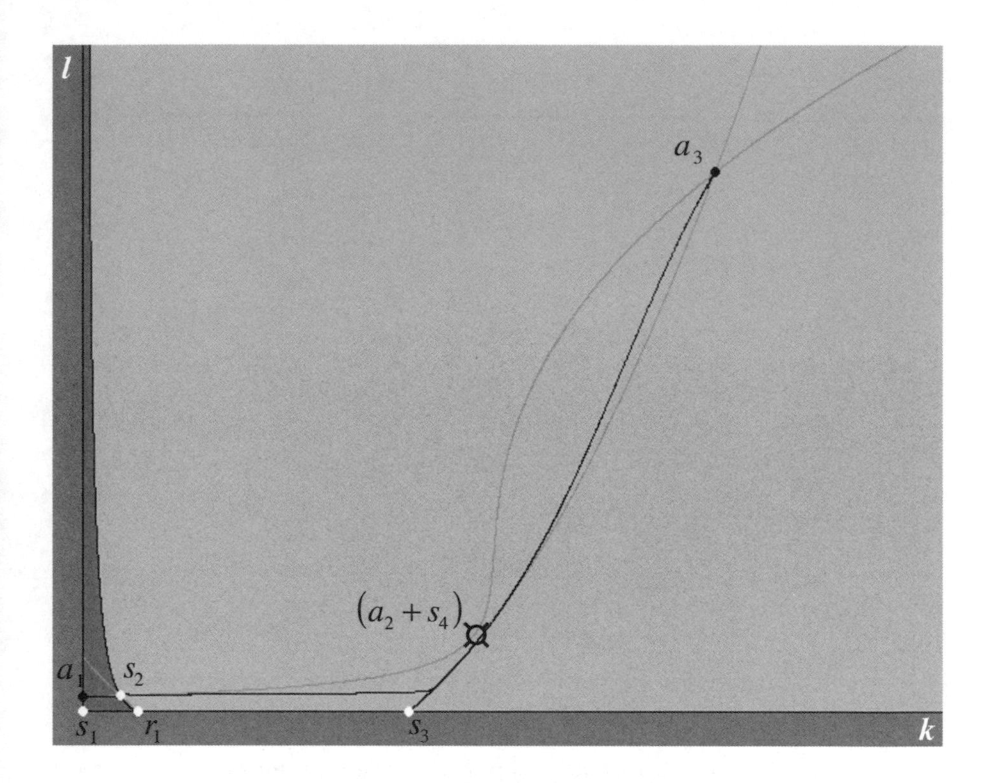

Fig. 13.9. Merger of middle and upper basins.

(13.46). Changes in the technological efficiency and scale parameters introduce more complex effects as they modify the curvature of, above all, the S-shaped zero population growth curve.

Figure 13.9 shows a case where the rate of capital depreciation has decreased and lowered the parabola so that it no longer intersects the S-shaped curve in the points a_2 and s_4. The picture shows the situation just after the moment when the saddle point and the attractor have merged and annihilated each other. The narrow waist between the curves shows where the evaporated equilibrium points were located.

As a consequence the two upper basins from Figure 13.8 merge. Note that, as the evaporated attractor is well in the interior of the merged basins, there is no doubt at all that the process will go to the highest attractor a_3.

But the attractor a_2 can also merge with the saddle s_2. This happens when the capital depreciation rate increases instead. In such a case it is the two lower basins that merge, and the process goes to the lowest attractor a_1. This scenario is displayed in Figure 13.10. The remains of the mutually annihilat-

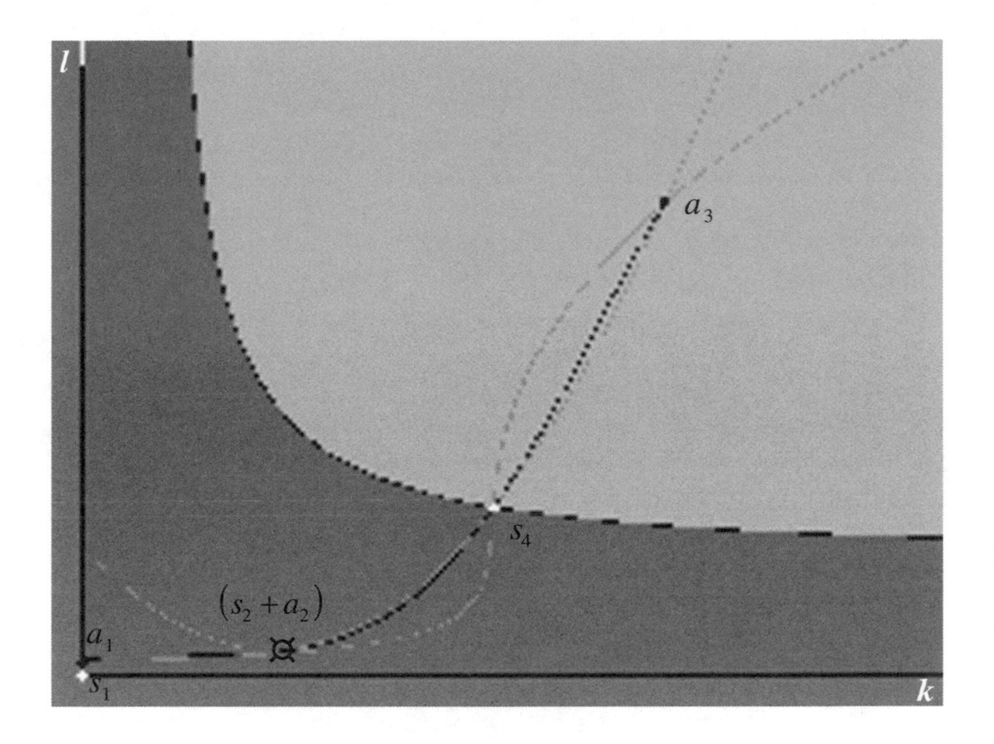

Fig. 13.10. Merger of lower and middle basins.

ing saddle and attractor can again be traced at the narrow waist. Note that in either case shown in Figures 13.9-13.10, had the system been close to any of the extreme equilibria a_1 or a_3, it would not have been affected by the disappearing of the intermediate attractor a_2.

We can summarize the bifurcations in Figure 13.11, where we just display the zero growth curves. The positions for the parabola are separated in three bands of different shade, and the boundaries of these bands are defined by tangencies between the zero growth curves for labour (black and immobile) and for capital (white and shifting). The tangencies mark bifurcation.

In the middle band we have the case shown in Figure 13.8, as there are four intersection points. In the upper band the two lowest intersections have vanished, whereas in the lowest band it is the two middle intersections that are gone. Accordingly, we have the cases shown in Figures 13.9 and 13.10 respectively.

We can also imagine that the parabola is located above the uppermost band. In that case the zero growth curves do not intersect at all, except at zero capital. Then it is only the lowest steady state that is left. On the other hand,

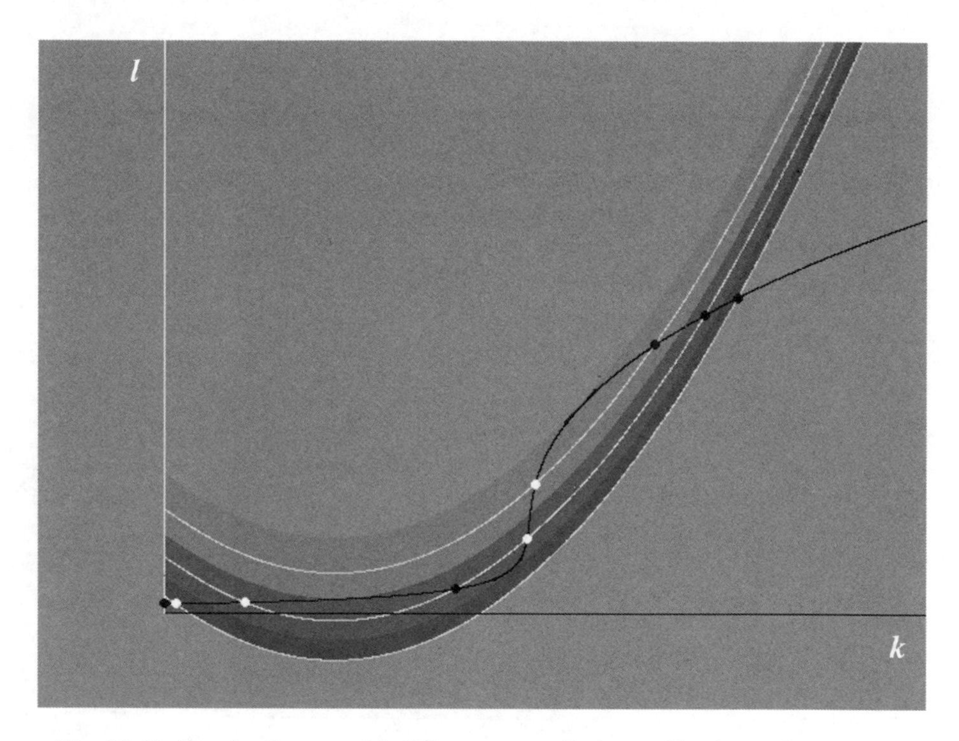

Fig. 13.11. Bands of cases with different constellations of basins and attractors.

if the parabola sinks still a little lower than the lowest band, then the nonzero capital steady state and the saddle which are already quite near in Figure 13.11 annihilate each other and the basin of the lowest attractor is destroyed. We are only left with the highest attractor and one single basin.

For the reader interested in numerical experiment we should mention that the coefficient values used were $\alpha = 3, \beta = 2, \gamma = 23$ throughout, whereas we varied $\kappa = 2,\ 3.5,\ 5$ to draw the curves in Figure 13.11 as well as in drawing the phase diagrams in Figures 13.8-13.10.

To find the bifurcation set is quite a complicated task. The equilibrium set of the system (13.43)-(13.44) is defined by (13.45)-(13.46) which is a nonlinear system, and far from easy to solve. The singularity set is as always obtained by putting the Jacobian determinant equal to zero. The Jacobian matrix is easily obtained by differentiating the right hand sides of (13.43)-(13.44):

$$\begin{bmatrix} 2\alpha\beta(2k+l)-9\alpha k^2-\kappa & 2\alpha\beta k \\ 2\alpha\beta l^2 & 1+2\alpha\beta(2k+3l)l-12\alpha l^3-2\gamma l \end{bmatrix} \quad (13.47)$$

Its determinant is a fifth order polynomial, so to solve the system (13.45)-(13.47) algebraically is out of question.

Note moreover that the system (13.43)-(13.44) is not a gradient system, and therefore we do not deal with a case of catastrophe theory.

By way of conclusion we should consider a few possible objections to a "theory of growth" like the one outlined. It may seem that the model would imply stationarity instead of change and this too would be contrary to all empirical experience. This, however, is not a true consequence of the model. The phase diagrams 13.8 through 13.10 everywhere have a very strong slow-fast direction dichotomy. The train of black line segments $a_1 s_2 a_2 s_4 a_3$ represents extremely slow movement, whereas transverse motion, more or less parallel to the hyperbolic basin boundaries is very fast. This means that from any initial condition we have a very fast approach to a neighbourhood layer of the slow route $a_1 s_2 a_2 s_4 a_3$, which we may take as a sort of (extremely congested) "turnpike" (a favourite term from growth theory) for nonlinear growth. In this boundary layer the process may linger any time in its approach to the attractor of the basin. The closer to the "turnpike", the slower the process is. The route $s_4 \to a_3$ for instance takes infinite time.

The fixed point is only the ultimate target of the system, and on its slow approach the system may even be caught by another attractor due to slow changes of parameters and resulting bifurcations, as discussed. As an asymptotic destiny a finite fixed point, in contrast to a point at infinity, is, however, the only reasonable choice. Moreover, the multiplicity of coexistent basins and attractors in a truly nonlinear model too is the only empirically reasonable choice. Conventional growth theories in economics miss both points.

We could have introduced diffusion of both labour and capital along with growth in the present model, though even diffusion should be introduced less mechanically than in Hotelling's original contribution, by linking it to economic forces. This would result in a model somewhat resembling the reaction-diffusion models in chemistry, where nonlinear functions for the interaction of chemicals are coupled with spatial diffusion. Especially in two dimensional space these models become quite interesting in terms of pattern formation, and also chaotic dynamics, but they also become very complex analytically. Recent software could provide means for research on these lines.

References

Abraham, R. H. and Shaw, C. D., 1992, *Dynamics - The Geometry of Beahavior* (Addison-Wesley)

Abraham, R. H., Gardini, L. and Mira, C., 1997, *Chaos in Discrete Dynamical Systems* (Springer-Verlag)

Agliari, A., Gardini, L. and Puu, T., *Chaos, Solitond & Fractals* (forthcoming)

Allen, R.G.D., 1956, *Mathematical Economics* (MacMillan, New York)

Arnold, V. I., 1983, *Geometrical Methods in the Theory of Ordinary Differential Equations* (Springer, Berlin)

Arthur, B., 1990, "Positive Feedbacks in the Economy", *Scientific American*, February

Barnsley, M., 1988, *Fractals Everywhere* (Academic Press, Boston)

Basalla, G., 1988, *The Evolution of Technology*, (Cambridge University Press)

Beckmann, M.J., 1952, "A continuous model of transportation", *Econometrica* 20:642-660.

Beckmann, M. J. 1953, "The partial equilibrium of a continuous space economy", *Weltwirtschaftliches Archiv* 71:73-89

Beckmann, M. J., 1976, "Equilibrium and stability in a continuous space market", *Operations Research Verfahren* 14:48-63

Beckmann, M. J. and Puu, T., 1985, *Spatial Economics: Density, Potential, and Flow* (North-Holland Publishing Company)

Beckmann, M. J. and Puu, T., 1990, *Spatial Structures* (Springer-Verlag)

Cartwright, M. L. and Littlewood, J. E., 1945, On Nonlinear Differential Equations of the Second Order, *Journal of the London Mathematical Society* 20:180-189

Chiarella, C., 1990, The Elements of a Nonlinear Theory of Economic Dynamics, *Lecture Notes in Economics and Mathematical Systems* 343 (Springer-Verlag)

Courant, R. and Hilbert, D., 1953 (1927), *Methods of Mathematical Physics*, (Wiley Interscience)

Cournot, A., 1838, *Récherces sur les principes mathématiques de la théorie des richesses* (Paris)

Dendrinos, D.S., 1990, *Chaos and Socio-Spatial Dynamics*, (Springer-Verlag)

Day, R. H., 1994-98, *Complex Economic Dynamics* Vol. I-II, (MIT Press, Cambridge, Mass.)

Devaney, R. L., 1986, *An Introduction to Chaotic Dynamical Systems*, (Benjamin, Menlo Park, Calif.)

Duff, G. F. D. and Naylor, D., 1966, *Differential Equations of Applied Mathematics* (Wiley, New York)

Feller, W.F., 1957-66, *An Introduction to Probability Theory and Its Applications* (John Wiley, New York)

Field, M. and Golubitsky, M., 1992, *Symmetry in Chaos* (Oxford University Press)

Gilmore, R., 1981, *Catastrophe Theory for Scientists and Engeneers* (Wiley)

Goodwin, R.M., 1951, The nonlinear accelerator and the persistence of business cycles, *Econometrica* 19:1-17

Gough, W., Richards, J. P. G. and Williams, R. P., 1983, *Vibrations and Waves* (Ellis Horwood Ltd.)

Guckenheimer, J. and Holmes, P., 1986, *Nonlinear Oscillations, Dynamical Systems, and Bifurcations of Vector Fields* (Springer-Verlag, Berlin)

Haken, H., 1977, *Synergetics. Nonequilibrium Phase Transitions and Self-Organization in Physics, Chemistry, and Biology* (Springer-Verlag, Berlin)

Haken, H., 1983, *Advanced Synergetics. Instability Hierarchies of Self-Organizing Systems and Devices* (Springer-Verlag, Berlin)

Harrod, R. F., 1948, *Towards a Dynamic Economics* (MacMillan, London)

Hayashi, C., 1964, *Nonlinear Oscillations in Physical Systems* (Princeton University Press, Princeton, N.J.)

Hicks, J.R., 1950, *A Contribution to the Theory of the Trade Cycle* (Oxford University Press, Oxford)

Hilborn, R. C., 1994, *Chaos and Nonlinear Dynamics* (Oxford University Press)

Hirsch, M. W. and Smale, S., 1974, *Differential Equations, Dynamical Systems, and Linear Algebra* (Academic Press, New York)

Hommes, C.H., 1991, *Chaotic Dynamics in Economic Models*, (Wolters-Noordhoff, Groningen)

Jackson, E. A., 1990-1991, *Perspectives of Nonlinear Dynamics* Vol. 1-2 (Cambridge University Press)

Johansen, L., 1972, *Production Functions*, (North Holland)

Jordan, D. W. and Smith, P. S., 1977, *Nonlinear Ordinary Differential Equations*, (Oxford University Press, Oxford)

Kapitaniak, T., 1991, *Chaotic Oscillations in Mechanical Systems* (Manchester University Press)

Keller, J. B. and Kogelman, S., 1970, "Asymptotic solutions of initial value problems for nonlinear partial differential equations", *SIAM Journal of Applied Mathematics*, 18:748-758.

Kevorkian, J. and Cole, J. D., 1981, *Perturbation Methods in Applied Mathematics* (Springer, Berlin)

Kuznetsov, Y. A., 1995, *Elements of Applied Bifurcation Theory* (Springer-Verlag)

Lancaster, K., 1971, *Consumer Demand. A New Approach*

Levi, M., 1981, Qualitative analysis of the periodically forced relaxation oscillations, *Memoirs of the American Mathematical Society* 32:244

Mandelbrot, B. B., 1982, *The Fractal Geometry of Nature* (Freeman, New York)

Mira, C., Gardini, L., Barugola, A. and Cathala, J.-C., 1996, *Chaotic Dynamics in Two-Dimensional Noninvertible Maps* (World Scientific)

Medio, A., 1992, *Chaotic Dynamics - Theory and Applications to Economics* (Cambridge University Press)

Millman M. H. and Keller J. B., 1969, "Perturbation theory of nonlinear boundary-value problems, *Journal of Mathematical Physics* 10:342-361.

Myerscough, C. J., 1973, A Simple Model of the Growth of Wind-Induced Oscillations in Overhead Lines, *Journal of Sound and Vibration* 28:699-713

Myerscough, C. J., 1975, Further Studies of the Growth of Wind-Induced Oscillations in Overhead Lines, *Journal of Sound and Vibration* 39:503-517

Nayfeh, A. H., 1973, *Perturbation Methods* (Wiley, New York)

Nusse, H. E. and Yorke, J. A., 1994, *Dynamics: Numerical Explorations* (Springer-Verlag)

Oliver, D. and Hoviss, D., 1994, *Fractal Graphics* (SAMS Publishing)

Parker, T. S. and Chua, L. O., 1989, *Practical Numerical Algorithms for Chaotic Systems* (Springer-Verlag)

Perlitz, U., and Lauterborn, W., 1987, Period-doubling cascades and devil's staircases of the driven van der Pol oscillator, *Physical Review A* 36:1428-1434

Peitgen, H. O., and Richter, P. H., 1986, *The Beauty of Fractals. Images of Complex Dynamical Systems* (Springer, Berlin)

Peitgen, H. O., and Saupe, D., (Eds.), 1988, *The Science of Fractal Images* (Springer, Berlin)

Peitgen, H. O., Jürgens, H., and Saupe, D., 1992, *Chaos and Fractals. New Frontiers of Science* (Springer-Verlag)

Peixoto, M.M., 1977, Generic Properties of Ordinary Differential Equations, *MAA Studies in Mathematics* 14:52-92

Phillips, A.W., 1954, Stabilization policy in a closed economy, *Economic Journal* 64:290-323

Popper, K. R., 1959, *The Logic of Scientific Discovery* (Hutchinson, London)

Poston, T. and Stewart, I., 1978, *Catastrophe Theory and its Applications* (Pitman)

Puu, T, 1989, 1991, 1993, 1997, *Nonlinear Economic Dynamics* (Springer-Verlag)

Puu, T, 1997, *Mathematical Location and Land-Use Theory* (Springer-Verlag)

Rand, D., 1978, Exotic phenomena in games and duopoly models, *Journal of Mathematical Economics* 5:173

Lord Rayleigh, J. W. S., 1894, *The Theory of Sound* (MacMillan, London)

Robinson, J., 1933, *The Economics of Imperfect Competition* (Cambride University Press)

Rosser, J. B., 1991, *From Catastrophe to Chaos: A General Theory of Economic Discontinuities* (Kluwer, Boston)

Samuelson, P. A., 1939, Interactions between the Multiplier Analysis and the Principle of Acceleration, *Review of Economic Statistics* 21:75-78

Samuelson, P. A., 1952, "Spatial price equilibrium and linear programming", *American Economic Review* 42:283-303.

Samuelson, P. A., 1947, *Foundations of Economic Analysis* (Harvard University Press, Cambridge, Mass.)

Schumpeter, J. A., 1954, *History of Economic Analysis* (Oxford University Press, Oxford)

von Stackelberg, H., 1938, Probleme der unvollkommenen Konkurrenz, *Weltwirtschaftliches Archiv* 48:95

Sprott, J. C., 1993, Strange Attractors. Creating Patterns in Chaos (M&T Books)

Stoker, J. J., 1950, *Nonlinear Vibrations in Mechanical and Electrical Systems* (Wiley, New York)

Strogatz, S. H., 1994, *Nonlinear Dynamics and Chaos* (Addison-Wesley)

Saunders, P. T., 1980, *An Introduction to Catastrophe Theory* (Cambridge University Press)

Thom, R. 1975, *Structural Stability and Morphogenesis* (Benjamin)

Thompson, J. M. T., 1982, *Instabilitis and Catastrophes in Science and Engeneering* (Wiley)

Thompson, J. M. T. and Stewart, H.B., 1986, *Nonlinear Dynamics and Chaos* (Wiley, New York)

Ueda, Y., 1991, Survey of Regular and Chaotic Phenomena in the Forced Duffing Oscillator, *Chaos, Solitons and Fractals* 1:199

Wegner, T. and Tyler, B., 1993, *Fractal Creations* (Waite Group Press)

Index

A

B

C

D

E

F

G

H

I

J

K

L

M

T

U

V